ARCTIC UNDERWATER OPERATIONS

Organizing Committee

President: Professor Louis Rey, Switzerland; Alaska
Vice-President: Surgeon Vice-Admiral Sir John Rawlins, United Kingdom
Secretary General: Professor Per-Ola Granberg, Sweden
Members: Johan Axelsson, Iceland; Bo Cassel, Sweden; Jakie Chappuis,
Switzerland; P Christopher, United Kingdom; Henrik Forsius, Finland;
Ron Goodfellow, United Kingdom; Bent Harvald, Denmark; Anders Karlqvist,
Sweden; D E Lennard, United Kingdom; G C May, United Kingdom;
Frej Stenbeck, Finland; Joergen Taagholt, Denmark; Leif Vangaard, Denmark

Proceedings of an international conference ('Icedive '84), organised jointly by
Comité Arctique International, Society for Underwater Technology, The Nordic
Council for Arctic Medical Research and the University of Alaska – Fairbanks, and
held in Stockholm on 3–6 June 1984.

Edited by **Louis Rey** (President, Comité Arctique International) with the assistance
of **Sir John Rawlins** (Past President, Society for Underwater Technology)
Donn K. Haglund (Professor of Geography, The University of Wisconsin –
Milwaukee) **Per-Ola Granberg** (President, The Nordic Council for Arctic Medical
Research)

ARCTIC UNDERWATER OPERATIONS

Medical and Operational Aspects of Diving Activities in Arctic Conditions

Edited by

Louis Rey

University of Alaska – Fairbanks

Comité
Arctique International

Society for
Underwater Technology

Nordic Council for
Arctic Medical Research

University of
Alaska – Fairbanks

Published by

Graham & Trotman

First published in 1985 by

Graham & Trotman Limited
Sterling House
66 Wilton Road
London SW1V 1DE

© Louis Rey, 1985
Softcover reprint of the hardcover 1st edition 1985

British Library Cataloguing in Publication Data
Icedive '84' (*Conference: Stockholm*)
  Arctic underwater operations: medical and
  operational aspects of diving activities in
  arctic conditions: edited proceedings of an
  international conference ('Icedive '84').
  1. Diving, Submarine—Arctic Ocean 2. Ocean
  engineering—Arctic Ocean
  I. Title II. Comité Arctique International
  III. Rey, Louis IV. Rawlins, John
  V. Granberg, Per-Ola VI. Haglund, Donn
  627'.72'091632   VM981

ISBN 978-94-011-9657-4   ISBN 978-94-011-9655-0 (eBook)
DOI 10.1007/978-94-011-9655-0

Typeset in Great Britain by Spire Print Services Ltd., Salisbury.

CONTENTS

PART II
DIVING OPERATIONAL MANAGEMENT

PART III
UNDERWATER OPERATIONS

Preface

Opening Speech of the ICEDIVE 84 Conference by
His Royal Highness
Prince Bertil of Sweden

I am very pleased to be invited to open the International Conference ICEDIVE 84, dealing with medical and technical problems of diving and related underwater activities in arctic conditions.

Until recent times, the arctic was considered a strange and remote area of minor importance. However, in a world with diminishing natural resources, arctic areas have become a region of global importance because of their enormous resources and strategic position. Certain experts believe that more than 50% of oil reserves are "sleeping" in these northern areas which are cold, harsh and hostile to man.

Operations in arctic areas are extremely difficult, expensive, and demand high levels of technical, scientific and physiological achievement. One should recall for example, that Alaskan oil investment only became economically viable after the 1973–1974 price explosion. Recent political/military troubles in the Gulf have increased interest in the development of polar resources.

This conference is unique as it is the first time that medical and technical specialists interested in the problem of diving in arctic conditions have met in an international forum.

Development of the arctic resources is a matter of international urgency, and it pleases me that scientists from the USA, Canada, the USSR, Australia and Europe have gathered here in Stockholm to present their experience and to discuss problems in this field.

I am glad that the conference came to Stockholm in early summer, a time when our city is at its most attractive. I wish you all a hearty welcome and I wish the organizations — Comité Arctique International, Society for Underwater Technology and The Nordic Council for Arctic Medical Research — good luck with the meeting.

I hereby declare ICEDIVE 84 opened.

Opening Address

Human Life in the Arctic

L. Rey, President, Comité Arctique International, Distinguished Professor of
Arctic Science and History, The University of Alaska, Fairbanks, USA

Man's involvement in the Arctic goes back almost to the beginning of civilization
and, since time immemorial, the forbidden boreal horizons have lured adventurers to
the challenging circumpolar regions.

However, about the middle of the Neo-Pleistocene period, these regions were not
always open to human adventure and, during each glacial period, expanding
ice-sheets compelled Neanderthal man to seek shelter deep in the mid-latitude taiga.
At the same time, a substantial part of the world's waters being imprisoned in
mighty ice-shelves, the mean ocean level dropped by nearly 200 m, uncovering vast
areas of dry, cold, subarctic tundra where primitive man deployed, following the big
game: mammoths, bears and bison.

The climate became so harsh that, at the peak of the Wurm–Wisconsin period,
some 25 000 years ago (the time of the Lascaux "optimum"), the Bering Strait itself
became a land bridge offering a wide new route to the wandering ancient hunters,
who, journeying from Siberia, set foot in North America.

Then, around 15 000 years BP (before present), the world's temperature rose
abruptly: ice-shelves melted and the oceans, swelling again, flooded the low
epicontinental plains and thereby restored the original maritime passages, including
the Bering Strait. Thus, 11 000 years BP, the North American Arcto-mongoloids
became completely severed from their Siberian origins. Initially landlocked in
Alaska and Yukon territories, they followed the receding ice-caps and spread over
the entire Arctic, where they diversified during the following millennia. Some, the
ancestors of the Athabascan and Algonkian Indians, remained inland. Others,
moving along the Arctic shores and hunting sea-mammals, became the predecessors
of the paleo-Eskimo, who migrated to Ellesmere Island and Northern Greenland
more than 4000 years ago. As such, they were the forefathers of the present-day
Inuit, and they also learned how to live by reindeer husbandry in the vast barren
lands of North America.

During the same period, along the immense Eurasiatic expanses, many other
ethnic groups settled on the Arctic rim. The people of the deer (Lapps, Samoyeds,
Ostiaks, Voguls, Tunguz, Iukaghieri and Chukchi) progressively freed themselves
from the powerful influence of the Scythian and Sarmatian Confederations and were

joined, a few hundred years later, by the people of the Arctic horse, the Iakuts, stemming from the ancient Turks. Demonstrating a remarkable and almost unbelievable adaptation to one of the harshest climates on earth, they settled there, in the far remote Septentrio, and developed highly specialized cultures. By living in close harmony with their environment and by always maintaining a proper balance between their own needs and the surrounding wild life, they managed to survive, to expand and even to create permanent settlements in this desolate universe of ice and rocks, blanketed by snow, swept by winds, elusive and challenging under thick layers of clouds or deadly ice-fogs in the long, depressing polar night but, also sometimes, shining white and blue under the continuous bright summer sun. So gentle was these people's impact on the environment that today, despite millennia of permanent presence, there is almost no trace of their passage and they still fuse perfectly with the sensitive boreal ecosystems.

Thousands of years have passed but the circumarctic populations are still there on the steppes and tundras trodden by their fathers and their fathers' fathers since the dawn of time. Thus, who can dispute their rights to consider these places as their aboriginal homelands?

Yet, they did not remain alone in the frozen north. Many "southerners" moved there also, sometimes centuries ago — as in the Norwegian Finnmark or along the Baltic coasts. It is even claimed by some scholars that, in south Greenland, the Vikings might well have been the original settlers. In any event, it seems fairly clear that, contrary to Antarctica, which remained virgin territory until the last century, the Arctic has been the permanent home of native settlers for thousands of years and their undisputed kingdom in the most recent historical past.

Today, however, approaching the end of the twentieth century, the overall pattern has changed substantially and the once forgotten, mysterious Arctic wastes have been brought into the spotlight of global politics. The discovery of huge natural resources, mainly of large fossil fuel deposits, within the northern lands, as well as in the adjacent continental shelves, has suddenly driven a vast cohort of Arctic entrepreneurs into an area which, in the meantime, has also become a critical zone in the delicate strategic balance between the Eastern and Western alliance blocks. As a consequence, more and more manpower is poured into the boreal circumpolar regions, billions of dollars are invested in accelerated industrial development of Arctic energy resources and in support logistics as well as in the completion of sophisticated early warning stations and operational military bases.

Thus, in many places, the traditional socio-cultural patterns of the local populations have been seriously endangered, if not dramatically changed or wrecked, and this is a matter of growing concern for governments, native leaders and Arctic entrepreneurs, who, from Siberia to northern Scandinavia, Greenland and Septentrional America, endeavor to promote a harmonious development of the North but at the same time prevent the thousand-year-old cultures from capsizing in the maelstrom of modern technology.

It is a delicate and complex problem, one which cannot be addressed in pure rational terms since it presents obvious political dimensions in addition to raising numerous emotional issues. It is equally a permanent challenge for those entrepreneurs who, bearing the responsibility of carrying out the operations, are stretched to the limit on completely isolated sites, under extreme climatic conditions and, often, alas, with casual, untrained personnel.

In fact, the effects of the Arctic environment on man are still virtually unknown. For instance low temperatures, a direct consequence of the circumpolar location of the Arctic regions, exert a constant pressure on men and play a major role in polar life. However, there are wide variations, especially inland, and a 90 °C difference between summer and winter is not uncommon. Moreover, the effect of cold is aggravated by the strong winds, which dramatically increase the chill factor and present a serious challenge to men and equipment. It is therefore compulsory to don heavy protective garments, hoods and appropriate boots, which considerably add to physical stress and reduce handling and operational ability due to increased numbness.

Cold air is also very dry (in fact, the Arctic is a true desert), and this imposes a serious strain on the upper airways and the lungs. An indirect consequence is the high degree of water loss through respiration and the associated risk of developing kidney stones due to unbalanced water regulation. More subtle effects on men can also be caused by the pumping action exerted by violent winds on closed vehicles and habitats which, indoors, generate low-frequency air vibrations accompanied by irritating, battering noises, often inducing psychotic disorders in wintering parties.

Recent studies have also suggested that steady prevailing air-streams impacting on regular periodical structures such as mountain ranges or sequenced individual hills could generate, on the lee-side, atmospheric pressure-waves associated with infra-sounds which could influence human reactions and add a new load to the already stressed organism.

Among the different environmental factors specific to polar regions, special emphasis should be placed on light and on the progressively shifting photo-periods, which result in a continuous out-phasing of circadian rhythms and biological clocks and often induce, in sensitive individuals, serious impairment of the sleep cycle and its structure. Finally, it is more than likely that the strong magnetic disturbances which occur around the geo-magnetic poles and within the wider auroral zone interact with brain activities and interfere with sleep.

The obvious conclusion is that the Arctic environment is very demanding and that it exerts a deep and diversified action on man, which might affect not only his physical condition but also his mental health, through increased anxiety, abnormal sleep patterns and disturbed metabolic and glandular functions due to the cyclic desynchronization of the biological and circadian rhythms. This will explain the many strange reactions which are known to occur in circumpolar regions, mainly in transition periods, such as the "ruska reaction" in the fall in Finland, or the "Spring crazy period" in Alaska and Canada. It is equally obvious that this may well have serious effects not only on behaviour, mood and affectivity, but also, with sensitive individuals, on their rate of activity, their memory, vigilance and aptitude to make rapid decisions in emergency situations, with all their subsequent consequences. To this should also be added the still unknown physiological effects of frequent and regular commuting to Arctic sites from southern latitudes, with the associated abrupt changes in life-style and time budget.

Indeed, life in the Arctic is a permanent challenge and, under such conditions, it can be readily understood that any additional load is bound to have profound repercussions on a man's performance and behavior. This is precisely what may occur when diving in Arctic conditions, since to the already existing major stress of the polar environment are added the stringent constraints of underwater operations. In that particular case, the situation is paradoxical since, in the heart of the cold Arctic regions, the polar sea behaves as a heat vent and a thermal buffer, where abrupt ambient temperature variations are converted into phase transitions within the overlying drifting ice-pack. Indeed, in contrast to the hostile, wind-buffeted, sterile Arctic lands, northern sea waters offer a stable, quiet environment where biological life abounds and where divers are protected from the bitter cold outside. At first glance, and within less than a 10 °C difference, underwater operations in the Arctic appear as any other underwater operation in a cold mid-latitude oceanic compartment.

In fact, this is not at all the case. In near-zero waters, heat losses are considerable, and many adverse physiological reactions do occur if divers are not properly insulated against cold. Moreover, in the Arctic, the sea is often at sub-zero temperatures over great depths and, accordingly, the water vapor released by human metabolism might freeze abruptly in the mask or within the diving apparatus itself, thus blocking gas circulation completely. Finally (and with the exception of saturation diving, where divers re-enter directly the bell or the transfer modules in the depths), Arctic divers who operate under the ice may face major difficulties in trying to reach open water when they surface again and are, in any event, bound to meet very low temperatures and a highly hostile environment outside. Curiously enough, the same situation occurs in lower latitudes at high-altitude natural or artificial lakes.

Furthermore, if we add the difficulties inherent to any Arctic or high mountain operation, in terms of logistics, transport, communication, security, rescue, we see that diving in ice-covered waters is of an order of magnitude more complex and delicate than any conventional operation in free water. Indeed, in Arctic conditions, a lost bell is always a dramatic issue and, in the cold, dark, turbid waters, moral distress and uncontrollable heat losses might drive the stranded divers into hopeless situations, which might turn into silent drama in a short time when the lost men enter the imprecise and mysterious frontiers which lie between deep hypothermia, clinical death and death. This is the reason why many Arctic operations favor submersibles, tethered or free, manned or remotely operated, which do not present the same risk factors and may be used in difficult or even hazardous conditions. Of course, their flexibility and operational capabilities are quite different, and they cannot always substitute direct human intervention; hence the development of lock-out submarines capable of bringing the divers directly on site and then back to the pressurized habitat of the mother ship.

Arctic underwater technology is still in its early stages, and many new interesting ideas and projects are being developed to achieve better and safer operations both with regard to the human operators and the environment. This was the common feeling of the Comité Arctique International, the Society for Underwater Technology and the Nordic Council for Arctic Medical Research in laying the basis for the Icedive Conference.

Part I

Medical and Physical Problems

1

Medical and Physiological Problems

L. A. Kuehn, Director of Science and Technology (Human Performance),
National Defence Headquarters, Ottawa, Canada

The physiological problems of hyperbaric exposure are reviewed briefly. Specific physiological responses in humans to diving are considered, together with risks associated with significant habituation and acclimitization. In particular, the rigours of polar diving are stressed.

Despite the substantial interest and technical development in diving afforded by many investigators and engineers over the years, the cost-effectiveness and success attained has been satisfactory only for diving at relatively shallow depths. That this is so is due as much to the intrinsic physiological limitations of the human body as to the rigours of the underwater environment itself (Berghage, 1978).

The principal physiological problems pertinent to hyperbaric exposure are twofold: those engendered by the requirement for decompression and those implicit in the need for thermal protection to offset the considerable heat drain of such environments. Before dealing with the peculiarities of the physiology of arctic diving, I will review briefly these underlying physiological concerns.

BASIC LIMITATIONS OF THE HUMAN IN HYPERBARIC ENVIRONMENTS

Excluding consideration of short-duration breath-holding dives, the requirement for respiration in longer dives necessitates that divers breathe oxygen/inert gas mixtures at the ambient hyperbaric pressure; consequently their body tissues absorb large amounts of inert gas until they are saturated to equilibrium tissue pressures (Kidd and Stubbs, 1969). (Such high levels of inert gas are necessary to prevent oxygen poisoning or oxygen toxicity, even though oxygen is a metabolizable gas.) On ascent from the dive, particularly if the body tissues have been saturated, the dissolved inert gas leaves the tissues by diffusion and perfusion. If the ascent is too rapid, the dissolved gas is precipitated by various means, in small bubbles which enlarge quickly thus causing the various cardiovascular, neurological, and respiratory ailments which are generally characterized by the term "decompression sickness" (Kidd and Elliott, 1969). Various other types of decompression ailments exist, apart from those presumed to have been caused by bubbles. These include barotrauma in its various forms, ranging from rupture of the eardrum to rupture of lung tissue, all caused by rapid expansion of gas in confined cavities. Another decompression concern is decompression arthralgia, that is, pain experienced in joints and limbs. All of these decompression maladies are ameliorated by sensible decompression practices, based mainly on slow step-wise ascents.

The other principal problem area of diving which limits deployment of humans underwater is that of thermal stress (Webb, 1974), due to the great thermal demand imposed by cold water on the human thermoregulatory system — a demand that is far greater than that for which this system evolved. The breathing of cold inert gases, which are highly heat-diffusive at great hyperbaric pressures, imposes a severe respiratory heat drain, dependent not only on the involuntary warming of this gas in the respiratory system but also on the involuntary humidification of the gas by the moist lungs and upper airways of the diver. Below depths of 100 meters it is essential that warming of the breathing gas supply, if not humidification, takes place to reduce the risk of diver hypothermia and subsequent unconsciousness at depth.

Another important avenue of heat loss is conductive/convective heat loss from the diver through his suit to the surrounding cold water. Despite significant

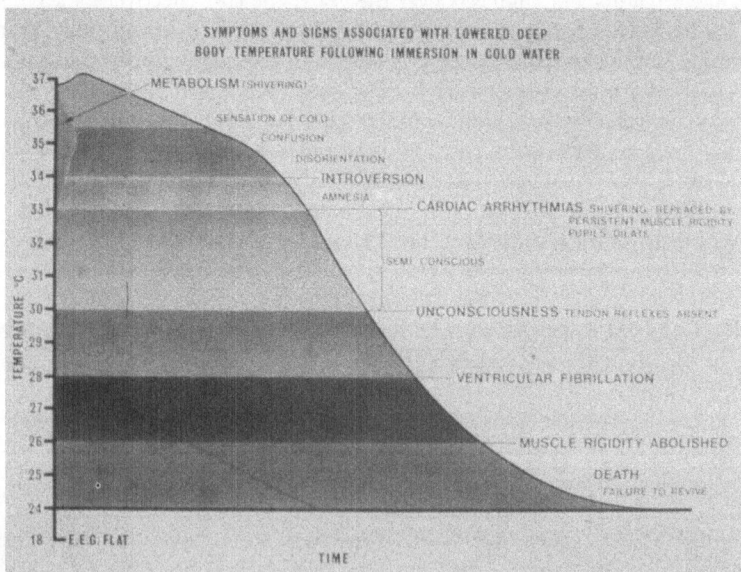

Figure 1 Hyperthermic response of the human body. Various physiological events at different body core temperatures are plotted against a non-dimensional time axis. (Reproduced permission of Frank Golden, RN)

improvements in suit design and material, no passive thermal insulation material yet developed successfully prevents diver hypothermia from convective heat losses. For diving deeper than depths of 100 (meters water depth), divers must have active thermal personal heating to ensure normal thermoregulation and safety. Although boiler-generated open-loop hot water systems have proven commercially successful, newer diver-carried closed-loop systems are being developed; however, protection of extremities remains a problem for all suit designs.

There are many other diver maladies that impact on diver performance and efficiency, such as the high pressure nervous syndrome, inert gas narcosis, oxygen toxicity, cardiovascular effects, and vestibular dysfunction, but these will not be treated specifically in this paper as they are not of paramount importance in arctic diving. Rather, the problems of decompression and thermal protection will be treated in much greater detail.

THE POLAR FACTOR

Arctic, or rather polar, diving (for diving is also performed in the Antarctic) is typified not only by excessively cold water, often covered by ice, but also by isolation and the consequent lack of access to technological resources (Jenkins, 1974). Before diving can even be undertaken, it is necessary to find the dive site (usually on ice), clear it of snow, prepare the through-ice diving hole, erect shelters for equipment and personnel protection, provide a supply route, and bring in support equipment and personnel. Once the dive begins, site maintenance, protection of support equipment from the environment, and provision of suitable facilities for the divers consumes much energy. Given that the ambient environmental air temperatures can be of the order of 10 °C or lower in the summer and much colder in the winter, with water as cold as -2 °C, these various tasks take on an added difficulty.

Although strong winds are not typical of the Arctic, they are common in coastal areas, where much diving is conducted. However, their effect is magnified by the low air temperature, the combination being a cooling power that is described in terms of windchill (Siple, 1945). This problem can be particularly acute for support personnel at the ice surface. Although precipitation is not great, and is somewhat variable, its main effect on diving is the lack of visibility at the ice surface in conditions that involve substantial blowing snow on windy days or whiteout of the surroundings during overcast periods with little or no wind.

Ice thickness alone can be a confounding factor in terms of the difficulty of providing a dive-entry point. Much of the Arctic is ice-covered for all or most of the year with thickness of the order of 2–3 meters. Fortunately, the water is exceptionally clear in polar areas (MacInnis, 1976), yet the long periods of darkness in winter months can cause the light level underwater and under-ice to be so low as to reduce the efficiency of underwater operations. Even sound transmission is reduced in the surface air environment under certain conditions in which soft snow absorbs the sound energy, making communication among surface personnel difficult.

SPECIFIC PHYSIOLOGICAL CONCERNS

Thermal physiology

The primary arctic diving problem is protection from the cold, both in the recognised as underwater and surface-support modes. Body fat has long been a major factor in the protection of diving personnel but new studies (Golden *et al.*, 1980) show that the metabolic heat production or the peripheral vascular response of an individual may be even more important. Habituation to cold water is another factor considered important. It is known that regular daily cold water diving will lead to habituation in approximately one week. It appears to be an acquired peripheral physiological response in that it depends specifically on the task involved. However, although it is possible to acquire significant habituation and even acclimatization to cold water, much of this can be lost by the adoption of improved thermal protection (Park *et al.*, 1969). Acclimatization and habituation are important in that these phenomena may alter physiological responses in complex ways, placing the diver at greater risk in terms of his self-appraisal of his physiological status.

This concern is especially warranted in terms of the progressive symptomless hypothermia in cold hyperbaric operations, now thought to be a factor in many diving fatalities in cold water, primarily due to the effects of insensible respiratory heat loss (Childs and Norman, 1978; Hayward and Keatinge, 1979). Although subjective thermal discomfort is related to a decrease in core temperature, with high respiratory heat losses a diver's assessment of comfort is not always related to physiological temperature change. Proper provision of respiratory heating to the diver would alleviate this problem.

Because of the lack of knowledge in the diving community as to the general physiological effects of cold exposure, and given that divers tend to be a self-selected population that terminates their exposures on subjective grounds, there is no consensus on recommended limits for cold exposure of divers. The question exists as to whether there should be one set of limits or a scale of limits set conservatively to apply to the majority of the population to which they are to be administered. Nevertheless, selection of divers with particular skin folds, or size, or body shape or previous diving history may be an important means whereby diver hazard and risk are significantly reduced in operational polar cold water environments.

Despite the lack of agreement on cold water exposure limits, standards for minimum inspired gas temperature have been established by several navies, detailing the effects of environmental hyperbaric pressures, gas composition, and respiratory rates on the cooling of the respiratory pathways (Piantadosi, 1980). Proposed new standards of the US Navy were based on the concept that a rectal temperature change of 0.25 °C per hour can be tolerated for a cumulative rectal temperature decrease of 1.0 °C in most diving operations. However, these standards are pertinent to seated subjects in hyperbaric chambers and more conservative limits may have to be developed for working helmeted cold water polar divers. Furthermore, control of hot water flow and temperature to a diver's suit and breathing gas heater to effect these standards is difficult.

Thermal effects on decompression

The thermal state of a diver has an important bearing on the efficacy of decompression schedules and tables, particularly in very cold water. Doppler ultrasonic studies (Dunford and Hayward, 1981) have shown that divers who are cold during their bottom exposures take up less gas in solution in body tissues than those who are warmer during the same diving exposures. Consequently, on decompression, those divers that were warm at the bottom phase of their dives and cold during decompression may have more decompression problems. Many anecdotes attest to the fact that divers accustomed to working at the bottom part of their dives without hot water suits are able to decompress faster than is the case when wearing hot water suits on similar dives.

Diver monitoring

The efficacy of physiological monitoring in operational diving is questionable, since most divers voluntarily end their exposure when experiencing the initial effects of hypothermia, but there have been numerous calls for such technology in the literature (Davis, 1979; James, 1981), particularly because of the concern for symptomless hypothermia that is conjectured to be the cause of many diving accidents. The state of monitoring technology is considered to be poor despite the elegant data transmission means that are now available. The parameters for monitoring include subjective verbal comments, deep body (core) temperature, individual or mean skin temperatures, heart rate and direct heat flow. No one physiological variable alone is an accurate and completely reliable indicator of a

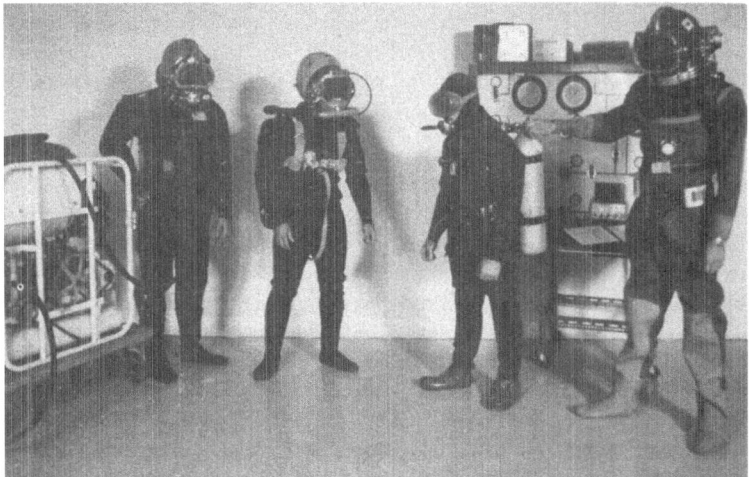

Figure 2 Four diving suits for protection against hyperbaric cold stress. *From left to right*: commercial hot-water suit with surface supplied water, wet suit, variable dry suit, Yokohama dry suit.

thermal problem but the single most useful and easily-implemented dive monitoring parameter is direct verbal contact (Pearson, 1981). Environmental monitoring, such as that of the hot water temperature at point of entry into a suit, should be used wherever possible to gather information on diver health and safety.

Cold water diver protection

Until recently, diver thermal protection was based on either passive neoprene foam-based insulation or expensive hot water heating systems, but significant advances have now been made in both areas, particularly by the US Navy (Nuckols, 1980). The passive thermal protection system is centered around a dry suit outergarment, which acts as a flexible water barrier with minimum insulation, worn over a thermal undergarment made of a hydrophobic material called Thinsolite, manufactured by the 3 M Company. It retains its insulation qualities even when wet and can provide thermal comfort for a diver operating in shallow 2 °C water over a six-hour mission duration.

Respiratory gas heating technology is not nearly so well developed; new minimal breathing gas heating limits cannot be met unless use is made of "active hot water insulation". The most critical area for technological improvement in this context is the interface of hot water heating to the respiratory gas supply at the helmet (Middleton, 1980). Other pertinent problems are the questions of what respiratory system limits there should be for thermal comfort and the avoidance of hypothermia.

It should be remembered that even though hot water heating systems have been successfully applied to the problems of the shallow-depth surface-supported polar diver, there is a limit to the usefulness of such expensive systems. As the depth increases, so must the surface hot water pressure, with the result that the connecting hot water umbilical becomes unwieldy. Consequently, bell-mounted, electrically-supplied closed-loop hot water systems have now been developed that provide divers with thermal energy sufficient for six-hour lockouts. Various new technologies have also been proposed to improve the generation of heat at the bell.

Concern for the lost bell diver

The most pressing thermal problem of recent years has been that of the lost bell diver (DMAC, 1980; Tonjum et al., 1983), a problem that would be aggravated if it were to occur at northern latitudes where the problems of locating the bell and obtaining resources to effect a rescue are extremely difficult. Various recommendations for extending survival time of lost bells have been implemented, some of which pertain also to improved protection of the polar diver. These include better insulation of the diver, isolation of the diver from the sides of the bell, adoption of a fetal position for reduction of heat loss, avoidance of any unnecessary physical activity and the use of emergency respiratory heat reclaimers/CO_2 scrubbers. Application of some or all of these recommendations has assured 24 hours of thermal safety, with only mild hypothermia, for trapped divers. Commercial survival technology systems for lost bell divers, based on a passive insulation garment ensemble to be used with a reclamative thermal generator/CO_2 scrubber, have been developed. However, there is a need for extension of such technology to afford three days of thermal protection (Kuehn et al., 1977), particularly for application in northern latitudes.

Figure 3 Stages in increased insulation afforded by donning various layers of Canadian Forces arctic winter clothing.

PARTICULAR POLAR TOPICS

Cold acclimatization

A primary concern for individuals diving in cold water under arctic conditions on many sequential dives is cold acclimatization or acclimation or, in other words, physiological changes in the individual in response to long-term stress of the cold. These changes are more significant and of greater permanence than those typified by habituation, or the psychological acceptance of the irritation and discomfort associated with cold exposure. Whereas effects of cold acclimatization are easy to demonstrate in animals, the evidence in the case of man is much more specious, unlike the case for heat acclimatization which is as easy to demonstrate in man as it is for other animals.

Metabolic rate increases in mammals exposed to cold, due partly to increased muscular activity in the form of shivering. However, measured at low temperatures under great environmental cooling power for long periods, metabolic rate can be maintained at a high level without gross shivering being apparent. This increase in metabolism is accompanied by an increased food intake, without much change in body weight. Objective evidence of acclimatization in man centers around increased tolerance time to given cold insults (Burton and Edholm, 1955), delayed onset of shivering after repeated exposures, as well as diminished reductions in body and skin temperature.

Generally speaking, modern diving suit technology may ameliorate the formation of a cold adaptive response. Divers such as the nearly-nude Korean Ama, who have been shown to have acquired significant cold acclimatization to cold water, have lost much of this physiological benefit on becoming accustomed to the regular use of modern wet suits (Park *et al.*, 1983). The wearing of wet suits routinely for three or four years completely reverses central and peripheral metabolic acclimatization.

Nevertheless, it has been shown that there is physiological benefit to the pre-acclimatization of personnel intended for deployment in arctic environments (Boutelier *et al.*, 1982). Young fit military troops fared better in cold arctic deployment after conditioning experiences of cold water baths for 30 min periods on the days leading up to the deployment.

Surface cold stress

As to the complaints of cold exposure voiced by diving personnel in the Arctic, fewer are received from the cold divers themselves than from the surface support personnel who act as their tenders. This is due to the fact that the latter personnel are exposed to the convective properties of cold winds whipping unobstructed along barren open ice fields. Invariably, the shelters set up for protection of these people are light mobile structures that are not impervious to cold drafts of blowing snow. The close association of the structures to open water in the diving hole in the ice, as well as the ice itself which serves as the floor or sub-floor, causes these shelters to be cold enclosures indeed. Unless the personnel are fairly physically active, cold discomfort or mild hypothermia can be a result.

Diving equipment freeze-up

The cold water itself is the source of problems of freeze-up within the respiratory breathing equipment (Jenkins, 1974) that can occasionally lead to life-threatening situations for the diver. Polar saline seawater is as cold as -2 °C and fresh water, either as residual vapour in the breathing gas supply or acquired inadvertently from the environment, can freeze in the first or second stages of the diving regulator, causing hazardous free-flow of breathing gas from the diver's tanks. Proper use of the breathing system, particularly on the surface in cold windchill conditions, as well as the avoidance of rapid purging, can reduce incidences of this phenomenon. This problem can also be obviated by design of the diving regulators so that such freezing is not possible. This can be accomplished by various types of insulation on the exposed respiratory supply lines, the most interesting and novel being a "hot water shroud" from the hot water supply line for heating of the diver's suit.

Novel under-ice technology

The ice-cover of the water, while acting as a secure and large platform for the surface support personnel and surface-positioned diver support equipment, can also act as a constraint to diver mobility under water by the necessity of a lifeline to the diving hole. The range of motion of a diver in this situation is limited by the mass and drag of the lifeline in the water. The entry and exit of divers through the diving hole is also physically demanding.

One solution may be the maintenance of divers in an underwater habitat positioned immediately below the ice, held up by its buoyancy, adjacent to the diving hole (English, 1975). One such structure that was developed several years ago is the eight-foot acrylic–aluminium sphere, Sub-igloo. The habitat storage depth was approximately 16 feet (4.3 m). The feasibility of use of such structures in arctic ice-covered waters has been demonstrated; however, the desirability of such application of shallow water saturation air diving must still be confirmed.

An extended variant of this idea would be the development of an under-ice crawler (Dotto, 1977), a mobile sub-ice habitat which would crawl underneath the ice, supported by its buoyancy also. It could be used to transport divers from site to site or to act as an underwater exploration vehicle. The mode of movement of such a vehicle would be from large lugged wheels, either mechanically or electrically powered, in contact with the ice surface. As the under-ice surface is very smooth and extended, travel in this mode could be more easily facilitated than travel above the ice surface, exposed to the vagaries of the cold atmosphere and the rough upper ice surface. Although such a vehicle has not been built as yet, various designs have been put forward.

Arctic diving has already been extended by technology in the form of a one-atmosphere submersible (Bachrach, 1982). Dives have been made to depths as great as 300 msw, without the problems of decompression, heat loss, communications, and diver life support attendant to ambient-pressure hyperbaric exposures. As has been demonstrated, many hours of deployment of such a vessel are possible, with different operators working in shifts and with only a few moments of turn-around time between shifts required. The need for extensive ambient-pressure diving experience is not mandatory for operation of such submersibles and various professionals of diverse backgrounds can use these vessels without undue hazard. This may indeed by the future for polar sea diving, particularly as the sophistication of mechanical manipulators for such vessels are improved.

CONCLUSION

For years the diver has been accustomed to cold water exposure, "a measure of his competence being his ability to withstand cold" (Rawlins and Tauber, 1972), but gradually there has been a successful evolution of technology to prevent his direct exposure to the effects of cold water, save for protection of the hands. Yet the rigours of polar diving are such that certain physiological problems cannot be ignored and "the evaluation of heating systems can[not] proceed on an engineering basis" only (Rawlins and Tauber, 1972).

REFERENCES

Bachrach, A. 1982. Performance and physiology in the atmospheric diving system (ADS) JIM-4, *Proceedings of the First Annual Canadian Ocean Technology Congress*, Defence and Civil Institute of Environmental Medicine, Downsview, Ontario, Canada, pp. 452–465.

Berghage, T. 1978. Man at high pressure: A review of the past, a look at the present and a projection into the future, *Marine Technology Society Journal*, 12, 18–21.

Boutelier, C., Livingstone, S., Bougues, L. and Read, L. 1982. *Physiological, psychophysiological and ergonomic aspects of the exposure of man to Arctic cold (Kool Stool II)*, M. Radomski, C. Boutelier and A. Buguet (eds) Defence and Civil Institute of Environmental Medicine, Downsview, Ontario, Canada.

Burton, A. and Edholm, O. 1955. *Man in a Cold Environment*, Edward Arnold, London.

Childs, C. and Norman, J. 1978. Unexplained loss of consciousness in divers, *Med. Aero. Spat. Med. Subaquat. Hyper.*, 17, 127–128.

Davis, F. 1979. Diving and hypothermia, *Brit. Med. J.*, 2, 494.

Diving Medical Advisory Committee (DMAC) 1981. *Thermal Stress in Relation to Diving*, DMAC Workshop Proceedings, Session Three, Chaired by D. Elliott and F. Golden, pp. 23–30.

Dotto, L. 1977. A vehicle that travels upside-down under the ice, *Globe and Mail*, 11,

Dunford, R. and Hayward, J. 1981. Venous gas bubble production following cold stress during a no-compression dive, *Undersea Biomedical Research*, 8, 41–44.

English, J. 1975. Shallow-air saturation dive in the high Arctic, *Proceedings of the IEEE Ocean '75 Conference*, pp. 264–267.

Golden, F., Hampton, I. and Smith, D. 1980. Lean long distance swimmers, *J. Roy. Nav. Med. Serv.*, 66, 26–30.

Hayward, M. and Keatinge, W. 1979. Progressive symptomless hypothermia in water — Possible cause of diving accidents, *Brit. Med. J.*, 1, 1182.

James, P. 1981. Human limitations and physiological monitoring in commercial diving, *Session B4, Proceedings of Divetech '81*, Society for Underwater Technology, London.

Jenkins, W. 1974. *A Guide to Polar Diving*, US Office of Naval Research, Arlington, Virginaia.

Kidd, D. and Stubbs, R. 1969. The use of the pneumatic analogue computer for divers, in *The Physiology and Medicine of Diving and Compressed Air Work*, P. B. Bennett and D. H. Elliot (eds), Balliére Tindall and Cassell, London, pp. 386–413.

Kidd, D. and Elliott, D. 1969. Clinical manifestations and treatment of decompression sickness in divers, in *The Physiology and Medicine of Diving and Compressed Air Work*, P. B. Bennett and D. H. Eliott (eds), Baliére Tindall and Cassell, London, pp. 464–490.

Kuehn, L. A., Smith, T. and Bell, D. 1977. Thermal requirements of dives and submersibles in arctic waters, in *Arctic Systems*, P. Amaria, A. Bruneau and P. Lapp (eds), Plenum Press, New York, pp. 801–831.

MacInnis, J. 1976. The Underwater Arctic: earth's most hostile frontier. *Proceedings of the 1976 Working Diver Symposium*, Marine Technology Society, Washington, pp. 196–214.

Middleton, J. 1980. Evaluation of the Kinergetics breathing gas heater for use in open circuit demand breathing apparatus, *Thermal Constraints in Diving: 24th Undersea Medical Society Workshop*, L. A. Kuehn (ed.), Undersea Medical Society, Bethesda, Maryland, pp. 167–184.

Nuckols, M. 1980. The development of thermal protection equipment for divers, *Thermal Constraints in Diving: 24th Undersea Medical Society Workshop*, L. A. Kuehn (ed.), Undersea Medical Society, Bethesda, Maryland, pp. 133–166.

Park, Y., Lee, I., Paik, K., Kang, D., Suh, D., Lee, S., Hong, S., Rennie, D. and Hong, S. 1981. Korean women divers revisited: Current status of cold adaption, *Undersea Biomedical Research*, 8 (1-Suppl.), A29.

Park, Y., Rennie, D. and Hong, S. 1983. Status of cold acclimatization in contemporary Korean women wearing wet suits, *Underwater Physiology VIII*, A. Bachrach (ed.), Undersea Medical Society, Bethesda, Maryland.

Pearson, R. 1981. Why do we need diver monitoring? *Session B4, Proceedings of Divetech '81*, Society for Underwater Technology, London.

Piantadosi, C. 1980. Respiratory heat loss limits in helium–oxygen saturation diving, in *Thermal Constraints in Diving: 24th Undersea Medical Society Workshop*, L. A. Kuehn (ed.), Undersea Medical Society, Bethseda, Maryland, pp. 45–54.

Rawlins, J. and Tauber, J. 1972. *Proceedings of the 1972 Working Diver Symposium*, Marine Technology Society, Washington, pp. 187–196.

Siple, P. 1945. General principles governing selection of clothing for cold climates, *Proc. Amer. Phil. Soc.*, 89, 177–199.

Tonjum, S., Pasche, A., Onarheim, J., Hayes, P. and Padbury, H. 1983. Cold exposure in heliox environment at 16 bars for 24 hours. *Underwater Physiology VIII*, A. Bachrach (ed.), Undersea Medical Society, Bethesda, Maryland.

Webb, P. 1974. Thermal problems in diving, *The Sixth Undersea Medical Society Workshop*, Undersea Medical Society, Bethesda, Maryland.

2

Thermal Balance

W. R. Keatinge, Department of Physiology, The London Hospital Medical College, London, UK

Measurements of regional and overall heat loss, made on people in a steady state of heat exchange, allow internal body insulations to be predicted directly from the individual's subcutaneous fat thickness when in the coldest water in which they can stabilise. Metabolic heat production in these conditions cannot be predicted so accurately, and this is the main limitation at present on our ability to predict the lowest water temperature in which an individual can stabilise body temperature, either with or without external insulation. Metabolic response to cold is affected by the pattern of skin temperature, as well as by many other factors which include prior exposures to cold and physical training. For example, uniform immersion in lukewarm water at 29 °C causes little metabolic response and so allows progressive core cooling, often to around 35 °C. Such cooling causes temporary but serious impairment of ability to learn new facts, and slows reasoning, and this appears to have been an important factor in accidents during deep diving with the warm water flooding system. Low core temperatures in those circumstances, and more notably during rewarming, may be associated with little or no sensation of cold. The safe lower limit for core temperature during demanding and potentially dangerous work is accordingly higher than the conventional "hypothermia" level of 35 °C, although the safe lower limit in other circumstances can be below 35 °C.

THE PROBLEMS OF THERMAL CONTROL IN SHIPWRECK SURVIVORS AND IN DIVING

The principal factors that determine the extent of heat loss and heat production by people immersed in cold water were established before 1970 (see Keatinge, 1969, for review), but the fine-tuning of the control mechanisms was not well understood at that time. Much of the subsequent experimental and theoretical work on the subject has been designed to establish the basis for this fine-tuning, and to enable detailed predictions to be made of heat exchanges under given conditions. In addition, the definition of what is an acceptable body temperature has itself had to be reassessed for particular situations. Most of the older work on thermal exchanges in water was concerned with the problems of shipwreck survivors. In these, survival depends essentially on preventing deep body temperature from falling to lethal levels before or after rescue, and substantial falls in temperature can be acceptable as long as the lethal level is not approached. The problem of maintaining a diver's temperature is a very different one. A man whose body temperature has fallen substantially may be in no immediate danger of death from hypothermia, but impairment of his mental performance could make it dangerous for him to carry out complex work underwater.

PREDICTION OF BODY HEAT EXCHANGES IN UNSTEADY STATES

Accurate quantitative information about internal insulation and heat production should permit accurate prediction of people's thermal changes during exposure to any thermal environment. Accordingly, extensive work has been carried out on computer models to predict the changes in deep body temperature that can be expected for persons of known physical characteristics, with particular external protection, when exposed to particular thermal environments. Some of the most notable models are those of Bullard and Rapp (1970), Wissler (1971), and Baker *et al.* (1979). Such programmes have been repeatedly modified, and it is much to the credit of those responsible for them that the most recent versions provide predictions for given situations which often approximate closely to experimentally observed changes in temperature in those situations. The most wide-ranging and flexible of these models require correspondingly wide-ranging and varied information about the parameters on which they are based. That underlying information often includes estimates of blood flow in different regions of the body under particular conditions of local and core temperature, and estimates of heat production in the same conditions, as well as the passive thermal characteristics of given body tissues and of any external protection provided.

The problem is that this underlying information is usually not available in an accurate form over the full range of possible conditions. Comprehensive programmes have therefore been very dependent on experimental checks of their predictions by direct exposure of volunteers to a wide range of thermal stress, and then on subsequent revisions of the basic parameters of the model to adjust for discrepancies. These revisions enable the model concerned to provide better predictions for the thermal conditions in question, but do not necessarily ensure that the basic parameters used are more correct or that the model will provide satisfactory predictions for very different situations. Some models of this kind in fact use basic parameters which are demonstrably incorrect, but also contain counterbalancing errors in other parameters, so as to give an adequate overall prediction of body temperature changes with particular types of thermal stress and internal physical characteristics. However, such models can be used for predictions only within areas in which each has been subjected to verification by direct experiment. As experimental data are accumulated, the models are becoming safe over a wider range of conditions. They must always, though, be used with an intelligent understanding of their existing limits of verification, and also with an understanding of the degree of variation that can occur in the response even of people of identical physical characteristics in identical situations. It has become clear that biological variation in people's responses can cause even the best possible formula to produce random errors in prediction which can be very substantial in degree. In particular, the metabolic response to cold can be affected by factors which are not predictable from physical characteristics or external thermal stress. The effect of this can be shown most clearly in prediction of the tolerable limits of external temperature during sustained thermal stress.

PREDICTION OF LIMITS OF STEADY-STATE TOLERANCE TO SUSTAINED THERMAL STRESS

One of the most important situations for which predictions are required is that of an individual in heat balance in the coldest and in the warmest water in which balance can be achieved. These upper and lower water temperatures then define the range of external temperature within which that individual can survive and function for long periods of time. This situation can be studied directly by immersing people in progressively colder water until the lowest water temperature in which core temperature can be stabilised is found, and then immersing in progressively warmer water until the highest water temperature at which core temperature can be stabilised is found. This requires repeated experiments, each of long duration, but once the subject has stabilised in the coldest or warmest water possible, very accurate measurements can be made of heat production and heat loss in a steady state of heat exchange at these limiting external temperatures. With measurements of local heat flux made at this time, from different skin areas, this provides direct measurements of all the overall and regional variables for heat transfer and heat production in these situations.

Early experiments of this kind (Cannon and Keatinge, 1960) showed that fat people could achieve heat balance in water as cold as 12 °C, while thin people could not do so in water colder than about 28 °C. This was associated with the ability of the fat people to produce a mean tissue insulation from body core to water about four times that which the thinnest could produce. However, in water colder than 12 °C even the fattest people suffered falls in insulation because of the appearance of cold vasodilatation in the limbs, which led to rapid heat loss and caused progressive body cooling in spite of increasing heat production from shivering as body core temperature fell. These studies established a lower limit of water temperature of a little below 12 °C at which unprotected fat people could stabilise and of 28–30 °C at which unprotected thin people could stabilise. As regards the upper limit for stability without serious overheating, both fat and thin people could stabilise in warm water at 38 °C but at the expense of substantial increase in their deep body temperature, so that the practical upper limit of water temperature for stability was about 38 °C.

More experiments of this kind have been made recently using the more accurate methods now available for measuring subcutaneous fat thickness and regional heat loss (Hayward and Keatinge, 1981). Subcutaneous fat thickness can be measured by both skinfold calipers and ultrasound, and a combination of the two methods eliminates most of the potential errors of each method. Heat flow devices, based on measurement of temperature difference across a thin insulating layer, all have the drawback that they themselves add insulation to the skin to which they are attached, and so modify the variable that they are designed to measure. This can be overcome by using methods of calibration and calculation that make allowance for the insulation of the device itself (Gin *et al.*, 1980). These additions made it possible to relate local rates of heat transfer to fat thickness in the heat and cold, and so enable maximum and minimum local insulations to regional skin areas to be predicted for people of specific fat thickness.

Even in the warm, tissue insulation was a little higher in fat than thin people. This was a surprising finding since the high cutaneous blood flow during vasodilatation might be expected to bypass and effectively to eliminate the insulation of the subcutaneous fat. In practice, it did not do so entirely, physical conduction of heat through the skin and subcutaneous fat still making a significant contribution to heat loss even during maximum vasodilatation in the warm. Another interesting finding was that exercise, even in the warm, caused a substantial increase in heat conduction from body core to skin surface. This again implies that cutaneous blood flow was not the only important factor in transfer of heat from the body core to the skin surface; muscle blood flow and local heat production also played a major part in transferring heat to near the body surface when the muscles were active.

The role of vasoconstriction in reducing body heat loss in the cold is a clear one. There has accordingly often been a temptation to regard thermal conductance from body core to a particular part of the skin surface as a measure of the degree of vasoconstriction in that part of the skin. However, other factors can greatly affect conductance in the presence of identical degrees of constriction. During immersion in moderately cold water, for example, at 15 °C, vasoconstriction in many skin regions may be virtually total, but the internal conductance can be very different in different regions of the body. Once blood flow has become negligible in the skin and underlying fat, the conductance through these tissues depends largely on the insulation provided by this fat, and hence on its thickness (Fig. 1). Accordingly, a fat person in cold water may have a much lower core to abdominal surface conductance than a thin person, although skin blood flow to the abdomen may be no different. In the limbs, a great deal of insulation can also be provided by muscle and bone deep to the subcutaneous fat; for a person at rest in cold water, body core to hand surface conductance can be less than 0.05 of core to abdominal surface conductance, with blood flow contributing little to either conductance (see Hayward and Keatinge, 1981). In cold limbs, exercise markedly increased conductance, since the blood flow in exercising muscle destroys much of the insulation that muscle otherwise provides. However, even over the trunk, where fat thickness is greatest, the total insulation provided by the tissues between body core and skin surface in the cold was always greater than the insulation that could be provided by the skin and the underlying layer of fat, even without blood flow in the skin and fat. This implies

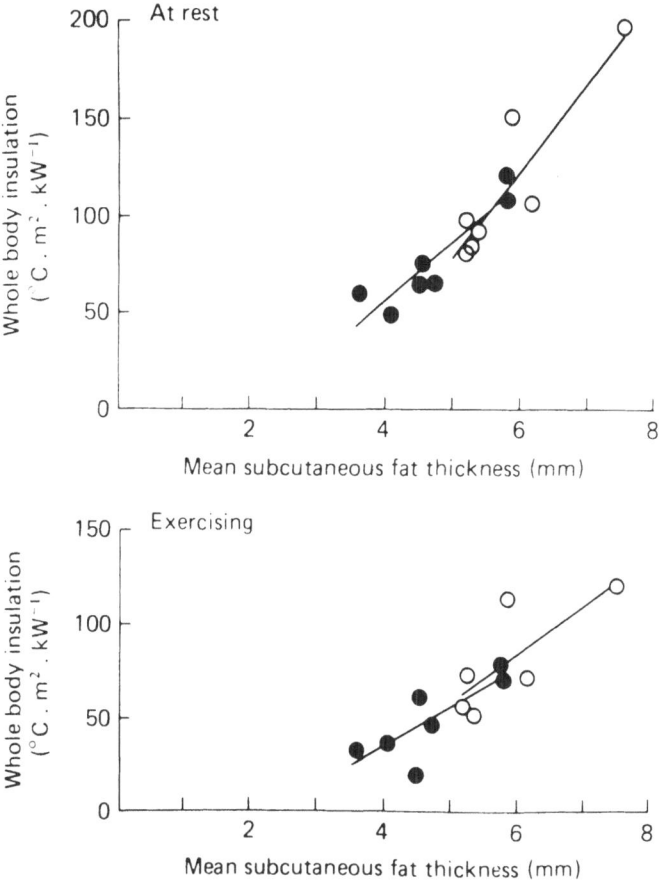

Figure 1 Total body insulation in relation to fat thickness, at rest (*top*) and during exercise (*bottom*). Solid circles men, open circles women. (From Hayward and Keatinge, 1981)

that the deeper tissues, such as muscle and bone, provided significant insulation even in the trunk.

In any event, the measurements made enable total internal insulation for each skin region in the cold to be predicted quite accurately from the subject's mean subcutaneous fat thickness (see tables given in Hayward and Keatinge, 1981). The regional insulations during exposure to heat could also be predicted reasonably accurately. This allows the minimum and maximum external temperature for heat balance to be calculated for an individual of any fat thickness, either unprotected or with given external insulation over different regions, provided that metabolic heat production is known. Metabolic rate during exercise is, of course, greatly dependent on effort, but even with the subjects at rest there was considerable variability of metabolic rate among individuals of similar fat thickness when they stabilised in the coldest possible water. One reason for this is that if two individuals of similar fat thickness give slightly different metabolic responses when immersed in moderately cold water at a given temperature, the one with the larger response will be enabled to stabilise in colder water; this colder water will further increase the metabolic response through greater cold stimulation of the skin, so that the initially larger response of the second than the first subject to cold will be exaggerated when their responses to immersion in the coldest possible water are compared. In one extreme case, one volunteer gave more than six times the metabolic response to cold given by another of similar fat thickness when they were both in the coldest water in which they could stabilise. Information about the factors which cause these differences is clearly of great importance to attempts to predict accurately the coldest water in which a particular person will be able to stabilise. Some of these factors are clear, though only in rather general terms. Repeated exposure to cold, particularly brief cold, usually reduces subsequent metabolic responses to cold (e.g. Carlson *et al.*, 1953; Keatinge and Evans, 1961; Golden *et al.*, 1979) though other patterns of exposure can have different effects. Also, people who are physically fit with a large mass of muscle are likely to have more capacity for prolonged shivering at a high rate than less fit people. These factors cannot be accurately quantified, but can give an indication whether a person is likely to give a larger or smaller metabolic rate than predicted by the usual metabolic rates in the heat and cold (these rates are given in Hayward and Keatinge, 1981). Another major factor which affects this response, and which came to light as a result of experiments prompted by problems in thermal control of divers in the North Sea, is the pattern of skin temperature as distinct from its mean level.

THERMAL CONTROL DURING DIVING

In shallow water diving, to about 30 m depth, the thermal problems are essentially the usual ones of people exposed to cold with varying amounts of protection. Protection is particularly difficult during diving because of compression of air under dry suits, and compression of the air bubbles in foam rubber suits, but otherwise there is little fundamental difference between problems of thermal balance in shallow diving and surface conditions. Shallow water divers often undergo skin cooling, but as soon as core temperature falls seriously, they feel cold, and can then come to the surface reasonably quickly when they feel the need to warm up. In any case, the need to recharge gas cylinders usually prevents scuba divers from staying down long enough to run into serious thermal problems. The only problem then is the distracting effect of cold skin (Baddeley *et al.*, 1975; Ellis, 1982). With deeper dives the problems multiply rapidly. The need to use "Heliox" instead of air increases the thermal conductance of the gas, promoting heat loss from the body surface, and the high heat-carrying capacity of highly compressed gas also causes serious heat loss in respiration. This calls for a heating system to prevent gross hypothermia in deep dives lasting 4 hours, which is the standard duration of commercial dives. The conventional heating system used in the North Sea, which is very successful within limits, is to pump down hot water from the surface to flood the interior of the diving suit, including the hands and feet. This hot water can also be used to heat the breathing gas. The diver is then essentially in a localised warm bath, and thermal problems would seem to be solved.

However, during the 1970s a number of unexplained problems appeared among commercial divers during the large-scale operations in the North Sea. Unexplained episodes of confusion and error by experienced divers were common, and in a high proportion of diving fatalities no complete explanation could be given for the sequence of events leading to death, whatever the final cause of death may have been (Childs and Norman, 1978). In 1979 we had an opportunity to obtain the body temperatures of divers as they finished commercial diving shifts at a depth of 130–145 m (Keatinge *et al.*, 1980), using the conventional hot water heating system. Body temperature was obtained when they returned to the diving bell on the seabed. Out of 13 dives, one diver had a body temperature as low as 34.7 °C and two others had temperatures between 35.0 and 35.5 °C as the dives ended. On the other hand, another was overheated at 38.3 °C. Temperature control was clearly poor, with some body temperatures so low that any further fall could be expected to produce confusion and loss of consciousness with little delay.

Later experiments in the laboratory indicated that, even without further cooling, the levels of temperature recorded in these divers could be an immediate hazard (Coleshaw *et al.*, 1983). Volunteers were cooled by immersion for an hour in water at 15 °C. They were then rewarmed in a bath at 41 °C. The immediate effect of the warm immersion was to accelerate the rate of fall of body core temperature, the well-known afterdrop. At about the lowest point of the afterdrop, the skin had been rewarmed and the subjects felt warm and comfortable, but their core temperatures were depressed by amounts up to 2.7 °C. There was therefore a low core temperature but no distracting effect of cold discomfort. Tests made at that time showed that people with low core temperatures had gross impairment of their ability to memorise facts. It was interesting that they had no impairment in ability to recall facts that they had previously learned when at normal core temperature. Nor was their ability to perform simple reasoning tasks impaired. However, they did show a significant increase in the time needed to perform more complex reasoning tasks when core temperature was low. The impairment in ability to learn, and slowing of reasoning, was severe with core cooling to about 35 °C, and must represent a serious hazard to a diver faced on the seabed with problems which he must resolve in order to complete his task and return safely. They must also represent a major loss of efficiency in divers trying to perform complex tasks.

There are two obvious ways in which major falls in divers' body temperatures can exist without the subject being aware of them. If a diver's heating system fails for a while and is later restored, he is in a similar position to our subjects being rewarmed after exposure to cold, undergoing an afterdrop in core temperature, but with warm skin and therefore no sensation of cold to give a warning that efficiency is impaired. Once the hot water is operating normally after a failure, neither the diver nor the operator has any signal to indicate that the diver will not be able to handle mentally situations that he could normally deal with readily.

The other reasons that a diver's body temperature might fall without sensory warning is less clearcut, but persistent operation of the hot water system at a lower than optimal temperature could, under some conditions, produce this. The deeper the dive, the longer the umbilical supplying the diver with hot water, and the greater the opportunity for the hot water to cool before it reaches him. This can be dealt with by raising the temperature of the water being pumped from the surface, but changes in the length of the umbilical, or in the extent to which it loses heat to surrounding mud or water, can make it difficult to maintain the outflow temperature at the best level. Laboratory experiments have shown circumstances in which prolonged operation of a hot water flooding system at a marginally low level could produce serious falls in body temperature without serious cold discomfort. Normally, during exposure to cold, the skin temperature of the hands and feet is lower than that of the trunk. Someone immersed in cool water, by contrast, has a uniform skin temperature. When volunteers were immersed in lukewarm water at 29 °C they all showed falls in body core temperature, on average of almost 1 °C after 3 hours, and at the end temperature was still falling (van Someren *et al.*, 1982). There was some cold discomfort but generally not of marked degree, and little or no shivering. Cooling of the hands and feet then caused shivering and increased metabolic rate and stopped the fall in core temperature. Clearly, the cooling of the hands and feet provided a sensory input that restored an adequate thermoregulatory response to cooling. It did not have an important counterbalancing effect by increasing heat loss from the hands and feet, since conductance to the hands and feet in the cold is low. Cooling the face can probably also halt insidious body core cooling in the same way, and this is likely to help prevent insidious cooling of this kind in divers with unheated or imperfectly heated breathing gas and unheated helmets, but the more effective the gas heating, the less stimulus will be available from this.

In any event, the problem is preventable with the hot water flooding system if the water is always run at a temperature which keeps the diver in a continuous state of thermal comfort, and if failure of the system is a signal to end the dive rather than to try to restore the heating after a delay. This is now general practice; measurements at the end of dives recently showed no low body core temperatures, and unexplained confusion as well as deaths among divers on the seabed appear to have become much less frequent.

REMAINING PROBLEMS FOR PREDICTION OF SAFE CONDITIONS FOR THERMAL BALANCE

The fact that mild core cooling can produce serious mental impairment requires revision of the acceptable limits within which core temperature should be maintained; 34–35 °C may be quite acceptable as a safe lower limit in many situations, but 36 °C or perhaps a little higher must be regarded as the minimum safe core temperature for people carrying out demanding and potentially hazardous tasks. The main problem is assessing the limits of the external conditions that will maintain safe core temperature is the difficulty in prediction of metabolic rate in the cold. This difficulty may be overcome by future work, at least partially, but it will remain important to keep in mind that not all cold situations are covered fully by experiments, and that no prediction should be applied to an entirely new practical situation without experimental verification in that situation.

REFERENCES

Baddeley, A. D., Cuccaro, W. J., Egstrom, G. H., Weltman, G. and Willis, M. A. 1975. Cognitive efficiency of divers working in cold water, *Hum. Factors,* 17, 446–454.

Baker, E. R., Ringuest, J. L. and Harnett, R. M. 1979. Human thermal models: application to the study of immersion protective devices, in *Aerospace Medical Association, Preprints of 1979 Annual Scientific Meeting,* Aerospace Medical Association, pp. 163–164.

Bullard, R. W. and Rapp, G. M. 1970. Problems of body heat loss in water immersion, *Aerospace Med.,* 41, 1269–1277.

Cannon, P. and Keatinge, W. R. 1960. The metabolic rate and heat loss of fat and thin men in heat balance in cold and warm water, *J. Physiol.,* 154, 329–344.

Carlson, L. D., Burns, H. L., Holmes, T. H. and Webb, P. P. 1953. Adaptive changes during exposure to cold, *J. Appl. Physiol.,* 5, 672–676.

Childs, C. M. and Norman, J. N. 1978. Unexplained loss of consciousness in divers, *Médecine subaquatique et hyperbare,* 17, 127–128.

Coleshaw, S. R. K., van Someren, R. N. M., Wolff, A. H., Davis, H. M. and Keatinge, W. R. 1983. Impaired memory registration and speed of reasoning caused by low body temperature, *J. Appl. Physiol.,* 55, 27–31.

Ellis, M. D. 1982. The effects of cold on the performance of serial choice reaction time and various discrete tasks, *Hum. Factors,* 24, 589–598.

Gin, A. R., Hayward, M. G. and Keatinge, W. R. 1980. Method for measuring regional heat losses in man, *J. Appl. Physiol.,* 49, 533–535.

Golden, F. St.C., Hampton, I. F. G. and Smith, D. 1979. Cold tolerance in long-distance swimmers, *J. Physiol.,* 290, 48–49.

Hayward, M. G. and Keatinge, W. R. 1981. Roles of subcutaneous fat and thermoregulatory reflexes in determining ability to stabilize body temperature in water, *J. Physiol.,* 320, 229–251.

Keatinge, W. R. 1969. *Survival in Cold Water. The Physiology and Treatment of Immersion Hypothermia and Drowning,* Blackwell Scientific, Oxford.

Keatinge, W. R. and Evans, M. 1961. The respiratory and cardiovascular response to immersion in cold and warm water, *Q. J. Exper. Physiol.,* 46, 83–94.

Keatinge, W. R., Hayward, M. G.. and McIver, N. K. I. 1980. Hypothermia during saturation diving in the North Sea, *Brit. Med. J.,* 1, 291.

van Someren, R. N. M., Coleshaw, S. R. K., Mincer, P. J. and Keatinge, W. R. 1982. Restoration of thermoregulatory response to body cooling by cooling hands and feet, *J. Appl. Physiol.,* 53, 1228–1233.

Wissler, E. M. 1971. Comparison of computed results obtained from two mathematical models — a simple 14-node model and a complex 250-node model, *J. Physiol. Paris,* 63, 455–458.

3

Environmental Stress

P. Lomax, Department of Pharmacology, School of Medicine and the Brain Research Institute, University of California, Los Angeles, CA, USA

In cases of fatal accidental hypothermia in which necropsy demonstrates drugs, such as ethanol, in the blood the concentrations are usually not those associated with severe toxicity. Studies in man and experimental animals have not demonstrated how the combination of such drugs and cold exposure causes a profound fall in body temperature. Similarly, the occurrence of accidental hypothermia in the elderly (some of whom may experience repeated episodes) cannot be explained purely in terms of the low ambient temperatures. There is fragmentary, but increasing evidence that some of the effects of CNS active drugs can be modulated by endogenous opioids. These (in particular β-endorphin) may have a regulatory role in maintaining normal thermal balance; also they can be released under conditions of environmental stress. Thus, it appears likely that the cold stress has a more complex effect in the pathogenesis of accidental hypothermia than simply constituting a favorable heat loss gradient. Neuroendocrine changes in aging could be a predisposing factor to the development of the syndrome in the elderly. Cold acclimation protects animals from cold exposure and this is associated with an increase in brown adipose tissue and non-shivering thermogenesis. High food intake in the rat will also mediate the development of brown adipose tissue. Endorphins have been implicated in the process of cold acclimation and the central control of food intake. This paper describes the development of animal models of accidental hypothermia which can be utilized to investigate the specific effects of environmental stress in the pathogenesis of the syndrome. The results of such studies could be of value in relation to the prevention and treatment of hypothermia in general. The current socioeconomic problems faced by large segments of the world population, compounded by the crisis in energy supplies, render an understanding of the concomitants of climatic stress of increasing importance. Furthermore, stratagems related to the adaptation of individuals to adverse environments, and the application of such knowledge during early development, might have important implications as exploration and settlement proceeds into areas dictated by the expanding utilization of global resources.

INTRODUCTION

Accidental hypothermia, in which the body temperature (T_b) falls to below 35 °C during exposure to a low ambient temperature (T_a), is not uncommon in northern

Europe and North America but there is a lack of reliable data as to the actual incidence. At one extreme it has been estimated that there may be up to 20 000 cases of urban hypothermia in the United Kingdom each year. At a recent US Senate Committee on Aging it was reported that there were around 400 deaths per year from hypothermia between 1970 and 1977; these involved all age groups. Ledingham (1980) and Hirvonen (1979) have reviewed the epidemiology of the condition.

Existing, although somewhat fragmentary, data seemed to point towards a defect in metabolic heat producing mechanisms as underlying the pathogenesis of the syndrome (Haight and Keatinge, 1973). Detailed studies have tended, generally, not to support this view; thus, altered heat production or pharmacokinetic considerations do not account for the thermoregulatory changes induced by the combination of administration of ethanol and adverse environmental stress (*vide infra*). Rather the results to date seem to point towards factors related to the adaptation of the organism to the environmental conditions as determinants of the ability to maintain thermal homeostasis.

The energy crisis and rising fuel prices have focussed attention on accidental hypothermia in general (a problem long recognized in countries such as Britain and Scandinavia) (see Cooper *et al.*, 1980) and this has led to the development of the concept that some individuals, particularly with aging, are less able to combat, physiologically, environmental stresses beyond those experienced in temperate climates (i.e. where economic or other conditions preclude the maintenance of a narrow thermoneutral microclimate). Concomitantly, recent investigations have cast doubt on the generally accepted concept of a direct relationship between food intake and metabolic heat production (see Elliott, 1980). Thus, genetically obese rats are unable to combat changes in environmental temperature by increasing heat production by non-shivering thermogenesis and have a high mortality rate when exposed to low temperatures. A dissociation between diet and obesity may exist in man, possibly related to the ability to expend excess caloric input by thermogenesis in adipose tissue.

These several considerations suggest that attention should be focussed primarily on the responses to environmental stress in explaining the syndrome of accidental hypothermia. The dramatic effect of cold exposure on the mortality of ethanol treated rats (Lomax *et al.*, 1981a; Lomax and Lee, 1982) exemplifies this suggestion.

The variable of a change in the environment exceeding that of the experimental laboratory has been little studied in the response of animals to centrally acting drugs. With the exposure of individuals to the temperature extremes experienced in the development of arctic or tropical areas, by modern military situations or to varying oxygen tensions with explorations at high altitudes or under excess atmospheric pressures (e.g. submarine), the physiological and behavioral responses to drugs such as ethanol take on considerable importance. Changes in body temperature and anoxia are associated with marked behavioral effects in man; lack of concentration, disorientation, coma and death may supervene. Many centrally active drugs have a similar spectrum of pharmacological actions which may be additive or synergistic to those induced by the ambient stress. These same factors may modify the receptor configuration or the metabolism of central neurotransmitters which, in turn, may be involved in the adaptive responses to environmental stress.

These considerations are exemplified by ethanol which selectively and reversibly inhibits the binding of enkephalins (δ agonists) to brain membrane preparations at concentrations which have little effect on opiate alkaloid binding. This suggests that δ receptors are specifically sensitive to the alcohol (Hiller et al., 1981). Acute administration of ethanol increased met-enkephalin and δ-endorphin levels in the pituitary (intermediate and posterior lobes) decreased 70% after chronic treatment with ethanol (Schultz et al., 1980). In a preliminary study Triana et al. (1980) suggested that endorphins may be involved in the epileptogenic effect of ethanol. δ-endorphin enhances ethanol-induced hypothermia and sedation in mice. Interference by ethanol with endorphin levels in the pituitary is suggested by functional abnormalities in the hypothalamo-pituitary-adrenal axis in chronic alcoholics (Merry, 1973) since ACTH and δ-endorphin are located in the same cells and derive from a common precursor molecule. Ethanol increases opioid immunoreactivity in the plasma of man (Naber et al., 1981).

Reports of effects of ethanol on the neuroamines have generally been confined to changes during long term administration or withdrawal.

Like the opiate alkaloids the opioid peptides cause thermoregulatory changes and these are generally reversible by naloxone (see Burks and Rosenfeld, 1979). A possible physiological role for endorphins in stress-induced hyperthermia in rats has been suggested (Bläsig and Herz, 1978) since there is a close relationship between stress and release of δ-endorphin (Guillemin et al., 1977). Indeed, stress-induced release of pituitary δ-endorphin may play a more important role in temperature regulation than in modifying nociception (Millan et al., 1981).

The thermoregulatory responses to opiates in rats have been shown to vary with age (McDougal et al., 1980a). Older (27 month) rats showed greater resistance to the acquisition of tolerance to the effect of morphine on body temperature (McDougal et al., 1981). Age-related differences in the hypothermic effect of ethanol are not seen under normal ambient temperature conditions but senescent animals developed a greater fall in core temperature when subjected to restraint stress at low ($\sim$18 °C) ambient temperature (McDougal et al., 1980b). In young animals food deprivation and mild cold stress prevented the increased heat production following administration of a pyrogen (Kleitman and Satinoff, 1981). Both exposure to cold and intraventricular injection of β-endorphin modify seizure activity in seizure-sensitive gerbils, suggesting that ambient temperature may exert an effect on the peptide release (Bajorek and Lomax, 1982).

Apart from those associated with non-specific aging processes, more defined age-related changes can result in disorders of homeostatic mechanisms. The clearest example is the series of physiological involutions occurring during the menopause (although the time frame is at the lower limit of the onset of the aging process in modern society). The pathophysiological basis of post-menopausal hot flushes has been demonstrated to be the result of a specific thermoregulatory disorder (Tataryn et al., 1980a) and there is considerable evidence that the attacks are triggered by the episodic release of LHRH in the rostral hypothalamic nuclei (Tataryn et al., 1980b). Other disorders of CNS function, such as sleep disorders (Erlik et al., 1981), may also be involved in the low frequency episodic release of LHRH in women (Ropert et al., 1981) and naloxone has been reported to inhibit flushing attacks in patients

(Lightman and Jacobs, 1979). Almost all patients with hot flushes sufficiently severe to cause them to seek treatment report anecdotally that ethanol increases the frequency and severity of the attacks.

To summarize: The cold stress itself may play a more significant role in accidental hypothermia than that attributable to the thermal gradient; stress releases neuropeptides; neuropeptides may modulate the thermoregulatory centers. On another front, diet may modify the peripheral effector systems regulated directly, or indirectly, by the thermoregulatory centers and may be an important determinant of the induced body temperature changes. Finally, age (very young or aged animals) may be another important variable affecting any of the above systems and interactions, not only because of nonspecific aging changes but also due to age-related disorders of neuroendocrine function.

THE EFFECT OF ENVIRONMENTAL STRESS ON ETHANOL-INDUCED ACCIDENTAL HYPOTHERMIA

Normothermic conditions

It has been shown in several species that ethanol causes a dose-related fall in body temperature under thermoneutral ambient conditions ($T_a \sim 20$ °C in the rat), with the regression lines not showing any significant differences between species (Lomax and Bajorek, 1980).

The minimal effective dose was 0.5 g.kg^{-1} i.p. and the ED_{50} was 1.5 g.kg^{-1} i.p. (a "maximum" response of -4 °C was taken since ethanol tends to lead to fatal irreversible hypothermia when the fall exceeds this level).

After an ED_{50} dose in the rat, blood ethanol concentrations at 30 min were 150 ± 0.009 ml.dl^{-1} (Lomax and Bajorek, 1980). These levels correspond to moderate intoxication in man and are similar to those found in necropsy samples from victims of fatal hypothermia after ethanol (Hirvonen, 1976).

Dose response curves relating the fall in core temperature to the dose of ethanol were compared in the same animals tested at a 24 h interval. There was no significant difference between the curves. Nor were the blood ethanol levels at 30 and 60 min different on day 1 and day 2 at the 1.5 g.kg^{-1} dose. Thus, there was no evidence of acute tolerance to the response nor were there any pharmacokinetic changes.

Behavioral studies revealed that rats avoided a heat source during the period the body temperature was falling after injection of ethanol, that is, the animals' behavior was such as to "allow" the fall, indicating that the drug had caused a downward shift in the thermoregulatory set point (Lomax *et al.*, 1980).

At ambient temperatures of 21 °C and 26 °C (at which tail skin blood flow would normally be expected to play a role in regulating core temperature in the rat) there was no change in tail skin temperature during the period in which core temperature was falling after administration of ethanol. Increased cutaneous radiant heat loss does not, therefore, play a significant role in the hypothermic effect of ethanol. Respiratory rate and tidal volume were unaffected by ethanol (Lomax *et al.*, 1981b), so that increased convective or evaporative heat loss do not occur. Oxygen consumption remained steady, at approximately 3 ml.min^{-1}.100 g^{-1} STPD, before and after injection of the alcohol so a decrease in normal metabolic activity does not account for the fall in temperature. It is unclear at the present time which peripheral effector mechanism(s) mediates the heat loss after the set temperature is lowered by the drug.

In summary, these results indicate that in the rat consistent effects of ethanol on body temperature occur under normothermic conditions (i.e. at environmental temperatures close to the thermoneutral zone) and the responses are graded over a dose range that does not cause serious CNS depression or nonspecific toxicity.

The behavioral studies indicate that the fall in core temperature induced by ethanol is due, at least in part, to a lowering of the thermoregulatory set point. However, the metabolism results and measurements reflecting cutaneous heat loss have failed to indicate the route(s) by which the body heat content is reduced.

The effect of environmental temperature

The problem of why certain individuals are prone to develop hypothermia when exposed to low environmental temperatures can be considered from two viewpoints: are there predisposing factors which render them susceptible, or do they fail to respond to the environmental stress as does the general population? Our first approach to these questions was to determine the effect of acclimation to low environmental temperatures on the hypothermic effect of ethanol. Although the physiological changes occurring during cold acclimation have been investigated extensively (see Janský, 1979), there does not appear to be any information concerning the effects of acclimation on the thermoregulatory changes induced by centrally active drugs. Such considerations are, however, clearly relevant to investigations into the etiology and pathogenesis of accidental hypothermia in man.

Groups of rats were maintained at T_a 20 ± 1 °C for 7 days. On the day of the experiment the animals of each group were placed in a constant temperature chamber maintained at 18, 15, 11, 6 or 0 ± 1 °C and the body temperature was monitored continuously. After a 1 h stabilization period, ethanol (1 g.kg^{-1} i.p.) was injected and the mean maximum fall in body temperature over the ensuing 45 min was recorded. As illustrated in Figure 1, the degree of hypothermia proved to be an exponential function of T_a over the range 0–18 °C. The calculated fall in core temperature at T_a 4 °C was almost twice (1.26 : 0.65) that at T_a 18 °C after ethanol (1 g.kg^{-1} i.p.). The ratio of T_b to T_a difference is, however, only 1.05 : 1.0 (on the basis of °Abs) so that it is unlikely that the cooling gradient alone could account for the increased heat loss at T_a 4 °C. This suggests that the environmental thermal stress potentiates the hypothermic effect of ethanol, or conversely, that ethanol interferes with the homeostatic thermoregulatory mechanisms which normally prevent a fall in core temperature during cold exposure. On the basis of proportional control theory of body temperature (see Hardy, 1961), on exposure to a low environmental temperature skin thermosensors signal the thermal stress to the rostral hypothalamic thermoregulatory centers and an upward shift in the set point occurs which results in the activation of heat conservation pathways to protect

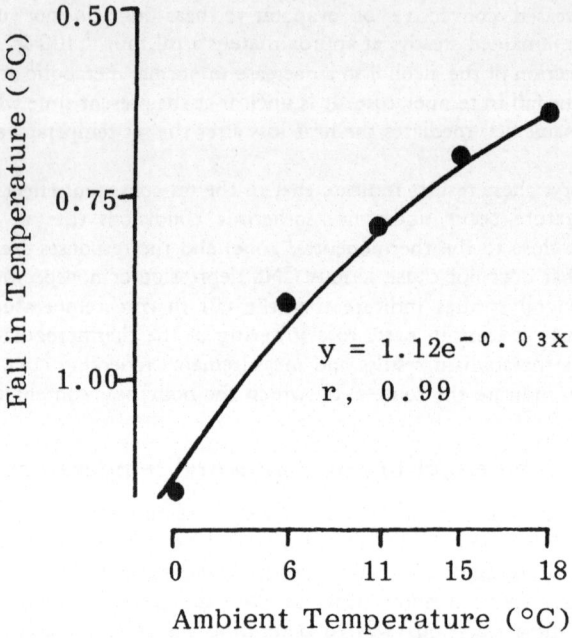

Figure 1   Effect of T_a on the hypothermic response to ethanol (1 g/kg^{-1} i.p.). Each point represents the mean from six rats.

against the potential fall in core temperature. As we have seen above, ethanol lowers the thermoregulatory set point under thermoneutral ambient conditions; the drug may additionally block the normal reflex response to cold exposure so that heat loss is not readjusted (or heat production is not initiated) to combat the environmental thermal stress. During cold acclimation the threshold levels for the several thermoregulatory effector systems become shifted, that is, the set point for these reactions adjusts to the chronic thermal stress (for review see Jansky, 1979), possibly due to an effect of noradrenergic pathways on reference units in the caudal hypothalamus (Zeisberger and Brück, 1971). Thus, such cold acclimation should reduce the hypothermic response to ethanol by limiting the induced lowering of the set point.

The effect of cold acclimation

Dose-response curves for the hypothermic effect of ethanol were determined in separate groups of rats acclimated to T_a 4 °C or 20 °C for 7 days. The maximum fall in body temperature within 45 min of administration of ethanol at T_a 4 ± 1 °C was recorded (Lomax and Lee, 1982). As seen in Figure 2, the fall in temperature was markedly attenuated in the animals acclimated to T_a 4 °C. At the highest dose nine of the unacclimated animals developed irreversible hypothermia and died within 120 min whereas all of the acclimated animals recovered completely. Blood ethanol

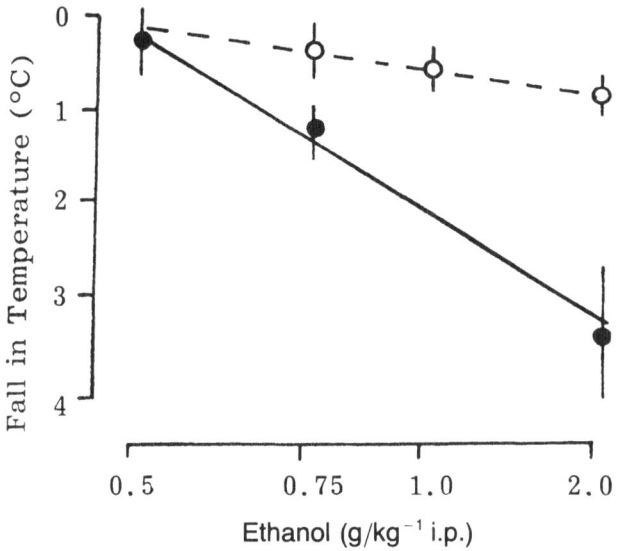

Figure 2 Effect of acclimation to 4 °C (broken line) compared to unacclimated rats (solid line) on the dose response relationships for ethanol-induced hypothermia. Each point represents the mean from ten rats.

concentrations were compared in control and cold acclimated rats 30 and 60 min after injection of ethanol (1 or 2 $g.kg^{-1}$ i.p.) at T_a 4 ± 1 °C. As seen in Table 1, there were no significant differences between the two groups.

These findings are in accord with the hypothesis developed in the previous section. Acclimation for as short a period as 7 days afforded considerable protection from the hypothermic and lethal effects of ethanol administered at a low T_a. Pharmacokinetic or metabolic changes are not implicated in this amelioration in

Table 1
Blood ethanol concentrations[a] in normal and cold acclimated rats at T_a 4 ± 1 °C

Group[b]	Ethanol (1 $g.kg^{-1}$ i.p.)		Ethanol (2 $g.kg^{-1}$ i.p.)	
	30 min	60 min	30 min	60 min
Normal	72 ± 3[c]	56 ± 7[d]	162 ± 11[e]	163 ± 17[f]
Cold acclimated	61 ± 6[c]	46 ± 5[d]	153 ± 20[e]	142 ± 17[f]

[a]$mg.dl^{-1}$ ± SEM. Determined by GLC with propanol as internal standard.
[b]Ten animals in each group.
Not significantly different: [c]$p = 0.1144$; [d]$p = 0.1144$; [e]$p = 0.3553$; [f]$p = 0.1970$.

response since the blood ethanol levels were similar in acclimated and control animals.

Time course of acclimation

In the adult rat, exposure to low environmental temperatures induces non-shivering thermogenesis which reaches its maximum by 20 days with the initial rate of development being greater (see Janský, 1979). Rats were maintained at T_a 4 ± 1 °C or T_a 18 ± 1 °C and the hypothermic effect of ethanol (1 g.kg^{-1} i.p.) was determined on days 1, 3, 6, 8, 16, 20 (Lomax and Lee, 1982). As seen in Figure 3, the hypothermic response gradually decreased during the period of cold exposure and by day 20 was not significantly different from controls (Δ_t: 0.27 °C, control; 0.33 °C, acclimated).

Oxygen consumption was measured on days 1, 3, 6, 8, 16 and 20 in a comparable group of animals housed at T_a 4 ± 1 °C. At T_a 4 °C $\dot{V}_{O_2}$ was significantly greater than at T_a 20 °C (4.08 ± 0.23 compared with 2.06 ± 0.08 ml.min^{-1}.100 g^{-1} STPD). During the early days of cold exposure there was an increase in $\dot{V}_{O_2}$ in cold exposed rats but this had returned to the initial level by day 16 as non-shivering thermogenesis became developed.

From these results it would seem apparent that the process of cold adaptation (which reaches a maximum by 20 days) parallels the attenuation of the hypothermic effect of ethanol. However, there is no associated change in the metabolism of ethanol during cold exposure (Lomax et al., 1981b). Thus, it might be concluded

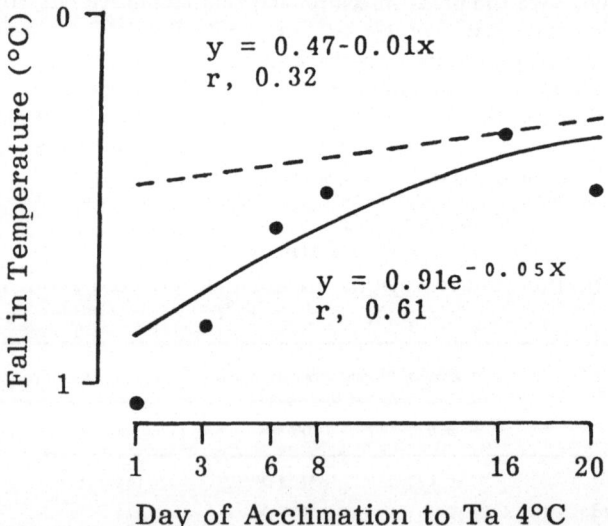

Figure 3 Regression lines for the fall in body temperature induced by ethanol (1 g/kg^{-1} i.p.) (solid line) or vehicle (broken line) during acclimation to T_a 4 °C. Each point represents the mean from six rats injected with ethanol.

that the underlying central mechanisms mediating the metabolic responses to cold exposure are also involved in suppressing the hypothermic response to ethanol. As discussed above, neural peptides, including the opioids, may be released by environmental stress and have been implicated in the ontogenesis of brown adipose tissue hypertrophy. These neuromodulators may constitute the common link in these several adaptive systems.

Diet-induced thermogenesis

Following the initial observations of Rothwell and Stock (1979) there has been considerable interest in the phenomenon of diet induced thermogenesis. Laboratory rats under normal conditions will eat just enough to fulfill their energy requirements. If offered an enriched diet (e.g. including ham, cookies, chocolate, etc.) over the standard chow, the animals will gain weight, but less than that calculated from their food intake. Furthermore, during the period of weight gain their metabolic rates increase — in short, it transpires that the animals are burning the excess calories and producing heat in brown adipose tissue. After 21 days animals on an enriched diet had double the amount of brown fat compared with the controls on a regular diet; the amount of brown fat was similar to that in animals acclimated to 5 °C. Recently, an increase in the respiration of brown fat has been demonstrated after a single meal (Glick et al., 1981). There is speculation that these findings could explain anomalous weight gain in humans. Rats on an enriched diet are resistant to cold induced (4 °C) hypothermia and Heroux (in press) has reported that the ability to maintain body temperature in the cold is related to 5-HT metabolism, or the norepinephrine/5-HT ratios in the brain. Finally, genetically obese (fa/fa) animals are deficient in brown fat and have little cold resistance. Pituitary β-endorphin may play a role in the development of the obese syndrome (Margules et al., 1978). Norepinephrine has been implicated in thermogenesis in normal and in cold acclimated rats (Schönbaum et al., 1963) and cold exposure (4 °C) causes a progressive increase in plasma norepinephrine concentrations (Picotti et al., 1981).

On the basis of these considerations we determined if diet-induced brown fat proliferation modifies the hypothermic effect of ethanol (Lomax et al., 1983). Groups of rats were housed at T_a 18 °C and fed standard laboratory chow or an enriched diet similar to that of Scalfani and Springer (1976) for 21 days. The hypothermic response to ethanol was determined on day 1 and day 21 for each group. On the days of ethanol testing they were placed in a constant temperature chamber at 4 ± 1 °C and allowed to stabilize for 1 h prior to drug administration. $\dot{V}_{O_2}$ for each group was also determined (at T_a 18 ± 1 °C) on days 1 and 21. The results are shown in Table 2. The gain in body weight was greater in the rats on the enriched diet (91.8 ± 1.68 g compared with 75.6 ± 1.78 g; $p = 0.031$); this weight gain was less than that reported by Rothwell and Stock (1979) (131 ± 8 g) for the same strain of rats on a similar diet for the same period but our animals had lower initial weights (250 ± 5 g compared with 315 ± 4 g). The difference in $\dot{V}_{O_2}$ between the groups is less than that in previous studies (Rothwell and Stock, 1979) (12% compared with 28%) but the differences in mean body weights and T_a (18 °C in our experiments versus 29 °C) could account for the discrepancy. The energy

Table 2
The effect of an enriched diet for 21 days on the hypothermic effect of ethanol (1.5 g.kg^{-1} i.p.), $\dot{V}_{O_2}$ and body weight in rats

Diet	Temperature change ($^\circ$C ± SEM)		$\dot{V}_{O_2}$ (ml.min^{-1}.100 g^{-1})		Weight (g ±SEM)	
	Day 1	Day 21	Day 1	Day 21	Day 1	Day 21
Control	2.47 ± 0.24	—	2.46 ± 0.21	2.28 ± 0.21	269 ± 6	345 ± 12
Enriched[a]	2.11 ± 0.36[b]	1.58 ± 0.28[b]	2.40 ± 0.12	2.56 ± 0.19	250 ± 5	342 ± 5

[a]Six animals in each group.
[b]$p = 0.13$.

intake on the enriched diet was approximately double that on the laboratory diet. Thus, the weight gain is less than that expected from the increased caloric intake, so that diet-induced thermogenesis had occurred. It is clear from these results that the hypothermic response to ethanol (Table 2) was certainly not prevented by diet-induced hypertrophy of brown adipose tissue.

CONCLUSIONS

One would conclude from these last results that the attenuation of ethanol hypothermia is not a direct result of the hypertrophy of brown adipose tissue or the onset of non-shivering thermogenesis *per se*. Indeed, it is most likely that the phenomenon is the direct effect of exposure to environmental thermal stress and results not from the adaptation to the stress (i.e. cold acclimation) but from the central nervous system responses to the stress. Certainly, none of the results from these studies would indicate that peripheral metabolic activity can be impugned as a significant factor in the reduced hypothermic effect. Nor does the pattern and time course of the decreased response seem to be analogous to the classical type of pharmacological tolerance mediated by changes in specific receptor function.

The tentative hypothesis which we would propose is that the environmental stress activates homeostatic responses which result in changes in CNS neuromodulator activity. This may involve one or more of the centrally active peptides, e.g. ACTH, endorphins, TRH, LHRH or possibly target organ hormones (e.g. we have recent data which demonstrate that the hypothermic response to narcotic analgesics is abolished by hyperthyroidism). Thermoregulatory effects of these neuropeptides (and others) have been described in several species (for review, see the several papers in the symposium: The newer putative central neurotransmitters: Role in thermoregulation, *Fed. Proc.*, 40, 2735–2768, 1981). Thus, it is envisaged that the environmental stress results in release of some neuromodulators (as a facet of the defence response) which also modify the activity of the thermoregulatory centers such that the set point is not now lowered by the presence of ethanol. Using appropriate methodologies this concept is eminently amenable to testing.

ACKNOWLEDGEMENTS

The research described has been supported by ONR contract N00014-75-C-0506 and by USPHS grant AAO 3513-01-05.

REFERENCES

Bajorek, J. G. and Lomax, P. 1982. Modulation of spontaneous seizures in the Mongolian gerbil: effects of β-endorphin, *Peptides*, 3, 83.

Bläsig, J. and Herz, A. 1978. Evidence for a role of endorphins in emotional hyperthermia, *Naunyn-Schmied. Arch. Pharmakol. Suppl.*, 302, R61.

Burks, T. F. and Rosenfeld, G. C. 1979. Narcotic analgesics, in *Body Temperature*. P. Lomax and E. Schönbaum (eds), Dekker, New York, p. 614.

Cooper, K. E., Brody, H., Lomax, P., Lyman, C. P., Paton, B. C., Rennie, D. W. and Stitt, J. T. 1980. *Accidental hypothermia*, NIH NIA Workshop, December 11–12, 1979, Vol. I & II.

Elliott, J. 1980. Blame it all on brown fat now, *Medical News*, 243, 1983.

Erlik, Y., Tataryn, I. V., Meldrum, D. R., Lomax, P., Bajorek, J. G. and Judd, H. L. 1981. Association of waking episodes with menopausal hot flushes, *J. Amer. Med. Assoc.*, 245, 1741.

Glick, Z., Teague, R. J. and Bray, G. A. 1981. Brown adipose tissue: thermic response increased by a single low protein, high carbohydrate meal, *Science* 213, 1125.

Guillemin, R., Vargo, T., Rossier, J., Mimick, S., Ling, N., Rivier, C., Vale W. and Bloom, F. 1977. β-endorphin and adrenocorticotropin are secreted concomitantly by the pituitary gland, *Science*, 197, 1367.

Haight, J. S. J., and Keatinge, W. R. 1973. Failure of thermoregulation in the cold during hypoglycemia induced by exercise and ethanol, *J. Physiol. (Lond.)*, 229, 87.

Hardy, J. D. 1961. Physiology of temperature regulation, *Physiol. Rev.*, 41, 521.

Heroux, in press.

Hiller, J. M., Angel, L. M. and Simon, E. J. 1981. Multiple opiate receptors: alcohol selectively inhibits binding to delta receptors, *Science*, 214, 468.

Hirvonen, J. 1976. Necropsy findings in fatal hypothermia cases, *Forensic Sci.*, 8, 155.

Hirvonen, J. 1979. Accidental hypothermia, in ·*Body Temperature*, P. Lomax and E. Schönbaum (eds), Dekker, New York, p. 561.

Janský, L. 1979. Heat production, in *Body Temperature*, P. Lomax and E. Schönbaum (eds), Dekker, New York, p. 89.

Kleitman, N. and Satinoff, E. 1981. Behavioral responses to pyrogen in cold-stressed and starved newborn rabbits, *Am. J. Physiol.*, 241, R167.

Ledingham, I.McA. 1980. Treatment of accidental hypothermia: a prospective clinical study, *Brit. Med. J.*, i, 1102.

Lightman, S. L. and Jacobs, H. S. 1979. Naloxone: non-steroidal; treatment for postmenopausal flushing? *Lancet*, II, 1071.

Lomax, P. and Bajorek, J. G. 1980. Comparative thermoregulatory effects of ethanol in rats, mice and gerbils, *Proc. West. Pharmacol. Soc.*, 23, 219.

Lomax, P. and Lee, R. J. 1982. Cold acclimation and resistance to ethanol-induced hypothermia, *Europ. J. Pharmacol.*, 84, 87.

Lomax P., Bajorek, J. G., Chesarek, W. A. and Chaffee, R. R. J. 1980. Ethanol induced hypothermia in the rat, *Pharmacol.*, 21, 288.

Lomax, P., Bajorek, J. G., Bajorek, T-A. and Chaffee, R. R. J. 1981a. Cold acclimation and the hypothermic effect of ethanol in the rat, *Proc. West. Pharmacol. Soc.*, 24, 33.

Lomax, P., Bajorek, J. G., Bajorek, T-A. and Chaffee, R. R. J. 1981b. Thermoregulatory mechanisms and ethanol hypothermia, *Europ. J. Pharmacol.*, 71, 483.

Lomax, P., Bajorek, J. G., Bajorek T-A. and Lee, R. J. 1983. The development of resistance to ethanol induced hypothermia during cold acclimation, in *Environment Drugs and Thermoregulation*, P. Lomax and E. Schönbaum (eds), Karger, Basel, p. 176.

Margules, D. L., Boisset, B., Lewis, M. J., Shibuya, H. and Pert, C. B. 1978. β-endorphin is associated with overeating in genetically obese mice (*ob/ob*) and rats (*fa/fa*), *Science*, 202, 988.

McDougal, J. N., Marques, P. R. and Burks, T. F. 1980a. Thermic responses to morphine in old and young rats, *Proc. West. Pharmacol. Soc.*, 23, 235.

McDougal, J. N., Marques, P. R. and Burks, T. F. 1980b. Age-related changes in body temperature responses to morphine in rats, *Life Sci.*, 27, 2679.

McDougal, J. N., Marques, P. R. and Burks, R. F. 1981. Reduced tolerance to morphine thermoregulatory effects in senescent rats, *Life Sci.*, 28, 137.

Merry, J. 1973. Hypothalamic-pituitary adrenal function in chronic alcoholics, in *Advances in Experimental Medicine and Biology*, vol. 35, Gross (ed.), Plenum Press, New York, p. 167.

Millan, M. J., Przewlocki, R., Jerlicz, M., Gramsch, C. R., Höllt, V. and Herz, A. 1981. Stress induced release of brain and pituitary β-endorphin: major role of endorphins in generation of hyperthermia, not analgesia, *Brain Res.*, 208, 325.

Naber, D., Soble, M. G. and Pickar, D. 1981. Ethanol increases opioid activity in plasma of normal volunteers, *Pharmacopsychiatria*, 14, 155.

Picotti, G. B., Carruba, M. O., Ravazzani, C., Cessura, A. M., Galva, M. D. and da Prada, M. 1981. Plasma catecholamines in rats exposed to cold: effects of ganglionic and adrenoreceptor blockade, *Europ. J. Pharmacol.*, 69, 321.

Ropert, J. F., Quigley, M. E. and Yen, S. S. C. 1981. Endogenous opiates modulate pulsatile luteinizing hormone release in humans, *J. Clin. Endocrinol. Metab.*, 52, 583.

Rothwell, N. J. and Stock, M. J. 1979. A role for brown adipose tissue in diet-induced thermogenesis, *Nature (Lond.)*, 281, 31.

Schönbaum, E., Sellers, E. A. and Johnson, G. E. 1963. Noradrenaline and survival of rats in a cold environment, *Can. J. Biochem. Physiol.*, 41, 975.

Schultz, R., Wüster, M., Duba, T. and Herz, A. 1980. Acute and chronic ethanol treatment changes endorphin levels in brain and pituitary, *Psychopharmacol.*, 68, 221.

Sclafani, A. and Springer, D. 1976. Dietary obesity in adult rats: similarities to hypothalamic and human obesity syndromes, *Physiol. Behav.*, 17, 461.

Tataryn, I. V., Lomax, P., Bajorek, J. G., Chesarek, W., Meldrum, D. R. and Judd, H. L. 1980a. Postmenopausal hot flushes: a disorder of thermoregulation, *Maturitas*, 2, 101.

Tataryn, I. V., Meldrum, D. R., Frumar, A. M., Lu, K. H., Judd, H. L., Bajorek, J. G., Chesarek, W. and Lomax, P. 1980b. The hormonal and thermoregulatory changes in postmenopausal hot flushes, in *Thermoregulatory Mechanisms*, B. Cox, P. Lomax, A. S. Milton and E. Schönbaum (eds), Karger, Basel, p. 202.

Triana, E., Frances, R. J. and Stokes, P. E. 1980. The relationship between endorphin and alcohol-induced subcortical activity, *Am. J. Psychiatry*, 137, 491.

Zeisberger, E. and Brück, K. 1971. Central effect of noradrenaline on the control of body temperature in the guinea pig, *Pflügers Arch. Ges. Physiol.*, 322, 151.

4

Cold-induced Changes

L. Vanggaard, Surg. Capt., Chief of Defence Royal Danish Navy, Vedbæk, Denmark

Normal thermoregulatory responses might be offset in divers. It is argued that maintenance of a diver's body temperature should be viewed from an engineering rather than a physiological viewpoint. The special nature of the temperature response in hands and feet should be given more attention. Much more effective heat exchange can be accomplished by supplying heat to a diver's hands and feet first and the trunk second.

The thermoregulation of a diver is different from that of a man in air. The diver cannot get rid of heat by sweating, which is the most efficient mechanism for moment to moment thermal regulation of the man in air. The diver is confined to an environment characterized by a close to constant temperature of the skin. When diving in cold water — and the mean temperature of the oceans is around 4 °C — the diver is facing a constant heat drain. In shallow diving, he will lose heat through the skin and to some extent by warming and humidifying the expired air; at great depths the heat loss from the lungs will surpass that across the skin and present the main thermal danger, thus active warming of the respiratory gases and auxiliary heating of the suit are required. The sudden collapse of divers during deep diving when respiratory gas heating is suddenly withdrawn seems more to be an effect of sudden cooling of the structures in the chest, lungs and heart, than a result of a sudden general lowering of body temperature, hypothermia.

Thermoregulation is normally considered as the physiological means of keeping the body temperature at a high and constant level around 37–38 °C. The body thermostat is said to be "set" at that temperature. This "set-point" temperature is believed to govern the thermoregulatory responses: vasoconstriction and, later, shivering when the body is threatened by a fall in temperature, and vasodilation and sweating when the body temperature tends to rise above the "set-point". In the thermoregulatory centre located in the brain stem in the hypothalamus, the thermosensory impulses from the skin, from internal structures and from the hypothalamus itself are weighted, the resulting "body temperature" is compared with the "set-point" temperature and the necessary correcting thermal responses are made. These thermoregulatory responses might be simplified into those elicited from the skin, where a falling skin temperature primarily leads to an increase in metabolic heat production, that is, shivering, and those resulting from a fall in deep temperature, leading to vasoconstriction in the skin and thus increasing the insulation of the body.

In the diver, however, these normal thermoregulatory responses might be offset. For example, in deep diving the heating of the expired gases might cool the body core rapidly, resulting in hypothermia even though the skin temperatures are high due to a heated suit. In this situation the normal thermoregulatory responses of the body will not be activated, and the diver might be unaware of the danger he is in.

Arctic diving presents special problems but these problems are more of a technical character, as they involve maintaining the auxiliary equipment at the surface in a functioning state. The diver in the high arctic waters of Greenland is not exposed to a thermal environment different from that of more southern locations. Even in the Arctic, water cannot defy the laws of nature and cannot be cooler than freezing point. Thus the thermal problems of divers are very similar all over the world. The basic thermal problems are those also encountered in shallow diving, that is, maintaining optimal function in a cold watery environment. In deep diving (below 100 m) the problems encountered in shallow diving are augmented by the extra thermal strain due to heat exchange between the diver and the respiratory gases.

If the commonly recognized aim of the thermal balance of the diver were shifted from preserving the constant body temperature and were seen as a mechanism that allows the diver to perform optimally in his environment, the basic concept of thermoregulation would be changed. It is an engineering rather than a physiological view point to regard thermoregulation as a means of maintaining deep temperature. A man's ability to perform in a work situation is, of course, related to deep body temperature and its maintenance, and it is equally important to maintain the temperature of his working muscles and nerves (i.e. the functioning of the extremities) during cold stress. And here the importance of thermoregulatory changes in skin circulation comes in and must be accounted for in terms of a concept of thermoregulation for the diver.

A working man is dependent upon a very fine and delicate adjustment of sensory and motor functions. This is especially important for the hands. The hand is one of nature's finest instruments. Distally, in the fingers, and in the hand are placed the structures that ensure a fine and very exact control. Here the muscles that govern the small adjustments in motor function are located. The stronger and more coarse muscles are placed along the forearm, and thus do not interfere with the function of the hand. This instrument and its proper and optimal function is not directly dependent upon the deep temperature of the body, but directly upon the local temperature. And the regulation of the local temperature of the hands (and of the feet) is due to the influence of the cutaneous circulation in the extremities.

CUTANEOUS CIRCULATION

The blood circulation in the skin serves different needs of the body. The skin itself must have a *nutritional blood supply*. The vascular pathways for the blood are the normal peripheral systems of the circulation, the arteries, branching out into arterioles and capillaries and the venous system, venules leading into veins of increasing size, returning the blood to the heart.

In the skin the blood flow is also supposed to have an influence upon the regulation of the heat exchange between the body and its surroundings. Thus vasoconstriction is mainly regarded as a means of increasing the insulation of the skin, thereby diminishing heat loss, whereas vasodilation, in which the blood flow is increased, augments heat loss from the body.

This simple and physically attractive explanation has long had its place in the textbooks on man's thermoregulation, but in reality it is a speculative concept, which today needs further questioning. It implies that, when a man is exposed to cold, the skin surface temperatures should fall rapidly and reach a final steady state value which is not just proportional to the difference in environmental temperature before and after cold exposure. But experimental evidence shows that the skin temperature of the body (i.e. over parts of the body where there is a subcutaneous layer of fat) falls only to that temperature which would be the resultant if the body core were covered with an inactive unchangeable insulation. There is no evidence that the skin blood circulation to any large degree changes the insulation of the subcutaneous layers of the body. It has been supposed that the blood in the skin is cooled to environmental temperatures or at least to skin surface temperatures, thus cooling the blood, and later participates in the regulatory cooling and lowering of the deep body temperature. But this does not seem to be the case. The blood in the skin is cooled on its passage to the surface, while it flows through the body shell. It might reach temperatures close to skin surface temperatures, but on its passage back towards the body core, the physical factors that determine the cooling now work to the opposite effect. When the blood reaches the core it has gained a higher temperature. In the skin there is evidence of an effective counter-current heat exchange between blood flowing to and returning from the surface. The temperature of the blood thus tends to follow the temperature gradient in the superficial part of the skin and it rests to be determined to what extent the blood flow to the skin participates in the regulation of the skin surface temperatures over the body. In this argument the blood flow to the extremities has not been dealt with.

For a warm man, it has been postulated that the skin blood flow is correlated to the need to transport to the surface exactly that amount of heat that was lost by evaporation of sweat. Anyone who has undressed after strenuous exercise will know that this is not true. After sweating, the skin is very cold, which means that the cooling of the skin by evaporation of sweat is not compensated by a higher level of blood circulation in the skin.

The concept of body core and body shell distribution of temperatures in a cold man is attractive. In the original concept, the formation of the body shell was a regulatory answer to a situation where the body core was threatened by a fall in temperature. The formation of an insulating body shell was regarded as an active defence mechanism, preserving the deep body temperature. Nobody would argue against a fall in surface temperatures upon exposure to cold — that is a physical necessity. But whether the formation of a body shell is due to a specific physiological reaction is questionable. This question is of considerable importance, as it should be remembered that the outer 2 cm of the body, due to its geometry, constitute nearly

50% of the total body mass. The heat preserved or lost in this volume greatly influences the total heat balance of the body. In emergencies such as hypothermia, where the deep body temperature is lowered, the heat exchange from the body core and the body shell might give rise to the feared afterdrop in temperature seen upon rewarming of the hypothermic victim.

I therefore postulate that the traditional concept of the regulated skin blood flow is not an effective means of thermoregulation in man, and that the surface temperatures of the skin over the body are due more to the conductive transfer of heat through the subcutaneous tissues to the surface than to the convective transport of heat with the blood.

But still, man does regulate some of the temperatures of the skin. When skin temperatures are measured at different points of the skin and averaged, there seems to be evidence that the temperature is actively regulated. Upon cooling, the average skin temperatures fall, and the overall calculations of the body heat exchange with the environment indicate an active regulation. The reason for this becomes obvious when this average skin temperature is broken down into its constituent parts. The temperatures of the extremities, especially hands and feet, do fall, and might after prolonged cooling reach very low temperatures. It is astonishing that, although skin temperatures have been recorded by numerous researchers, most of them have used the different temperatures obtained only to calculate the average skin temperature, while it is quite clear that something special is going on in the hands and feet.

LOCAL HEAT REGULATION IN HANDS AND FEET

When man is exposed to *cooling*, for instance when he has been transferred from a warm room to a cooler one or immersed in water, the temperatures in his extremities will soon begin to fall. It has been shown that they follow a cooling curve. If a blood pressure cuff is placed over one arm, and is inflated to a pressure above arterial pressure at the time of cold exposure, the temperature in the hand where all blood supply, and thus also all convective heat transport, is abolished falls at the same rate as that in the other hand, although this has an unobstructed blood supply.

This shows that the changes in blood flow to the extremities during cold exposure are diminished to an extent where it does not influence the local temperatures. Even in a thermoneutral situation, the same effect can be seen. Here the temperatures of the fingers vary with time. The excursions are small, but the rate of change even during these small temperature fluctuations follow the same course as found when all blood supply to the hand is obstructed. These changes show the picture of an "on–off" regulation.

It thus seems as if thermoregulatory changes in the skin of man are restricted to the extremities, whereas changes in the skin of the trunk of the body are of a passive nature.

When man is *warm*, or even when he is in thermoregulatory balance, his extremities are always warm. It is even so that the warmest parts are the fingertips (and the toes). This is astonishing, when the geometry of the extremities is considered. The fingers and toes are most distal. And, in hands as well as in feet, the surface to volume ratio increases outwards, along the extremity, it was to be expected that the most distal parts were the coldest, but this is not so. Everyone can try placing palm and fingers over the opposite forearm and feel how the hand is warmer than the more proximal forearm. But upon cooling, when exposed to a cold environment, when the thermal balance is upset and the deep body temperature is threatened by a fall, this situation suddenly changes. Now the extremities and especially the distal parts will cool down. Eventually they will reach the temperature of the environment. In an experiment where subjects were exposed to a sudden cooling, the temperatures of the distal parts of the extremities fell to those of the environment within a few hours.

These changes in local temperatures in the extremities profoundly change the heat exchange from the body to the surroundings. The extremities account for around 50% of the body surface and close to 30% of the total body volume. When cold, a man's active heat dissipating surface is reduced by close to half the area of a warm man, and the body volume to be defended is reduced by 30%. This could be visualized as a "physiological amputation" of the extremities in order to keep the deep body temperatures normal, especially those of the heart and brain. During cooling, heat of course is lost from the extremities, but this does not influence the heat content of the regulated central part (the body core) as this heat loss takes place from a volume that, physiologically speaking is "outside" the body. On rewarming, there might be problems, if blood circulation is suddenly restored without applying heating to the extremities. This is especially important in the treatment of the hypothermic victim. During slow rewarming the extremities should be kept cold, not massaged or treated in a way that will stimulate circulation. In rapid rewarming, where the victim is immersed in hot (42–44 °C) water, the extremities should be immersed together with the body, as the warm water will heat the extremities and thus the blood circulating in them. Here the extremities might function as effective heat exchangers; this point will be discussed later.

THE SPECIAL VASCULAR ARRANGEMENT IN HANDS AND FEET

In hands and feet there is a nutritional blood supply to the different structures, and this nutritional blood flow through arterioles, capillaries and venules corresponds to that found in other parts of the body. But in hands and feet, the nutritional blood supply only accounts for around 10% of the maximal blood flow found in a warm man. The greater part of the blood to the hands flows through the *arteriovenous-anastomoses*, which are abundant in the distal parts of fingers and toes but lacking in other parts of the skin. These anastomoses are rather large blood vessels; they have an inner diameter when open of about 50–70 μm that is, they are just invisible to the naked eye. In comparison, the nutritional capillaries of the skin have an inner diameter of 5–10 μm. The arteriovenous-anastomoses shunt blood directly from the arteries to the veins. They have a thick muscular wall, which ensures an "on–off" function. They are either open or closed, and they are centrally regulated, presumably through the sympathetic nervous system. They function completely synchronously in fingers and toes, thus causing the mentioned drop in local temperatures in hands and feet in a cold man; in a warm man they are open, thus causing the very warm fingertips mentioned earlier.

The arteriovenous-anastomoses lead fully oxygenated blood from the arteries to the veins. In a warm man the blood drawn from a superficial vein in the forearm is bright red and has an oxygen saturation close to 100%. The anastomoses are placed on the volar side of the hand (and the foot). They drain to the superficial veins of the dorsum of the fingers and hand and further proximally along the superficial venous rete of the forearm. In the forearm there are few communicants across the fascia to the deeper structures. This arrangement of the venous return close to the surface gives very good possibilities for heat exchange from a warm man to the environment, a heat exchange that might be further increased during thermal sweating, which is pronounced at the dorsum of the hand and from the forearm skin.

The arteriovenous-anastomoses in hands and feet thus have an important role in the heat dissipating mechanisms of the body in a warm man, and by their closure in cold man increase his chances of survival when threatened by hypothermia, by performing a "physiological amputation", diminishing the heat losing surface and the volume of the body.

THE ROLE OF ARTERIOVENOUS-ANASTOMOSES IN EXTREMITY FUNCTION

As mentioned, the arteriovenous-anastomoses in hands and feet play a role in thermoregulation, but their main role seem to be that of ensuring optimal functioning of hands and feet. When open the anastomoses convey warm (36 °C) blood to those parts of the body that otherwise would be threatened by their large surface to volume ratio leading to low local temperatures. In this way optimal sensory and tactile function is ensured. The returning warm venous blood in the superficial venous rete forms a physiological glove around the muscles, nerves, joints and other structures of the forearm, thus ensuring a high and optimal working temperature. Everyone knows how the finer functioning of the hand disappears when the hand is cold. The function of the arteriovenous-anastomoses determines the local temperature of the hand and forearm. When a man gets cold, the anastomoses close, and the local temperatures fall to values close to that of the environment. In water this cooling will be rapid due to the high heat capacity of water. Warm hands can be taken as a sign of thermoneutrality of the body. Temperature measurements, at air temperatures of −40 °C, in the Greenlandic sledge dog, musk-ox and polar bear show the same feature. These animals all have warm feet, but the arctic animals are protected from the cold ground by a heavy insulation in the parts directly in contact with it. Of course, the arctic animals have the same need as man to keep their extremities at an optimal and high temperature.

The way the arteriovenous-anastomoses function explains why it is impossible ever to construct gloves or boots that will keep a cold man's hands and feet warm. Good insulation will increase the time before dangerously low temperatures are encountered. Good insulation might postpone the moment where the deep body temperature begins to drop. But when cold, the man's hands and feet will fall in temperature, and cold injury might occur, if ambient temperature is low enough. In this connection it is important to know that sensory as well as motor nerve activity is abolished at local temperatures around 7–9 °C. Thus the last thing a man feels before local cold injury occurs is that he feels nothing. In the Arctic a sleeping sensation should always lead to vigorous attempts at rewarming.

The concept of body core and body shell formation has been mentioned earlier. It was stated that the concept was not valid in its classical form. Another concept, based upon the function of the anastomoses would be more appropriate: the body shell formed during cold stress is that body volume whose local temperature is dependent upon the function of the arteriovenous-anastomoses.

The anastomoses are governed by the central thermal demands, but they are also influenced by the local temperature. Even in a warm man, the anastomoses will close if the fingers are suddenly exposed to a cold temperature, as, for instance, being immersed into cold (less than 15 °C) water. In a cold man, even in a hypothermic man, the anastomoses will open, if the hands are exposed to a warm environment. This gives the rationale for having the arms in the hot water during rapid rewarming in a warm water bath. But it also points to the dangers that might arise if the anastomoses open by accident, due for instance to rough handling or local heating without real capacity for effective heating — this will lead to an afterdrop in deep body temperature, as the relatively warm blood of the body core will be cooled in the extremity, and the cold blood will return to the body, leading to a further fall in deep body temperature.

SKIN CIRCULATION AND DIVING IN COLD WATER

The diver in cold water will sooner or later reach a thermal situation where he starts cooling. This leads to closing of the arteriovenous-anastomoses and subsequent cooling of hands and feet. If he is wearing insulating gloves this cooling might be slowed down sufficiently for him to perform the job he was sent out to do, but, if no auxiliary heating of hands and forearms are supplied, physical impairment will follow. The situation might be aggravated by the fact that the diver very often cannot feel this cooling of the hands. Cold water is uniformly cold and thus thermal sensations will not be evoked, as they are usually triggered by rapid changes in local skin temperature.

Auxiliary heat to the diver should be supplied to those areas of the body where a convective transport of heat by blood from the skin surface to the body core might be established. This means that the currently used heated suits, those that supply heat (warm water) to the trunk and only to a minor degree to the extremities, should be replaced by a warming system that works by heating the extremities first and the trunk second. With such a "reverse" heating, the temperatures of the muscles and nerves of the extremities could be kept at a reasonably high temperature, and optimal function thus ensured.

It is a question whether one should attempt to heat the diver via the skin. The heated suit is more to be regarded as an insulated suit, where the warm water creates a thermal barrier to heat transport from skin to the surrounding water. Actual heating via the skin over the trunk of the body leads to a time lag of around 15–20 minutes in regulation. In the extremities a very effective heat exchange via the superficial veins of the forearms could be established, but the demand is that the fingers then should be warmed to around 38–40 °C to bring about the necessary blood flow through the arteriovenous-anastomoses. The capacity for heat uptake from the forearm with open anastomoses and a water temperature around the arm of 44 °C is 1 kcal/min, which is close to basal metabolic rate. In the Danish Navy cold-exposed victims are treated by immersing only the hands and feet in hot water. Even in the hypothermic person local heat will open the anastomoses, and thus create an avenue for heat transport to the body core. After treatment the hands and forearms should be heavily insulated, as the forced vasodilation might lead to a heavy heat loss after rewarming has been performed.

There are other skin areas where active heating could be accomplished, and we need to concentrate more research upon the heat exchange properties of the different skin areas. It is important that future diving suits are constructed in such a way that they support the human body's normal thermal reactions. Active heating seems to be one way. But if a non-compressible material could be found, resembling the blubber of the arctic marine animals, this might be sufficient for keeping the diver warm, perhaps with a smaller heat input to the hands and arms to support their function.

In deep diving it should be recognized that the heat loss to the respiratory gases is the main problem, combined with an insulating non-compressible diving suit.

CONCLUSION

In the development of thermal support to the diver, it is important to change the normally accepted concept of human thermoregulation. Man is not just 70–80 kg of tissue kept at a temperature of 37 °C. Man's normal and natural responses to his thermal environment should be recognized and supported if the diver is to benefit. It has been my intention to show that there might be new ways to develop our concept of human thermoregulation.

5

The Diving Reflex in Free-Diving Birds

P. J. Butler, Head of Department of Zoology and Comparative Physiology,
University of Birmingham, UK

The works on involuntary submersion of various species or birds are reviewed.
Experiments on free-diving birds are described in which the oxygen-conserving
response is invoked only when the birds are caught unawares in a life-threatening
situation or when they have to swim long distances under water. Different animals may
show different physiological adjustments to diving, and one animal may show different
adjustments at different times of year.

Some of the earliest work on the physiology of diving was performed on birds and the
results set the stage for the development of the hypothesis which has been applied to
all air-breathing vertebrates that dive. In 1870, Paul Bert published the results of
his famous and oft-quoted experiment of placing the head of a domestic duck
water and feeling the pulsations of the heart through the breast (Bert, 1870). There
was a gradual and sustained reduction in heart rate until the head was surfaced and
the animal began ventilating again. Almost 30 years later, another French
physiologist, Claude Richet (1899), argued that because a duck weighing 1.5 kg has
a total oxygen store of 90 ml, normally consumes oxygen at a rate of 30 ml min^{-1}
and yet can survive head submersion for 20 min, there must be a decrease in the
rate of oxygen consumption during diving (quoted by Scholander, 1940). The
relationship between Bert's observation and Richet's conclusion was not apparent
until 1940. Scholander demonstrated in ducks, penguins and seals that not only is
there a reduction in oxygen consumption and heart rate during involuntary head
submersion, but there is also an increase in blood lactate, which rises even further
when the animal surfaces and begins to ventilate (Fig. 1.).

This work, as much as anything, supported the hypothesis of Irving (1934, 1939)
that diving homeotherms survive breath-hold asphyxia by reducing the supply of
oxygen to all those tissues and organs that can withstand oxygen lack in the short
term. They metabolize anaerobically, producing lactic acid, while the oxygen-
dependent tissues, the CNS and heart, have their oxygen supply maintained.
These regional changes in metabolism are the result of a redistribution of blood.
Associated with the reduction in heart rate (bradycardia) is vasoconstriction in

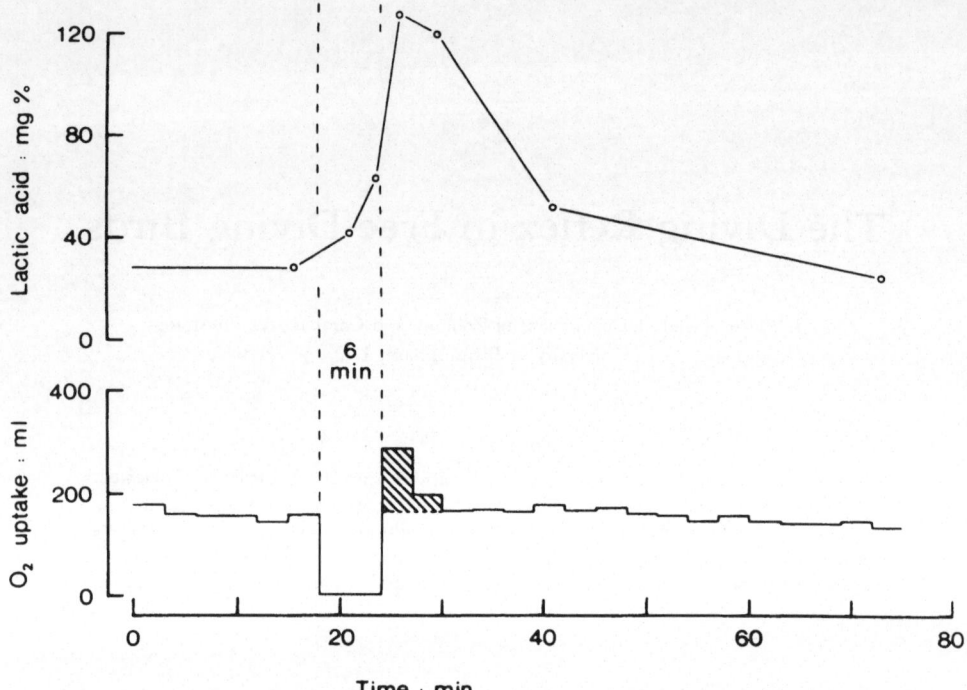

Figure 1 Oxygen uptake, ml $(3 \text{ min})^{-1}$, and lactic acid concentration in the blood of a domestic duck, before, during and after submersion of the head for 6 min. The extra oxygen uptake upon surfacing, above the pre-submersion level (cross-hatched), is thought to represent the amount used during the period of submersion (redrawn from Scholander, 1940).

virtually all vascular beds except the brain, heart and adrenal glands (Johansen, 1964; Butler and Jones, 1971; Jones *et al.*, 1979), but central arterial blood pressure is maintained (Fig. 2). As a result of such a lack of perfusion, the functioning of the kidney ceases in the duck (Sykes, 1966) but more importantly it is the regional hypoperfusion which reduces the supply of oxygen causing the anaerobiosis. There is a considerable reduction in total cardiac output, which approximately matches the increase in peripheral resistance (thus maintaining arterial pressure) and which is almost directly proportional to the decline in heart rate (i.e. stroke volume changes little or decreases). This diving bradycardia is the typical element of the cardiovascular, and hence of the metabolic, adjustments to involuntary submersion and is often taken as an indicator of the remainder of the response taking place (Påsche and Krog, 1980). The total response to involuntary submersion has been termed the "classical" response to submersion asphyxia (Butler, 1982).

The efficacy of the cardiovascular adjustments has been demonstrated by pharmacologically reducing the level of peripheral vasoconstriction or abolishing the bradycardia. In each case there is a more rapid decline in the partial pressure of

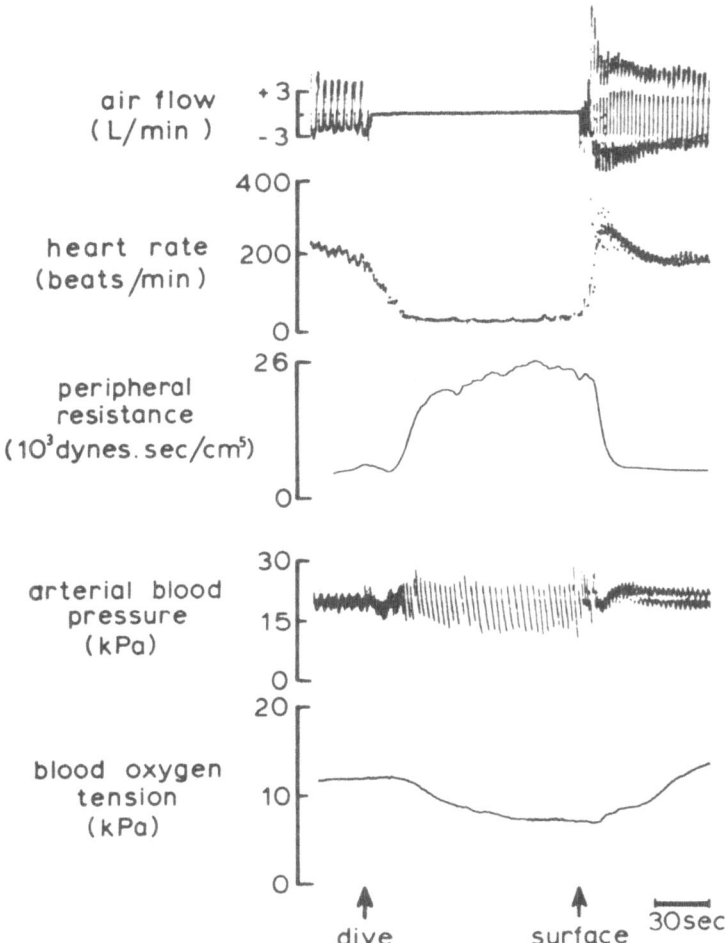

Figure 2 Cardiovascular responses to involuntary submersion of the head of a domestic duck (*Anas platyrhynchos*). "Peripheral resistance" refers to the resistance to blood flow through the sciatic vascular bed of one leg (Butler and Jones, 1982).

oxygen (pO_2) in arterial blood during submersion (Butler and Jones, 1971) and during the latter condition there is a more rapid increase in cerebral NADH (Bryan and Jones, 1980a). The brain of the duck is no more able to tolerate hypoxia than that of the chicken, so the duck's ability to withstand longer periods of submersion than the chicken is the result of its more effective cardiovascular adjustments (Bryan and Jones, 1980b). The pharmacological agents used in the above studies indicate that the vasoconstriction is mediated by α-adrenergic receptors whereas the bradycardia is entirely the result of an increase in cardiac vagal activity (Butler and Jones, 1971).

Denervation or perfusion of the carotid bodies with hyperoxic blood prevents much of the bradycardia in response to involuntary submersion in mallard ducks, *Anas platyrhynchos*, and their domesticated varieties (Hollenberg and Uvnas, 1963; Jones and Purves, 1970; Butler and Woakes, 1982a). In fact, stimulation of the peripheral chemoreceptors accounts for 85% of the bradycardia (providing initial heart rate is <190 beats min^{-1}) and 67% of the increased vascular resistance in domesticated ducks, whereas the central chemoreceptors, sensitive to pCO_2 in the blood, contribute about 30% of the total change in vascular resistance (Jones *et al.*, 1983). Thus, in mallard, and its domesticated varieties, peripheral and central chemoreceptors are predominant in producing the cardiovascular responses to involuntary submersion. That is why these responses are relatively slow in their onset (Fig. 2); they depend on the progressive accumulation of CO_2 and depletion of O_2. It was, in fact, this relative slowness of the oxygen-conserving response, taking anything from 20 to 40 s, that prompted me to look at naturally diving birds. With the exception of the larger penguins, birds do not usually remain under water, when diving voluntarily, for much longer than 60 s (Butler and Jones, 1982), so that under these circumstances a more rapid attainment of the maximum adjustments might be expected. Data from freely diving birds, however, do not lend support to this premise.

The small passerine bird, the dipper, *Cinclus mexicanus*, has a similar heart rate after 5 s (the usual duration) of spontaneous diving as it does after 5 s of involuntary submersion, whereas it falls to a much lower level if submersion is prolonged (Murrish, 1970). Freely diving gentoo and Adélie penguins, *Pygoscelis papua* and *P.*

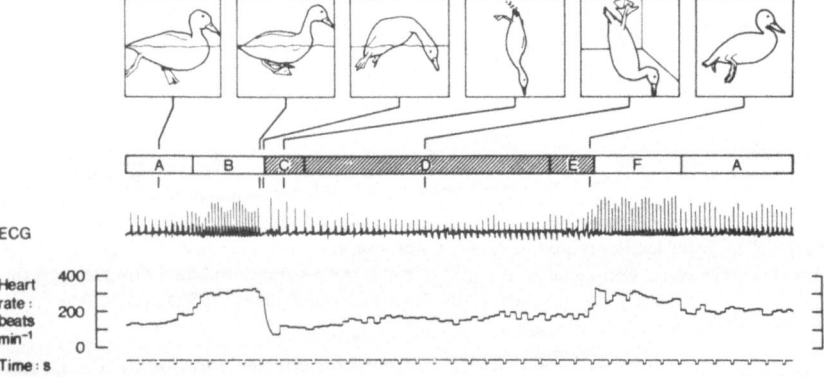

Figure 3 Relationship between the behaviour of a tufted duck, *Aythya fuligula*, and changes in heart rate during a voluntary dive in a glass-sided tank, 1.7 m deep. From above, downwards: tracings of duck from cine film showing, from left to right, swimming, preparing to dive, moment of submersion, descending, feeding on bottom, surfacing (10 cm from surface); time periods of A — swimming on surface, B — cardiac acceleration before submersion, C — descent, D — feeding on bottom. E — surfacing, F — cardiac acceleration following surfacing; ECG; heart rate; time marker (s). The lines between the pictures of the duck and the time boxes join coincident points in time. (Modified from Butler and Woakes, 1982b).

adeliae, also show a bradycardia and reduced blood flow through the femoral artery, both of which are less severe than the changes elicited by involuntary submersion (Millard *et al.*, 1973; Kooyman *et al.*, 1980). On the other hand in pochards, *Aythya ferina*, tufted ducks, *A. fuligula*, double crested cormorants, *Phalacrocorax auritus*, and Humboldt penguins, *Spheniscus humboldti*, there is no maintained bradycardia during voluntary dives (Butler and Woakes, 1976, 1979, 1982a, 1984; Kanwisher *et al.*, 1981).

Ducks and penguins dive repeatedly for extended periods without becoming exhausted. Preceding the first dive of a series there is an elevation of heart rate and an increase in respiratory frequency in pochards and tufted ducks (Butler and Woakes, 1976, 1979). Upon submersion, which appears to require much effort as the bird arches itself into the water, there is a transient bradycardia, but heart rate increases

(a)

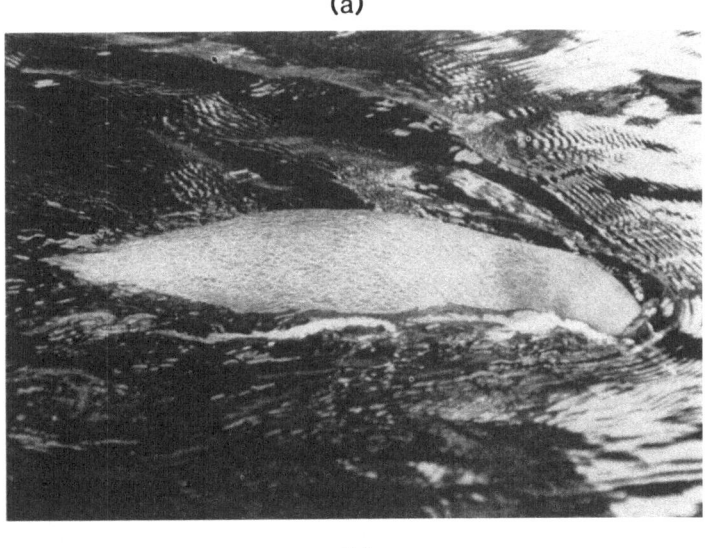

(b)

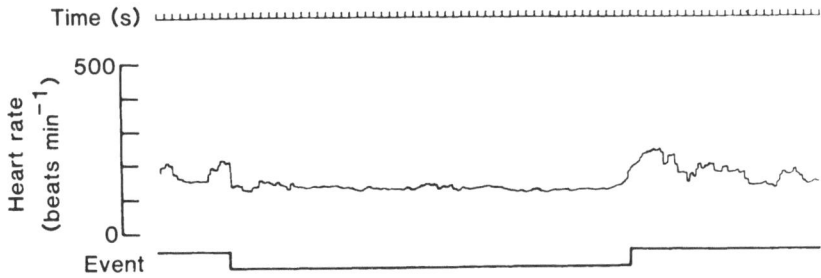

Figure 4 (a) Humboldt penguin, *Spheniscus humboldti*, swimming on the surface of the pond with its eyes under water before submerging completely. (b) Heart rate of Humboldt penguin (4.5 kg) before, during and after the longest recorded, voluntary dive (50 s). Duration of dive is indicated by the downward deflection of the event marker (Butler and Woakes, 1984).

somewhat during the first few seconds of the dive and then maintains a more or less steady rate (Fig. 3). This steady rate is certainly not lower than that seen before the pre-dive increase and may even be higher. The ducks have to continue paddling when under water and surfacing occurs passively once leg beating has ceased. In other words, the animals are positively buoyant. There is often cardioacceleration before the bird breaks the surface. By contrast, penguins glide under water, almost effortlessly, and do not exhibit an obvious tachycardia before the first dive of a series (Fig. 4); neither is it necessary for them to perform perceptible locomotor activity to remain submerged. They must be almost neutrally buoyant. Like the ducks, however, there is no sign of bradycardia during the dive (Butler and Woakes, 1984). These findings from naturally diving birds put a completely different complexion on the concept of differential blood supply, reduced aerobic metabolism and increased anaerobiosis during diving, for bradycardia is an integral part of this response.

Oxygen uptake has been estimated in tufted ducks at mean duration of voluntary dives and during surface swimming (Woakes and Butler, 1983). The birds were enclosed by an open circuit respirometer box which allowed them sufficient room to manoeuvre and to dive for food to the bottom of a deep (1.7 m) glass-sided tank (Fig. 5). Oxygen uptake was monitored continually by a mass spectrometer and heart rate was monitored by an implanted radiotransmitter. The same respirometer box was used when the birds swam at different velocities on a water channel. The estimated rate of oxygen consumption at mean dive duration (14.4 s) was 3.5 times the resting value and not significantly different from measured oxygen uptake at the maximum sustainable swimming speed (Fig. 6).

On the basis of the estimated rate of oxygen consumption during voluntary dives (0.57 ml s^{-1} STPD) and the size of the oxygen storage compartments in tufted ducks (Keijer and Butler, 1982), aerobic metabolism could continue for 44 s with the duck actively swimming under water. During the anticipatory period (which lasts, on average, for 6.8 s) before the first dive of a series, oxygen is extracted by the ducks at a rate of 1 ml s^{-1} STPD (6.3 times resting). Although some of this extra O_2 may be metabolized, it is estimated that at least an extra 4 ml of O_2 are taken in by the ducks and stored before the first dive. This would allow the animals to remain under water and metabolize aerobically for an extra 7 s (i.e. for a total of 51 s). This means that birds diving to depths of 6 m in the wild do so aerobically, for mean dive duration is 28 s. Even after denervation of the carotid bodies, which significantly increases duration of voluntary dives, they are still well within the aerobic limit (Butler and Woakes, 1982a).

When the ducks were swimming on the water channel, there was a good linear relationship between steady state oxygen uptake and heart rate (Woakes and Butler, 1983). If mean oxygen consumption and heart rate at average dive duration are plotted on the same graph, heart rate, at the same level of oxygen consumption, is significantly lower during diving than during surface swimming (Fig. 7). It is 1.5 times the resting value and 2.7 times the value after 15 s of involuntary head submersion.

As a result of these data, it is suggested that the circulatory adjustments during voluntary diving in ducks are similar to those during exercise in air in as much as the locomotory muscles, as well as the heart and CNS, receive an enhanced blood supply

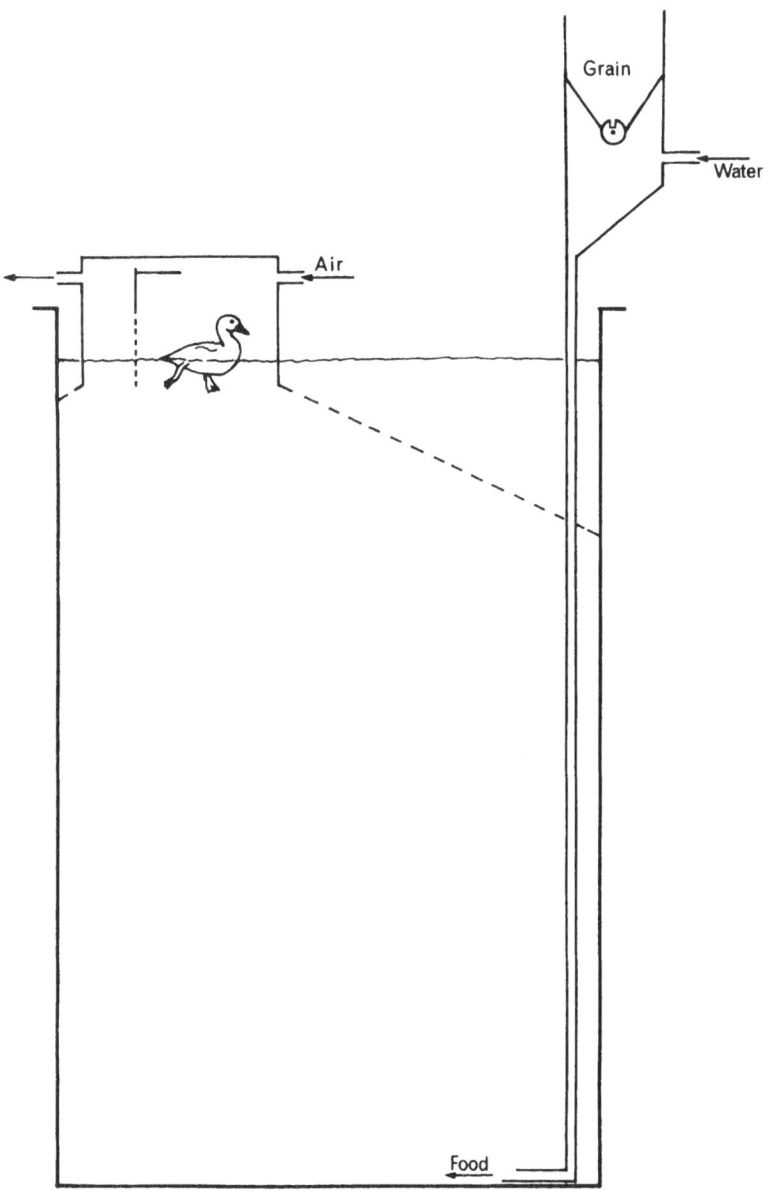

Figure 5   Diagram to show a duck in an open circuit respirometer on the top of a glass-sided tank 1.7 m deep. The duck learns to dive to the bottom of the tank and press on a red lever which causes a few pieces of grain to be delivered (see Woakes and Butler, 1983, for details).

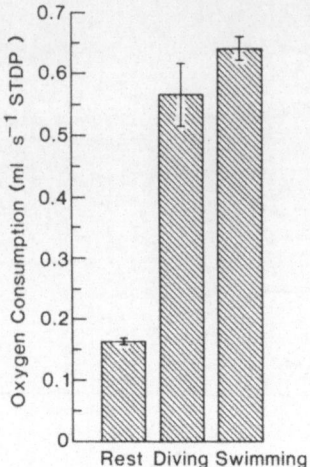

Figure 6   Mean oxygen uptake (±SE of mean) of 6 tufted ducks, *Aythya fuligula*, while at rest on water, at mean dive duration (14.4 s) while diving spontaneously on a glass-sided tank and while swimming at maximum sustainable velocity (0.78 m s^{-1}) on a water channel (data from Woakes and Butler, 1983).

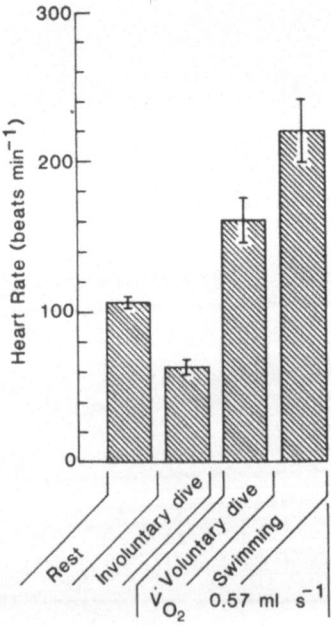

Figure 7   Mean heart rate (±SE of mean) for 6 tufted ducks, *Aythya fuligula*, at rest on water, 15 s after involuntary submersion of the head, voluntary dives of 14.4 s duration (10 ducks used)) and while swimming. Oxygen consumption (vO_2) at mean dive duration and while swimming was the same, 0.57 ml s^{-1} STDP (Woakes and Butler, 1983).

and sufficient oxygen for aerobic metabolism. The inactive muscles, viscera, kidneys and skin may well receive a reduced supply (Butler, 1982). The lower heart rate during diving could indicate, however, that the selective constriction is more intense and that oxygen extraction by the active muscles is greater when the ducks exercise under water than when they exercise in air. Certainly, the respiratory muscles are inactive when under water, in contrast to the situation during exercise in air, and probably receive a reduced blood supply. Heart rate is higher at the end of relatively long (>20 s) voluntary dives after denervation of the carotid bodies (Butler and Woakes, 1982a) which suggests that in intact ducks there is partial expression of the cardiovascular adjustments that are seen during involuntary submersion. During voluntary dives there would appear to be, in tufted ducks, a balance between the cardiovascular responses to involuntary submersion and to exercise in air, with the balance towards the latter (Fig. 7).

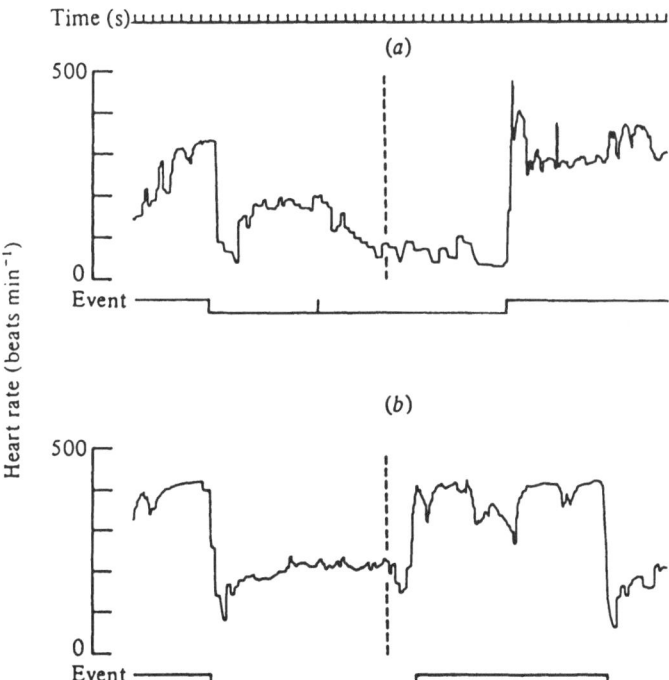

Figure 8 Traces showing heart rate of a tufted duck, *Aythya fuligula*, (a) when prevented from surfacing after a voluntary dive, (b) during unimpeded diving behaviour. In both cases the duck was in a glass-sided tank 1.7 m deep. Downward deflection of the event marker indicates the point of submersion and upward deflection the point of surfacing. In (a) the duck began to surface at the time indicated by the vertical line on the event marker, but was prevented from gaining access to air. The vertical dashed lines in (a) and (b) indicate identical times after submersion and illustrate the differences in heart rate under the two conditions (Butler and Woakes, unpublished).

The balance can be tipped in the opposite direction. Attempts were made to extend voluntary dives of tufted ducks to approximately 30 s duration in the glass-sided tank by briefly preventing access to the surface air following submersion. As soon as the duck was aware that access to air had been prevented, it displayed a progressive bradycardia which reached its lowest level within 10–15 s and this level was similar to that during involuntary submersion (Fig. 8). The bird continued to swim around and showed no supplementary locomotory activity, such as wing beating, which sometimes occurs during "play" (Butler and Woakes, 1982b). If, as has previously been assumed, such bradycardia is an indication of large-scale peripheral vasoconstriction and anaerobiosis, then the animal switched to that

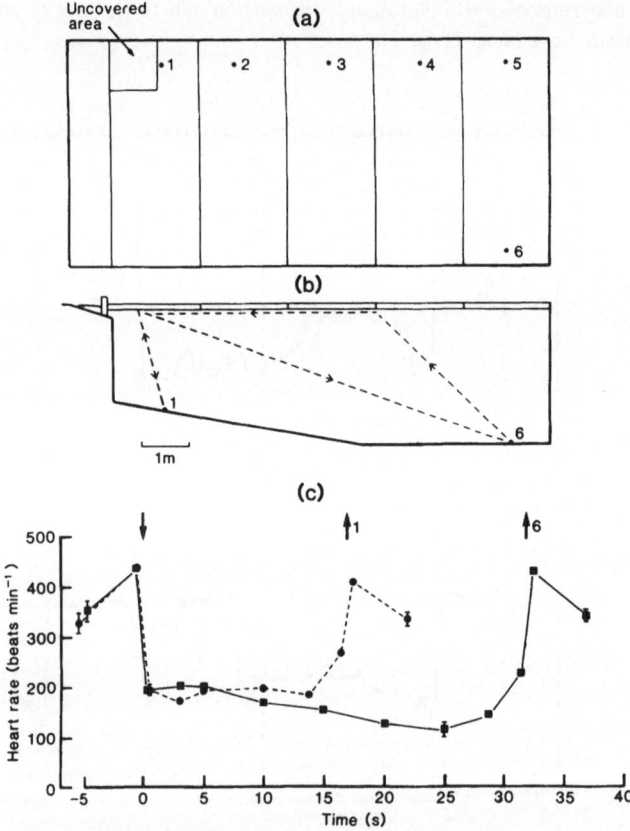

Figure 9   Plan (a) and side-view section (b) of an outside pond. The surface is almost totally covered except for one small area. Food is placed on the bottom of the pool at 6 different distances from the uncovered area (1–6 in (a)). (b) shows the route taken by the ducks to and from positions 1 and 6, and (c) shows the mean dive duration and mean (±SE of mean) heart rate for 6 tufted ducks while making these journeys. It should be noted that for the longer, diagonal journeys, the ducks have to swim actively during the return as well as during the outward leg (Stephenson and Butler, unpublished).

pattern of response when access to air was temporarily denied. This is interesting when viewed in the light of the earlier observation that heart rate is higher than during "normal" voluntary dives when ducks dive to avoid being caught by a hand net (Butler and Woakes, 1979). In this instance the birds knew that they could take a quick breath when required. Also, if ducks have to swim varying distances under a cover to obtain their food, there is no initial lowering of the heart rate when long distances have to be travelled, although heart rate does fall progressively during the longer dives after the first 5 s (Fig. 9). These observations must be pertinent to ducks feeding under ice in the winter. It would seem that only if caught unawares does the animal immediately invoke its oxygen-conserving response. It is likely, therefore, that when water is covered with ice, ducks make a number of short exploratory dives initially in order to reduce their chances of becoming disorientated. If they have to swim long distances under water they may progressively shift to the "classical", oxygen-conserving, response.

Using a similar set-up to that used with the ducks in the glass-sided tank, oxygen uptake of voluntarily diving Humboldt penguins was estimated on a large (9 × 4.6 × 2.78 m deep) outside pool (Butler and Woakes, 1984). Both heart rate and oxygen uptake during submersion were not significantly different from the resting values. On some occasions there was an indication of one deep inspiration followed by an expiration before diving, but as already mentioned, there were no anticipatory changes in heart rate before the first dive of a series, as seen in tufted ducks. Based on our data for oxygen consumption and values calculated by Kooyman *et al.* (1980) for usable oxygen stores, it has been deduced that Humboldt penguins could remain aerobic under water for 2.3 min (Butler and Woakes, 1984). This is interesting in that observations on freely diving chinstrap penguins, *Pygoscelis antarctica*, indicate they dive to less than 45 m and for an average of 1.6 min (Lishman and Croxall, 1983). Of course, active swimming to such depths may cause aerobic metabolism to be higher than that estimated by Butler and Woakes (1984), but it is likely that these dives were completely aerobic. It certainly seems that penguins are able to dive and remain aerobic for longer periods than ducks. This is probably because they are more efficient at underwater locomotion, partly as a result of their being almost neutrally buoyant (Butler and Woakes, 1984).

Biochemical studies suggest that some of the larger, more deeply diving penguins, particularly the emperor, king and Adélie, may resort more regularly to anaerobiosis during feeding (diving) than the royal, rock-hopper and gentoo penguins (Baldwin *et al.*, 1984). It would be interesting to know whether anaerobiosis occurs progressively during dives, as may happen in tufted ducks during longer voluntary dives, or whether it makes a varying contribution, early on in the dive, depending on the expected dive duration. On the basis of work on Weddell seals (Kooyman *et al.*, 1980) I would predict that anaerobiosis is used infrequently, even in the emperor penguin. The little penguin, *Eudyptula minor*, on the other hand, appears to be equipped predominantly for aerobic metabolism (Mill and Baldwin, 1983).

From the information that we have so far it appears that oxygen is used at the rate required to maintain a high level of aerobic metabolism in the active tissues. Inactive tissues may become anaerobic and produce some lactic acid, but this could be oxidized elsewhere in the body, for example, in the heart, so that there is little or no overall lactate accumulation. Upon surfacing, the oxygen stores are quickly replaced and another dive performed. Such a behaviour pattern, with relatively short periods at the surface, is more economical in terms of percentage of time actually spent feeding during a group of dives, than if longer dives are performed, but proportionately longer is spent at the surface metabolizing the lactate (Kooyman *et al.*, 1980). Also, animals in such an exhausted state would be more vulnerable to predators.

The ability to obtain physiological data from freely diving animals has opened up the whole area of diving physiology. What at first appeared to be a predictable, stereotyped reflex response of the cardiovascular system to asphyxia is clearly complex and labile. Different animals with different feeding behaviour may show subtle differences in their physiological adjustments to diving. Even the same animal may respond differently at different times of the year and the adjustments when escaping from predators or when unexpectedly trapped under water may be different again from those when feeding.

ACKNOWLEDGEMENTS

The author's work on this topic is supported by SERC.

REFERENCES

Baldwin, J., Jardel, J-P., Montague, T. and Tomkin, R. 1984. Energy metabolism in muscles of diving penguins. (In preparation).

Bert, P. 1870. Leçons sur a Physiologie Comparée de la Respiration, Bailliere, Paris.

Bryan, R. M. and Jones, D. R. 1980a. Cerebral energy metabolism in diving and non-diving birds during hypoxia and apnoeic asphyxia, *J. Physiol. Lond.*, **299**, 323–336.

Bryan, R. M. and Jones, D. R. 1980b. Cerebral energy metabolism in mallard ducks during apnoeic asphyxia: the role of oxygen conservation, *Am. J. Physiol.*, **239**, R353–357.

Butler, P. J. 1982. Respiratory and cardiovascular control during diving in birds and mammals, *J. exp. Biol.*, **100**, 195–221.

Butler, P. J. and Jones, D. R. 1971. The effect of variations in heart rate and regional distribution of blood flow on the normal pressor response to diving in ducks, *J. Physiol. Lond.*, **214**, 457–479.

Butler, P. J. and Jones, D. R. 1982. Comparative physiology of diving in vertebrates, in *Advances in Physiology and Biochemistry*, Vol. 8, O. E. Lowenstein (ed), Academic Press, New York, pp. 179–364.

Butler, P. J. and Woakes, A. J. 1976. Changes in heart rate and respiratory frequency associated with natural submersion in ducks, *J. Physiol. Lond.* **256**, 73–74.

Butler, P. J. and Woakes, A. J. 1979. Changes in heart rate and respiratory frequency during natural behaviour of ducks, with particular reference to diving, *J. exp. Biol.*, **79**, 283–300.

Butler, P. J. and Woakes, A. J. 1982a. Control of heart rate by carotid body chemoreceptors during diving in tufted ducks, *J. appl. Physiol.*, **53**, 1405–1410.

Butler, P. J. and Woakes, A. J. 1982b. Telemetry of physiological variables from diving and flying birds, *Symp. Zool. Soc. Lond.*, **49**, 106–128.

Butler, P. J. and Woakes, A. J. 1984. Heart rate and aerobic metabolism in Humboldt penguins, *Spheniscus humboldti*, during voluntary dives, *J. exp. Biol.* (In press).

Hollenburg, N. K. and Uvnas, B. 1963. The role of cardiovascular response in the resistance to asphyxia of avian divers, *Acta physiol. scand.*, **58**, 150–161.

Irving, L. 1934. On the ability of warm-blooded animals to survive without breathing, *Scient. Mon.*, **38**, 422–428.

Irving, L. 1939. Respiration in diving mammals, *Physiol. Rev.*, **19**, 112–134.

Johansen, K. 1964. Regional distribution of circulating blood during submersion asphyxia in the duck, *Acta physiol. scand.*, **62**, 1–9.

Jones, D. R. and Purves, M. J. 1970. The carotid body in the duck and the consequences of its denervation upon the cardiac responses to immersion, *J. Physiol. Lond.*, **211**, 279–294.

Jones, D. R., Milsom, W. K. and Gabbott, G. R. J. 1982. Role of central and peripheral chemoreceptors in diving responses of ducks, *Am. J. Physiol.*, **243**, R537–R545.

Jones, D. R., Bryan, R. M., West, N. H., Lord, R. H. and Clark, B. 1979. Regional distribution of blood flow during diving in the duck (*Anas platyrhynchos*), *Can. J. Zool.*, **57**, 995–1002.

Kanwisher, J. W., Gabrielsen, G. and Kanwisher, N. 1981. Free and forced diving in birds, *Science, N.Y.*, **211**, 717–719.

Keijer, E. and Butler, P. J. 1982. Volumes of the respiratory and circulatory system in tufted and mallard ducks, *J. exp. Biol.*, **101**, 213–220.

Kooyman, G. L., Wahrenbrock, E. A., Castellini, M. A., Davis, R. W. and Sinnett, E. E. 1980. Aerobic and anaerobic metabolism during voluntary diving in Weddell seals: Evidence of preferred pathways from blood chemistry and behavior, *J. comp. Physiol. B.*, **138**, 335–346.

Lishman, G. S. and Croxall, J. P. 1983. Diving depths of the chinstrap penguin, *Pygoscelis antarctica*, *Brit. Antarct. Surv. Bull.*, **61**, 21–25.

Mill, G. K. and Baldwin, J. 1983. Biochemical correlates of swimming and diving behavior in the little penguin *Eudyptula minor*, *Physiol. Zool.*; **56**, 242–254.

Millard, R. W., Johansen, K. and Milsom, W. K. 1973. Radiotelemetry of cardiovascular responses to exercise and diving in penguins, *Comp. Biochem. Physiol.*, **46A**, 227–240.

Murrish, D. E. 1970. Responses to diving in the dipper, *Cinclus mexicanus*, *Comp. Biochem. Physiol.*, **34**, 853–858.

Påsche, A. and Krog, J. 1980. Heart rate in resting seals on land and in water, *Comp. Biochem. Physiol.*, **67A**, 77–83.

Scholander, P. F. 1940. Experimental investigations on the respiratory function in diving mammals and birds, *Hvalråd. Skrift.*, **22**, 1–131.

Sykes, A. H. 1966. Submersion anuria in the duck, *J. Physiol. Lond.*, **184**, 16P.

Woakes, A. J. and Butler, P. J. 1983. Swimming and diving in tufted ducks, *Aythya fuligula*, with particular reference to heart rate and gas exchange, *J. exp. Biol.*, **107**, 311–329.

6

Pulmonary Function Changes

W. A. Crosbie, Employment Medical Adviser, Health and Safety Executive, Employment Medical Advisory Service.

The professional diver has larger than normal lungs due to an increase in the size of the alveoli. This enlargement is not matched by an increase in the size of the airways, hence there is present some limitation on expiratory airflow. The long-term effect of these lung changes is unknown but information is being gathered in order to answer this question.

The professional diver works in a unique environment which exposes him to physiologic stress not seen in any other occupation. He is required to breath gases at higher than normal pressure for several hours in a day, and in a saturation diving complex it may be for a continuous exposure period of 2–3 weeks at a time. It naturally follows that such individuals require to have good lung function so that they may cope with these demands. Anyone with poor lung function is unlikely to continue to work in this environment even if his disability is not detected by the medical examination required in certain parts of the world. There are many questions to be answered concerning the long- or even short-term effects which may result from working in such atmospheres. Do divers become divers because they know they function well when in the water, or does working under these conditions produce alterations in body structure and function which adapts the individual to this environment? If this is the case, what are the environmental factors which lead to these changes, what are the time scales of these changes, and do they lead to harmful results in the long term?

The lungs are an obvious target for study in these circumstances since they are the largest organ and surface area of the human which is exposed to the environment. Although there is considerable knowledge (Lamphier, 1975) on the performance of the cardio-pulmonary system under hyperbaric stress, there is a dearth of facts on the characteristics of the lungs themselves. The unique and detailed studies of the diving Ama carried out by Hong and co-workers in the 1950s showed that the active divers had significantly larger vital capacities compared with their non-diving sisters. (Hong *et al.*, 1963). This could be due to a selective system where the fitter subjects knowingly carried out the underwater work while the less fit individuals remained on shore to do the routine chores. It is of interest that these investigators report an increase in the vital capacity but state that the residual volume was not significantly different from the non-diving controls. Of course, the type of diving carried out by the Ama is of the breath-hold type which is completely different from the diving done by the modern professional diver, who breathes normally through his life support system while in the water or just normally while in a saturation chamber. Further information on this subject was obtained by Carey *et al.* (1956), who studied the changes which occurred in naval diving instructors who worked regularly in submarine escape training tanks. They showed that the instructors had larger than predicted lungs and were able to follow the changes which occurred in the lungs of these men during a period of duty in the tanks. They found that the increase in lung size developed in the men following a period of exposure to increased pressure and these changes regressed after a period of no diving exposure. Hence here was evidence that working under water was associated with an increase in the volume of the lungs. No further information on this subject was reported until 1974.

The discovery of large deposits of gas and oil under the North Sea led to the aggregation of large numbers of professional divers in this area, since their skills were vital to the development of the offshore oil industry. The regulation of

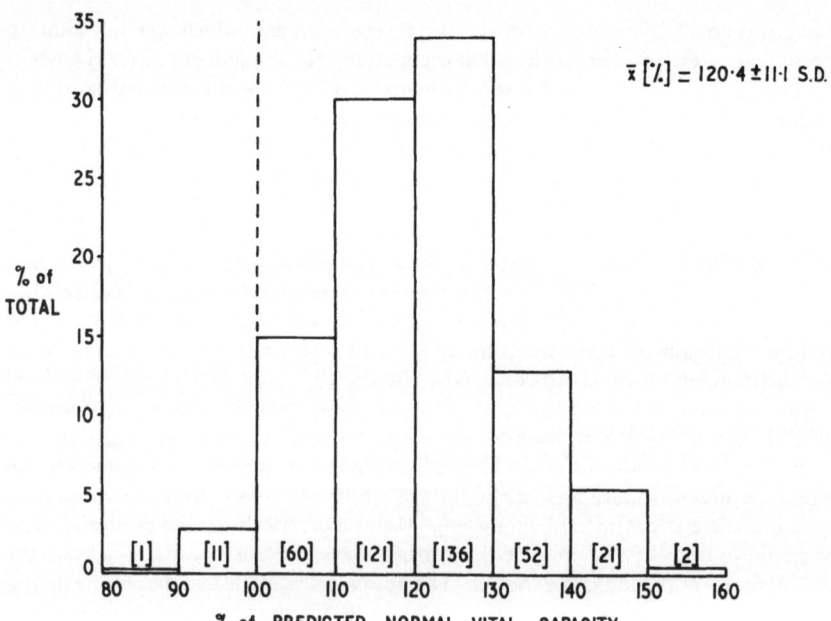

Figure 1   The distribution of the measurements of forced vital capacity (*FVC*) of 404 commercial divers. [   ] = number in each group.

employment of these men in waters under the jurisdiction of the United Kingdom gave rise to a system of medical surveillance of the workforce. These men were required to undergo an annual medical examination by specified approved doctors and had to be certified fit before they were allowed to work under water. The recording of simple lung ventilatory function became part of the medical examination and thus was gathered a unique and invaluable bank of information on the lung function of divers. The initial exploration of the North Sea took place in the southern basin; the major centre for servicing the diving industry at that time was Great Yarmouth. The North Sea Medical Centre developed as a medical service to the offshore industry which included diving. This group of doctors were able to collect the data from the flow of divers who were examined at the Centre, using the same techniques and equipment over a 1–2 year period. This information was collated, analysed and reported by Crosbie *et al.* (1977).

The first important finding is shown in Figure 1 where the distribution of values for the forced vital capacities (*FVC*) of the 404 divers is demonstrated. This histogram quite clearly shows that there is a generalised increase in the size of *FVC*, with a mean of 120% of the predicted value. This finding is of considerable interest since it is in accord with the findings reported by Hong and Carey and their co-workers. The second important finding is shown in Figure 2 where the comparable results of the forced expired volume in one second (*FEV*$_1$) is shown. Here again, it is apparent that there is a general increase in the size of this measurement

DISTRIBUTION OF FORCED EXPIRED VOLUME IN 1 SECOND OF 404 COMMERCIAL DIVERS OPERATING IN THE NORTH SEA.

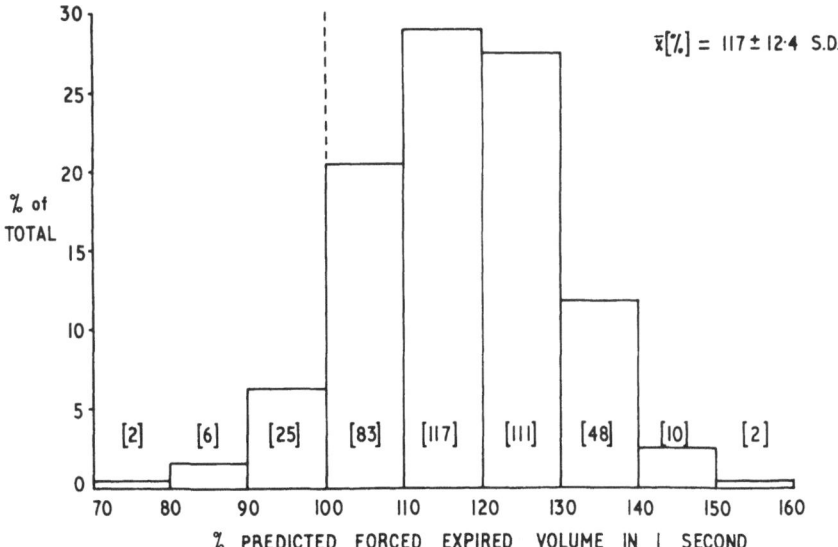

Figure 2 The distribution of the measurements of forced expired volume in 1 s (*FEV*$_1$) of 404 commercial divers. [ ] = number in each group.

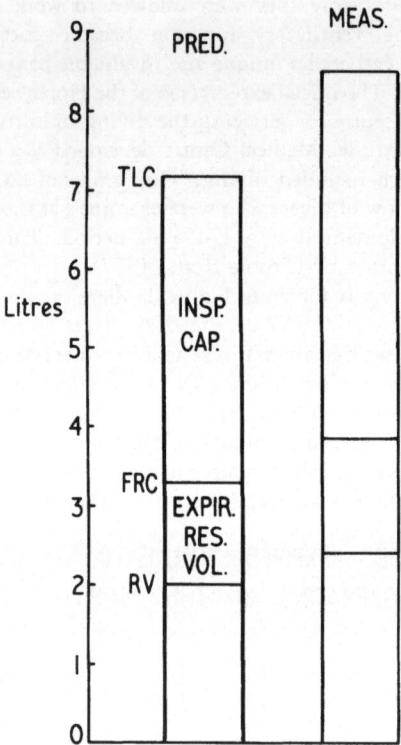

LUNG VOLUMES OF 26 COMMERCIAL DIVERS

Figure 3   The mean value of the measurements of the subdivisions of lung volume of the 26
divers in comparison with the predicted volumes.

but two other points also emerge: the mean value is less than the mean for the *FVC*
and the range of values is wider, hence the shape of this distribution curve is
different. These observations seem to indicate that the increase in lung size may be
different in the alveolar tissue from that seen in the airways. Further investigation of
these findings was required and a more detailed analysis of the function of divers
lungs was made in group of 26 of these men (Crosbie *et al.*, 1979). This smaller
group was chosen because they all had an $FEV_1/FVC \times 100$ of 75% or less. This
value was chosen because the regulatory authorities used this level as the lowest
acceptable value considered normal for this measurement.

We had shown that one lung volume was enlarged, but did this apply to the other
lung volumes? The answer is shown in Figure 3, where it is seen that the mean value
for all the other subdivisions of the lung volume, except for the expiratory reserve
volume, were significantly larger than the predicted values, Crosbie *et al.* (1979). This

finding is different from the information given in Hong and Carey's papers; they state that the residual volume of their groups were not significantly greater than their control populations. Perhaps the most interesting fact was the finding that the RV/TLC ratio remained normal in our group. In addition, the functional residual capacity (which is the volume where normal respiration is carried out) also was increased in proportion to the other lung volumes. These findings seemed to indicate that the balance of expansive and retractive thoracic forces were set at higher than normal level but the relative values remained the same. This means that any increase in the chest wall expansion was balanced by an increase in the lung recoil force of the lung. The latter could be increased because there is more tissue present in the lung or because it was stretched more than normal. Does this mean that the lungs of a diver contain an increased number of alveoli, or are there a normal number of alveoli present which are larger than normal? To help answer this question the information gained from the measurement of the gas transfer capacity (TL_{CO}) of these lungs is shown in Figure 4. Here it is again clearly seen that there is a significant increase in the mean value of the TL_{CO} compared with the predicted value. This finding indicates that either there is an increase in surface area over which gas exchange takes place, or there is more blood in the lung capillary bed compared with normal. We have no evidence for the latter explanation since these men all had normal

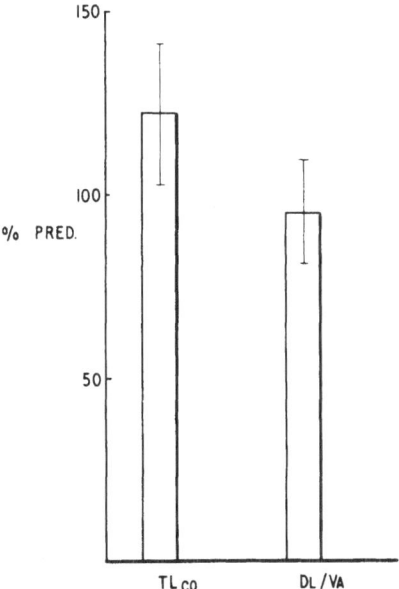

Figure 4 The means and standard deviations of the diffusion capacity of the lung (DL_{CO}) and diffusion capacity expressed as a proportion of the alveolar volume of diffusion DL/VA.

haemoglobin levels and there was no indication of the presence of left heart failure. Further information on this aspect is shown in the same diagram. The second column shows the mean value for the group of the gas transfer capacity, expressed as a function of the alveolar volume in which the gas exchange took place. Here it will be seen that this measurement is not increased compared with the predicted value or the gas transfer capacity itself. Combining the evidence shown in this figure, it seems that there is an increase in the total surface area over which gas exchange takes place without an equal increase in the volume of gas in contact with this surface. How can these findings be related to the structural and functional characteristics of the lungs themselves? The increase in volume of the lungs can either be due to an increase in the number of alveoli present, or there is an increase in the size of the alveoli present. If there is an increase in the number of alveoli, there should be a similar increase in both the TL_{CO} and TL/VA. If there has been an increase in the size of the alveoli, then the total surface area for gas exchange will increase but the volume of gas in contact with this surface will show a greater increase. This state comes about since the alveoli are basically spherical units and the increase which occurs in the surface of a sphere is smaller in proportion to the volume of a sphere when it enlarges. Hence, from this finding we suggest that the increase in size of the lungs of a diver is due to an increase in the volume of the alveoli rather than an increase in the number of alveoli present.

Is this state of benefit to the diver? It would seem to be obvious that a larger than normal lung would be of benefit to a diver. The increased ability to absorb oxygen and eliminate carbon dioxide must be an asset where heavy physical work is a necessary part of the work. A large vital capacity should mean that large volumes of gas can be ventilated throughout the lungs, and a normal RV/TLC ratio should lead to good distribution of the gas in the lungs. The enlargement of the surface area over which gas exchange takes place should be a physiological improvement and this is shown by the increased values for gas transfer. Such would be our reasoning, but there is one important element of lung structure and function which is fundamental to the whole process of respiration and that is the part played by the airways. Having large lungs and an increased surface for gas exchange may be of little avail unless the gas is able to reach, mix and be expired through the airways. The behaviour of the airways can be assessed by looking at an index of maximal expiratory flow. A measurement of this function is the FEV_1. In our studies we have the measurement of this index in 404 divers. If we examine the relationship between the $FEV_1/FVC \times 100$ and the FVC expressed as a percentage of the predicted value, then we can get some indication of the behaviour of the airways with increasing lung size. Such a relationship is shown in Figure 5. It is at once apparent that there is a statistically significant relationship between these two variables. It is obvious that $FEV_1/FVC \times 100$ reaches a mean maximal value about 110% of FVC, but thereafter falls off in a linear manner. t-testing of this slope shows the correlation to have a P value less than 0.0001, which is highly significant in statistical terms. This means in simple everyday terms that, as the lungs increase in size, there is a steady decline in the rate at which air can be maximally expired.

The larger the lung, the more difficult it is to empty. Why should this occur? The work of Pride and co-workers (1967) has shown us that the control of expiratory

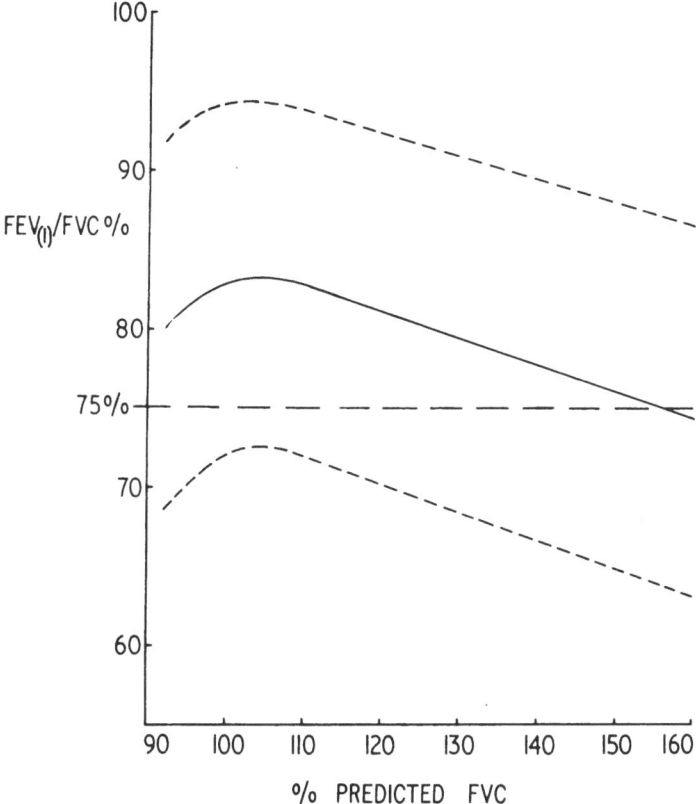

SHOWING RELATIONSHIP BETWEEN $FEV_{(1)}/FVC\%$ and $FVC\%$
PREDICTED on 404 COMMERCIAL DIVERS.

Figure 5 The relationship between the mean values of 5% incremental subgroups of the
FVC and the $FEV_1/FVC\%$ of 404 commercial divers. The solid line joins the means and the
broken lines shows the 2–50 range either side of the mean.

flow from the lungs depends on the interplay between airway size and number
and the elastic recoil force in the lung. At a specific lung volume, once a certain
flow has been reached further effort will not augment this flow. Applying this
theory to our findings, it means that, as the lungs become larger, the size
number and elastic recoil force of the lungs would have to increase in proportion to
maintain normal maximal expiratory flow rates. If this does not happen, then the
maximal expiratory flow rates will decrease. Since the 26 divers in our second study
(who all had an $FEV_1/FVC \times 100$ of less than 75%) had a normal RV/TLC ratio, we
feel that the elastic recoil force in the lungs is normal. It therefore seems that the
reduced expiratory flow is most likely due to a failure of the airways to dilate or

increase in number to respond to the increase in alveolar volume. In terms of function this means that, as the size of the lungs increase, there is a reduction in the ability of the airways to cope with the increased volume of gas which is being respired. Hence the ability to ventilate the lungs is determined by the limitation of expiratory flow produced by the relatively narrowed airways. Since the stimulus to enlargement of the lung in a diver occurs by the time the lung is fully mature, there is little chance of new lung units or larger airways developing in these lungs. In contrast, the lung parenchyma, which is mainly composed of alveolar tissue, can enlarge by stretching without the need for growth of new tissue. Hence the thoracic cavity can be increased in volume by hypertrophy of the muscles of respiration, which will produce an increased force of expansion in lungs. The alveoli, being made up of distensible tissue, will thus enlarge in response to the new balance of intrathoracic forces which will occur at a higher than normal resting lung volume, and this is seen in our group by the size of functional residual capacity. The airways are capable of only a lesser enlargement circumferentially and may even be elongated as the lungs enlarge. The result will be a lung capable of increased gas transfer but the ventilation will be limited by the ability of the airways to cope with large flows. The long-term effects of a lung showing alveolar distention with a relative expiratory airflow obstruction is not yet apparent.

It will be at once apparent that this combination of structural change is very like those seen in pulmonary emphysema. We have no evidence at present that this occurs but scientific observation of these men should be carried out so that the development of this pathological state may be detected at an early stage. The present system of medical surveillance of the professional diver in the United Kingdom should enable such information to be gathered. At present there are nearly 300 doctors approved under the United Kingdom Diving Operations at Work Regulation, 1981, to carry out the medical examination of divers who want to work in waters under our jurisdiction. They examine over 6000 divers each year and the information obtained is centralised in London. This system should enable us to answer some of the questions raised so far about the effects of diving on the lungs.

REFERENCES

Carey, C. R., Schaefer, K. E. and Alvis, H. J. 1956. Effect of skin diving on lung volumes, *Journal of Applied Physiology*, 8, 519–523.

Crosbie, W. A., Reed, J. W. and Clarke, M. C. 1979. Functional characteristics of the large lungs found in commercial divers, *Journal of Applied Physiology*, 46, 639–645.

Crosbie, W. A., Clarke, M. B., Cox, R. A. F., McIver, N. K. I., Anderson, I. K., Evans, H. A., Liddle, G. C., Cowan, J. L., Brookings, G. H. and Watson, F. G. 1977. Physical characteristics and ventilatory function of 404 commercial divers working in the North Sea, *British Journal of Industrial Medicine*, 34, 19–25.

Hong, S. K., Rahn, H., Kang, D. H., Song, S. H. and Kang, B. S. 1963. Diving pattern, lung volumes and alveoli gas of the Korean diving woman (ama), *Journal of Applied Physiology*, 18, 457–465.

Lamphier, E. H. 1975. in *The Physiology and Medicine of Diving and Compressed Air Work*, 2nd edn. P. B. Bennett and D. H. Elliott (eds), Balliere Tindall, London.

Pride, N. B., Permutt, S., Riley, R. L. and Bromberger-Barnea, 1967. Determinants of maximal expiratory flow from the lungs, *Journal of Applied Physiology*, **23**, 646–662.

7

Diving Response of Mammals and Birds

Arnoldus Schytte Blix, Department of Arctic Biology and Institute of Medical
Biology, University of Tromsø, Norway

An evaluation is made of how much of the bradycardia produced in force-dived seals
and ducks can be accounted for by stress and fear. The concept of "diving" bradycardia
as part of a natural defence mechanism against asphyxia is defended.

INTRODUCTION

In 1970 it had been known and appreciated for 100 years that ducks respond to
forced diving with a profound bradycardia (Bert, 1870). It consequently caused great
confusion when, shortly after this jubilee, it was reported that ducks diving
voluntarily displayed very little, if any, bradycardia (e.g. Woakes and Butler,
1975). This new discovery ultimately resulted in the development of the bold idea
that "stress-induced artifacts (can) account for a large part of the 'diving'
bradycardia reported in laboratory studies" (Kanwisher *et al.*, 1981).

Unfortunately, the promulgation of this idea without reference to the extensive
literature at hand can easily lead those who are not themselves physiologists to the
idea that the old and celebrated concept of "diving" bradycardia as part of a natural
defence mechanism against asphyxia is totally false. Moreover, since none of the
radio-biologists have found it important to define how much "a large part" is, I have
taken upon myself to evaluate just how much of the diving bradycardia of
force-dived seals and ducks can be accounted for by emotions, or (for those who enjoy
indulging in a more dramatic language) by stress and fear, and to explain why there
is so little bradycardia in free *versus* forced dives in ducks.

DIVING RESPONSE IN DUCKS

It has been known for quite some time that when ducks are forced under water their
"diving" bradycardia develops in two stages. First, there is an immediate 40%
reduction in heart rate accompanied by a decrease in peripheral blood flow. Second,

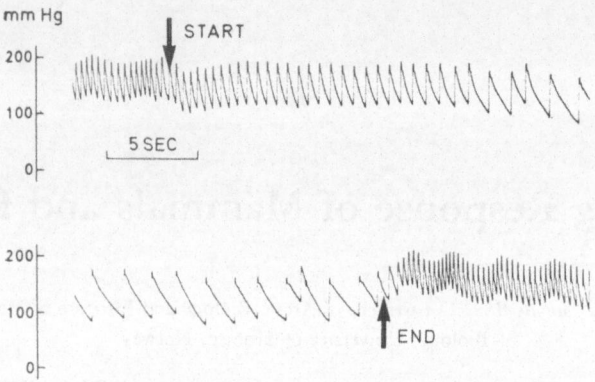

Figure 1 The normal cardiovascular response to forced diving in ducks. The figure includes arterial blood pressure recordings from the beginning (*upper tracing*) and the end (*lower tracing*) of a dive lasting for 40 s. (From Blix and Berg, 1974)

after 30–50 s, this "initial" response is greatly reinforced into the full response, usually amounting to a 90% reduction in heart rate (Fig. 1), and a profound reduction of muscle blood flow (e.g. Blix and Folkow, 1984).

Importance of arterial chemo-receptors

It was shown by Jones and Purves (1970) that the cardiovascular adjustments to forced diving in ducks were markedly reduced after surgical denervation of the arterial chemo-receptors. Later, Blix and Berg (1974) demonstrated that activation of the arterial chemo-receptors prior to forced dives, by exposure of the duck to a simulated altitude of 6000 m, resulted in the elicitation of the full response immediately upon submergence (Fig. 2). Moreover, Blix (1975) demonstrated that when force-dived ducks are kept normoxic in spite of a persistent apnoea while under water, *only* the "initial" 40% reduction of heart rate took place regardless of the

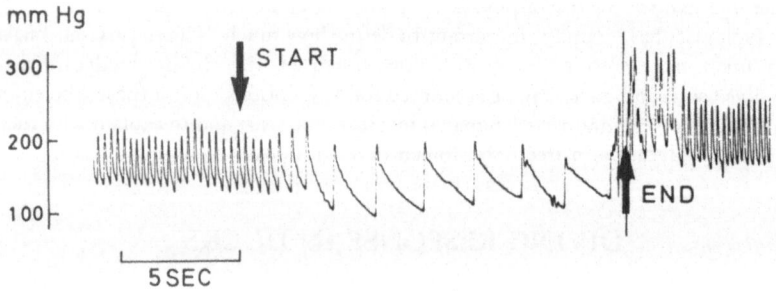

Figure 2 The arterial blood pressure response to forced immersion in a duck which entered the dive at an altitude of 6000 m, showing the effect of chemo-receptor stimulation prior to immersion of the head. (From Blix and Berg, 1974)

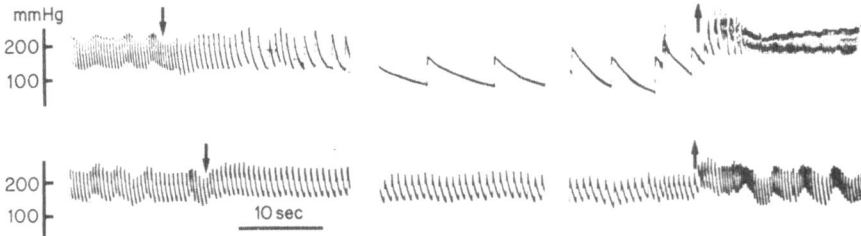

Figure 3 Typical blood pressure responses to 2 min forced immersion in ducks. The records were made at the beginning, after 60 s, and at the end of the dive. *Upper tracing*: the normal response (without any kind of ventilation). *Lower tracing*: the response of the same animal during a dive not involving arterial asphyxia. (From Blix, 1975)

duration of the underwater exposure (Fig. 3). These experiments clearly demonstrate that the secondary, delayed, reinforcement of the "initial" diving response does indeed depend on chemo-receptor stimulation in ducks. Thus, this fraction of the fully developed response can hardly be ascribed to emotional activation of the animal; rather, both the bradycardia and the simultaneously developed peripheral vasoconstriction should be regarded as perfectly normal autonomous reflexes aimed at the distribution of cardiac output, and, hence, the blood oxygen store to the brain and the heart with a largely maintained blood pressure (i.e. a defence mechanism against asphyxia). This implies that any emotionally induced component of the diving response must be hidden among the "initial" 40% of the response.

The "initial" diving responses — an orienting response?

It was suggested by Goodman and Weinberger (1970), working on an aquatic salamander, the mud-puppy (*Necturus maculosus*), that a relationship existed between the diving responses and the "orienting response" (e.g. Sokolov, 1963). Later, Blix *et al.* (1976), being at the time unaware of this report, suggested that the "initial" diving responses of ducks could indeed be regarded as an "orienting response" to the novel experience of being forced under water. It is generally agreed upon that orienting responses fade away upon repetition of exactly the same stimulus, that is, the animals habituate to the stimulus.

In order to uncover a possible orienting response component of the "initial" diving response in ducks, a colleague of mine (Gabrielsen, unpublished) subjected ducks to 60 consecutive dives lasting for 20 s, each separated by a recovery period of 40 s. In the first of these dives the "normal" 40% reduction of heart rate was recorded (after 10 s of submergence), but after 60 dives in a row, heart rate was reduced by only 17% after 10 s of submergence. Thereafter, further dives no longer resulted in further modifications of this response. This indicates that about 23% of the "initial" 40% reduction in heart rate can truly be ascribed to the emotional account and claimed to be caused by surprise or, for that matter, fear. This leaves about 17% of the fully developed response unaccounted for. The "residual" 17% reduction of heart rate compares very well with the bradycardia normally seen in free-diving ducks (e.g. Woakes and Butler, 1975), which usually dive for only about 10 s. A dive of 10 s duration is, of course, too short to cause any activation of the arterial chemo-receptors, and, since a voluntary dive is in itself a planned action, it can hardly be expected to produce an "orienting response", let alone any fear in a professional diver like the duck. Thus, ducks respond to voluntary diving with about a 15–20% reduction of heart rate.

Feigl and Folkow (1963) and Andersen (1963a) independently studied the effect of forced dives, during which the ducks were artificially ventilated through a tracheotomy. These workers found that heart rate in such dives was reduced by about 33%. If this value is compared with the 40% reduction obtained by Blix (1975), who subjected ducks to normoxic dives without ventilation, it may be concluded that silencing of the stretch receptors of lungs and/or thorax, contributes about 7% to the diving bradycardia. This leaves about 10% to be shared between trigeminal and glossopharyngeal influences (Andersen, 1963b; Bamford and Jones, 1984; Blix *et al.*, 1976) and central interactions between the inhibited respiratory "center" and the cardiovascular "centers" at medullary level.

This indicates that about 50% of the "initial" diving response can be accounted for by emotional factors, which amounts to about 25% of the fully developed response displayed in force-dived ducks.

DIVING RESPONSE IN SEALS

Unlike ducks, seals often display the fully developed diving responses within a few seconds of submergence. This implies that the full development of the response is independent of chemo-receptor activation in such creatures. It has been demonstrated, however, that maintenance of the response during prolonged dives is dependent on the arterial chemo-receptors, at least in anesthetized animals (Daly *et al.*, 1977). It follows that the immediate onset of the full diving response in seals could be regarded as a stress-induced phenomenon. It is noteworthy, however, that free-diving seals, contrary to ducks, usually respond to voluntary submergence with a drastically reduced heart rate, although the response seen in forced dives is slightly larger (Elsner, 1965). Recently, Kooyman *et al.* (1980) reported that seals occasionally, and usually in dives of short duration, do not exhibit any appreciable diving response. Furthermore, it has been noted by several authors that seals frequently show reduced heart rate in anticipation of a dive and frequently show tachycardia prior to surfacing (Jones *et al.*, 1981; Casson and Ronald, 1975) (Fig. 4). It has also been shown that seals can be conditioned to reduce their heart rate upon command while out of water (Ridgway *et al.*, 1975).

These results clearly indicate that seals, unlike ducks, have the ability to control their heart rate and other cardiovascular functions voluntarily, making it possible for them to adjust their responses according to the anticipated challenge of the dive to come. This could explain why seals, when they are forced to dive, usually respond with a maximal response whereas the response is variable in voluntary dives. When forced under water they do not know for how long the dive is going to last and,

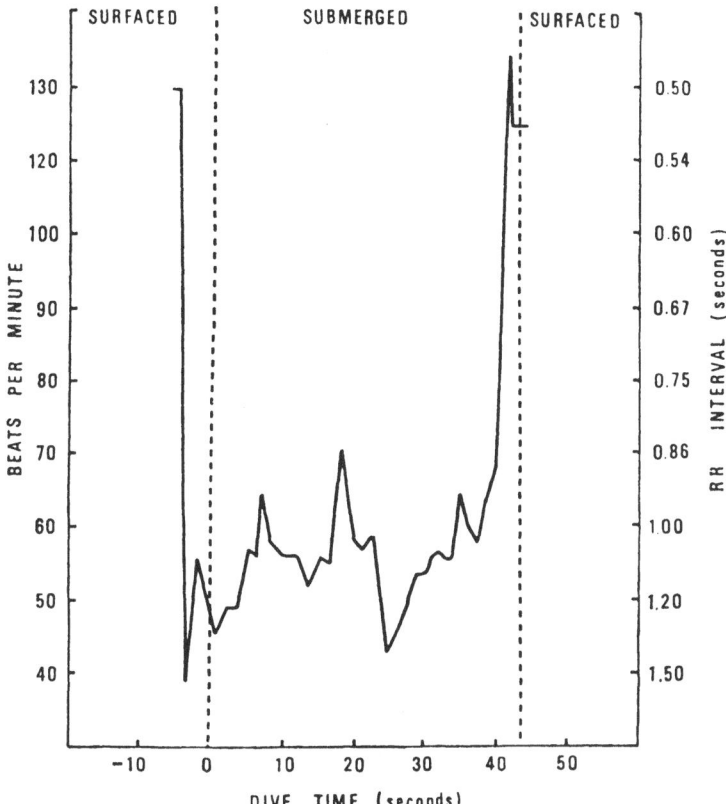

Figure 4 Typical heart rate recording from a harp seal (*Phagophilus groenlandicus*) diving voluntarily, showing pre-dive "anticipatory" bradycardia, pronounced diving bradycardia and pre-surfacing tachycardia. (From Casson and Ronald, 1975)

consequently, they "turn on" the full response for maximal O_2-economy from the very beginning of the dive. Whether this reflects stress and fear, or simply common sense, is open for discussion.

ACTIVE AND/OR PASSIVE AVOIDANCE IN DIVING

When seals and ducks are at first unaccustomed to experimental, restrained, diving they often struggle to get loose during the dive. This behaviour, reminiscent of the flight response, is associated with tachycardia; but when the struggle temporarily ceases a very pronounced bradycardia is usually seen. This indicates that divers, like many terrestrial species, such as ptarmigan (Gabrielsen *et al.*, 1984) and opossum (Gabrielsen and Smith, unpublished), may use two different strategies when exposed to novel and/or frightening stimuli. First, they might display active avoidance associated with tachycardia, high blood pressure and increased blood flow to skeletal muscles; second, when this does not get them out of the frightening situation, they might venture on a passive avoidance strategy, associated with bradycardia and peripheral vasoconstriction. Thus, it is possible that the initiation of the profound cardiovascular adjustments displayed by seals upon forced submersion could, in part, represent a passive avoidance response. It should be emphasized, however, that the autonomic component of such a response is, in principle, identical with the responses displayed by seals in prolonged voluntary dives; the result of both being the reservation of the blood oxygen store for the brain, and to some extent the heart (Zapol *et al.*, 1979; Blix *et al.*, 1983).

REFERENCES

Andersen, H. T. 1963a. Factors determining the circulatory adjustments to diving. II. Asphyxia, *Acta Physiol. Scand.*, **58**, 186–200.

Andersen, H. T. 1963b. The reflex nature of the physiological adjustments to diving and their afferent pathway, *Acta Physiol. Scand.*, **58**, 263–273.

Bamford, O. S. and Jones, D. R. 1974. On the initiation of apnoea and some cardiovascular responses to submergence in ducks, *Respir. Physiol.*, **22**, 199–216.

Bert, P. 1870. *Lecons sur la Physiologie Comparée de la Respiration*, Balliére, Paris, pp. 526–553.

Blix, A. S. 1975. The importance of asphyxia for the development of diving bradycardia in ducks, *Acta Physiol. Scand.*, **95**, 41–45.

Blix, A. S. and Berg, T. 1974. Arterial hypoxia and the diving responses of ducks, *Acta Physiol. Scand.*, **92**, 566–568.

Blix, A. S. and Folkow, B. 1984. *Cardiovascular Responses to Diving in Mammals and Birds. Handbook of Physiology. The cardiovascular system IV.* American Physiological Society, Washington, D.C. (In press).

Blix, A. S., Rettedal, A. and Stokkan, K.-A. 1976. On the elicitation of the diving responses in ducks, *Acta Physio. Scand.*, **98**, 470–483.

Blix, A. S., Elsner, R. and Kjekshus, J. K. 1983. Cardiac output and its distribution through capillaries and A-V shunts in diving seals, *Acta Physiol. Scand.*, **118**, 109–116.

Casson, D. M. and Ronald, K. 1975. The harp seal, *Pagophilus groenlandicus* (Erxleben, 1977). XIV. Cardiac arrythmias, *Comp. Biochem. Physiol.*, A50, 307–314.

Daly, M. De B., Elsner, R. and Angell-James, J. E. 1977. Cardiorespiratory control by carotid chemoreceptors during experimental dives in the seal, *Am. J. Physiol.*, **232** (*Hear Circ. Physiol.* 1), H508–H516.

Elsner, R. 1965. Heart rate response in forced versus trained experimental dives in pinnipeds, *Hvalrådets Skr.*, **48**, 24–29.

Feigl, E. and Folkow, B. 1963. Cardiovascular responses in "diving" and during brain stimulation in ducks, *Acta Physiol. Scand.*, **57**, 99–110.

Gabrielsen, G. W., Blix, A. S. and Ursin, H. 1984. Orienting and freezing responses in incubating ptarmigan hens, *Physiology and Behaviour* (submitted).

Goodman, D. A. and Weinberger, N. M. 1970. Possible relationships between orienting and diving reflexes, *Nature*, **255**, 1153–1154.

Jones, D. R. and Purves, M. J. 1970. The carotid body in duck and the consequences of its denervation upon the cardiac responses to immersion, *J. Physiol. London*, 211, 279–294.

Jones, D. R., Fisher, H. D. McTaggart, S. and West, N. H. 1973. Heart rate during breath-holding and diving in the unrestrained harbor seal, *Can. J. Zool.*, **51**, 671–680.

Kanwisher, J., Gabrielsen, G. and Kanwisher, N. 1981. Free and forced diving in birds, *Science*, 211, 717–719.

Kooyman, G. L., Wahrenbrock, E. A., Castellini, M. A., Davis, R. W. and Sinnett, E. E. 1980. Aerobic and anaerobic metabolism during voluntary diving in Weddell seals: evidence of preferred pathways from blood chemistry and behavior, *J. Comp. Physiol.*, 138, 335–346.

Ridgway, S. H., Carder, D. A. and Clark, W. 1975. Conditioned bradycardia in the sea lion *Zalophus californianus*, *Nature London*, 256, 37–38.

Sokolov, E. N. 1963. Higher nervous functions: the orienting reflex, *Annu. Rev. Physiol.*, 25, 545–580.

Woakes, A. J. and Butler, P. J. 1975. An implantable transmitter for monitoring heart rate and respiratory frequency in diving ducks, *Biotelemetry*, 2, 153–160.

Zapol, W. M., Higgins, G. C., Schneider, R. C., Qvist, J., Snider, M. T., Creasy, R. K. and Hochacka, P. W. 1979. Regional blood flow during simulated diving in the conscious Weddell seal, *J. Appl. Physiol. Respirat. Environ. Exercise Physiol.*, 47, 968–973.

8

The Diving Response in Man

J. W. Kanwisher, Research Laboratory of Electronics, Massachusetts Institute of
Technology, Cambridge, MA, USA
and
G. W. Gabrielsen, Department of Arctic Biology, University of Tromsø,
Norway

It is argued that the diving reflex theory has shortcomings which are exposed when
studies are made of voluntarily diving birds. Consideration of the physiology of marine
mammals is made, particularly of the mechanisms for functioning under temporary
anaerobic conditions. The implications for human diving are considered.

How do seals and whales do so well living in the open sea under conditions which
create dangers and difficulties to man? Such animals are able to stay underwater most
of the time, making long feeding excursions to deep water between their brief
periods of surface breathing. As one example, we know that sperm whales dive
deeper than 1 km, and can stay below for more than an hour. Only in this way have
they been able to tap an immense food resource of giant squid in the deep sea.
Independently, in another group of marine mammals, the pinnipeds, Weddell seals
are observed to go nearly as deep and as long while feeding beneath the Antarctic ice.

The widely accepted physiological explanation of such impressive aquatic
performances such as these was first suggested by Scholander (1940) nearly half a
century ago. He carried out a series of clever and original observations during forced
immersion of seals and ducks. They were lashed to a board and pushed under the
surface while he monitored physiological functions. Such animals always showed an
immediate large decrease in heart rate, as much as 16-fold or more. The reduced
cardiac output that resulted was channelled principally to the animal's heart and
brain. This was the result of a simultaneous peripheral vasoconstriction which
blocked blood flow to other major areas such as the skeletal muscles and the viscera.
Scholander aptly named this massive cardiovascular adjustment the "diving reflex".

He explained this reflex as a mechanism by which the animal's limited oxygen
supplies available during a dive were selectively saved for vital anaerobic-intolerant
tissues, such as the central nervous system. The rest of the body, starved of blood
flow, was seen to go anaerobic, accumulating a mounting oxygen debt in the form of
lactic acid. This could only be paid off over a period of time, after the animal was
allowed to breathe again. The biochemical sequence was similar to an athlete
recovering from intense exercise such as a sprint.

Many following physiologists repeated and expanded such forced immersion studies, believing they represented a reasonable model of a diving animal. The concept of a diving response became one of the basic elements of comparative physiology. Whenever an animal's heart slowed down, the diving response was likely to be called forth. This included also situations as when a fish was lifted out of water (Scholander, 1963).

Unfortunately, the physiological details of diving in marine mammals and birds are not that simple. Telemetry monitoring of voluntary diving animals shows that a diving reflex to stretch the oxygen supply is rarely, if ever, present. In addition, both aquatic and terrestrial animals show no apparent diving reflex when no water is involved. So we welcome this opportunity to suggest some different explanations of diving. We shall use these newer data, mostly acquired during the last decade, and combine them with an evolutionary point of view that stresses the similarity rather than the differences of these aquatic animals from their ancestral antecedents on land.

Let us begin by citing some graphic shortcomings to the conventional forced diving experiments from which has emerged the diving reflex theory. It was the accumulation of such personal doubts that originally motivated our own telemetric observations of unrestrained diving. One early experiment we observed, in particular, initiated what became a lengthy and agonizing reappraisal of our own views on diving physiology.

In this experiment a huge elephant seal was lashed to a cargo platform and lowered over the side of the ship by a hydrowinch. One could hardly suppress a feeling of sympathy for the unwilling subject who was forced to endure such mistreatment in the cause of physiology. It had no way of knowing when, if ever, it would be allowed to breathe again. Such episodes, typical of much past diving research, seemed to us more a study of drowning rather than diving. Yet most laboratory physiologists are habitually reluctant to take behavioral considerations into consideration during the choosing of an experimental design. Animals are usually considered to be mechanisms ruled by a finite and relatively simple set of reflexes, most of which are physiological in nature. These are believed to be largely intact, even when the animal is securely restrained and physically mistreated. Only by such assumptions can a busy investigator maintain a punctual schedule. Yet to us that elephant seal looked mighty unhappy as it was lowered into the sea. Even the most hard-bitten experimenter should sense the possibility of emotional stress affecting the physiology of such a brutally treated animal.

So it should come as little surprise that unrestrained animals diving voluntarily in the open sea tell a considerably different physiological story. Such experiments are laborious and time-consuming, and sometimes fail completely when the diving animal of interest does not cooperate. Yet, when they work the data that result have a greater credibility because of the biologically more realistic conditions of the experiment.

Voluntarily diving animals rarely, if ever, display evidence of a "diving reflex" when they submerge. Yet the same animal, when frightened either in or out of the water, will frequently display the deep bradycardia previously associated with diving. In our telemetry observations, it soon became clear that all diving is not with the heretofore expected reflex (bradycardia). Also, all precipitous heart slowing is not the consequence of an animal that is diving. We now realize that such heart slowing can occur under a variety of conditions, most of which indicate that it is a behavioral rather than a physiological response. We have yet to see deep bradycardia as part of an oxygen-conserving response in any animal which is diving voluntarily. Yet when the animal is threatened, either in or out of the water, it shows a passive defense response, a fear response that fits the picture of a dive reflex.

The remarkable persistence of the classical diving reflex down to the present, in spite of increasing contradictory evidence such as this, can serve only to illustrate the staying power of a rationally satisfying and broadly accepted theory. Countless subsequent investigators since Scholander have held seals and ducks underwater and observed their abrupt decrease in heart rate. And Scholander supplied the simple explanation necessary to make these easily performed experiments seem a fundamental physiological component of diving mammals and birds. Unfortunately, as we and others have found, a classical diving response may be rarely, if ever, present when marine animals go underwater naturally. It is equally unfortunate that ideas that have ascended to the level of textbook dogma do not really go away. If the past is reliable evidence, a changing view may emerge only with a new generation of physiologists.

It is for the naive, and thus uncommitted, in diving theories that we supply this brief review in the current confusion in physiological and behavioral thinking concerning marine mammals and birds. The most recent review (Blix and Folkow, 1984) illustrates how unchanged much of the mainstream thinking has remained. Diving physiologists still tie down unfortunate seals and ducks and place them under medical X-ray machines. Here they can observe ever more detailed features of blood flow changes when the animals are forced to hold their breath. And the concept of a "diving reflex" still holds a central position in the explanations offered for these cardiovascular changes.

The medical school location of most physiology departments offers a partial explanation for the persistence of this paradigm. Rarely stated clearly, but implicit in the argument of these researchers, is the deeply held belief that the brain of a porpoise or a seal is like that of man. When oxygen to the human brain is removed, consciousness is lost in seconds and irreversible brain damage is believed to follow after a few minutes. This concept that all mammals were created in the model of man may have emerged from our Christian heritage and the Genesis version of our origin, but more likely it is the result of the medical training of many physiologists, such as Scholander. The central role of man in such studies makes it tempting to project one's learning to other mammals by implied, but rarely proved, analogy.

A few words are necessary in defense of Scholander, whose ideas and personality breathed so much fresh air into modern comparative physiology. In his original paper he recorded the absence of a bradycardia reflex when his seal could "dive" voluntarily. This came when the animal was held at a shallow depth where it could stretch its nose to the surface for air whenever it chose. And he also graphically described the bradycardia that occurred when a seal out of water was frightened by a threatening motion of the experimenter's hand. His own prophetic words were "It thus seems very suggestive that physical factors are capable of inducing bradycardia"

(Schólander, 1940). But this dual contradictory evidence was largely disregarded by both himself and his followers. Accepting it, of course, would have denied the attractively simple concept of a one-to-one correspondence between an animal's diving and its bradycardia. This abandonment of a popular and simple explanation was apparently too high a price.

We feel it is best to disregard much of the evidence from forced diving studies, unless one is interested in fear and drowning. We will center our following explanation on telemetry data from voluntary, unrestrained divers. Also a certain amount of physiological common sense combined with long hours of unrestrained animals has provided us with additional insight. We will also take a comparative view. Vertebrate taxonomy shows that biology has independently played the evolutionary game of aquatic adaptation for warm-blooded air breathers at least twice with mammals and many times with birds. We will begin by considering the nature of the physiological problems to be solved for such an animal to have the capacity for living most of its time underwater.

THE NATURE OF THE PHYSIOLOGICAL PROBLEM

Imagine you are diving among a school of porpoises or seals. After the excitement has worn off a bit, the analysis of what you are seeing can begin. One is initially impressed by how all the marine mammals move about underwater with such grace and dexterity. They only occasionally come to the surface for brief periods of breathing. They appear to complete this as rapidly as possible so they are free to go below again. A resting gray seal might spend two minutes in this surface breathing cycle, and then have 10 min below before a necessary return to the surface. In a very real sense such animals are seen to live mostly underwater, and only occasionally rise up to the surface for air. They do not usually live on the surface, and then dive down, as is generally perceived. This is important in assessing the nature of the physiological problems they face. Following such observations of animals in nature we can now begin to sense what kind of adaptations would be most useful.

We soon realize that, if a marine mammal is to have maximum freedom in its midwater life, it should minimize the time spent breathing at the surface. This implies the natural evolutionary tendency for both rapid respiratory gas exchange and an enlarged oxygen store that can be drawn on while beneath the surface. This increased store is realized in part by a greatly enlarged blood volume, along with a somewhat increased oxygen-carrying capacity of the blood itself. There is also a sizeable amount of oxygen temporarily combined with myoglobin, a protein similar in function to the hemoglobin in the red cells. This is dispersed in the muscles, giving them a characteristic dark red color. There is enough myoglobin in some species to increase by half again the amount of oxygen carried by the blood (Robinson, 1939).

The speedy reloading of this oxygen store, depleted from an extended period submerged, depends on a vascular system which can rapidly pump all of the animal's blood through its lungs. For this purpose there is a more powerful heart, and a grossly altered circulatory system. Within it are such features as large venous sinuses, that we are only beginning to consider functionally.

There is little, if any, tendency in marine mammals to increase their oxygen supply by an enlargement in their lung volume. Buoyancy problems while trying to dive would probably inhibit such an evolutionary trend. During the long evolution of marine mammals the lung gases have become a decreased part of the animal's total oxygen store. This is particularly true of pinnipeds, which have some of the largest blood volumes, sometimes exceeding 20% of their weight. After a breathing cycle, a seal will frequently submerge with a final expiration, apparently not needing the additional increment of oxygen in its lungs.

It is important to realize, therefore, that in all of these marine mammals many lungfuls of air are needed in order to reload the blood and myoglobin with oxygen. Equally important in the animal's respiratory gas exchange is its need to discharge a nearly equivalent volume of carbon dioxide. The buffering of this gas by the bicarbonate system sets an effective upper limit on its concentration in the expired breath. This in turn determines the minimum volume of lung ventilation that is needed to dispose of it. So even though these mammals can, and do, draw the oxygen down close to zero in a given breath-hold, they must eventually move an amount of gas through their lungs that is determined by the total carbon dioxide production which must be exhaled.

One can observe that both seals and porpoises come to the surface for a series of quick breaths. During this brief time all the blood is rapidly circulated through the lungs. The temporarily increased circulation is reflected in a high heart rate, a breathing tachycardia.

This shows us some fundamental differences from the respiratory and cardiovascular physiology of terrestrial animals, such as man, where relatively smooth blood flow passes sequentially through the pulmonary and systematic capillary beds, alternating loading and unloading oxygen (and carbon dioxide conversely). Regular breathing in humans keeps a relatively constant alveolar gas composition, which is "seen" by the blood circulating through the lungs.

The intermittent breathing of a marine mammal requires a fundamentally different circulatory system. All of the blood must flow rapidly through the lungs, but it is largely unknown where much of this oxygenated blood is then stored. We would also like to know how smoothly, both in the time course during a dive, and also specifically to which organs, this oxygenated blood is subsequently delivered. Whether all of this elevated blood flow during the brief breathing period also passes through the animal's systemic tissue capillary bed is not clear. Some of the flow presumably goes to the muscles to reload the depleted myoglobin.

The anatomy of the circulatory system in both pinnipeds and cetaceans is complex and very little understood. There are large venous sinuses in which a major fraction of the blood can be temporarily stored. And in the thorax region there appear to be arterial venous shunts. These may allow some temporary systematic by-passing during the brief breathing period. We have almost no hard data with which to evaluate such a suggestion.

EVOLUTIONARY LIMITATIONS OF AEROBIC BREATH-HOLDING

What are the evolutionary limits to the time a seal or porpoise can stay underwater? We could well ask why a marine mammal does not have twice the percentage of blood and muscle myoglobin. In this way it would double its oxygen supply, and thereby the time it could stay below the surface during its routine aerobic diving. But there must be constraints, in the form of a reduced performance in other directions, which would penalize such an animal.

Consider the admittedly absurd situation of an animal that is 99% blood. It would have little space left over for all other needs, such as muscles for locomotion, as well as fat storage for the animal's thermal insulation and its future metabolic fuel supply. Equally important, such an animal would have somewhat the physical properties of a balloon filled with water. We could not expect such an anatomical arrangement to have very elegant hydrodynamic properties even if there was sufficient muscle power to drive it through the water. Observations on swimming seals indicate that, even at 15% blood, they may already be paying a large penalty from insufficient stiffness of form. They, as well as porpoises, show pronounced wrinkling of the skin when they swim fast. Part of this may also be due to the relatively flabby surface blubber layer.

Myoglobin, by being dispersed in the muscles, does not create such liquid difficulties of form. It does not have to be in a fluid form like blood because it does not move in the animal's vascular system. This makes it at least twice as efficient on a weight basis as an oxygen store over that of the hemoglobin in blood. In the best divers there can be enough of this oxygen-combining pigment in the muscle cells to increase the animal's total oxygen reserves half again over that carried in the blood. But the increasing amount must at some level eventually reach a level where, by displacing actin and myosin, it represents an invasion of the basic contractile function of the muscle in doing work.

Myoglobin has a low oxygen-loading tension in order for it to draw the gas from the hemoglobin circulation in blood. This is equivalent to saying that it loads up at a much lower partial pressure of oxygen than the hemoglobin. This stronger binding inhibits its oxygen from re-entering the circulatory system during a dive. So we should think of the myoglobin as a purely local source of oxygen for the muscle.

Thus the animal's myoglobin oxygen represents a supply which is already in place at the beginning of a dive, within diffusing distance of where it will subsequently be used at the mitochondria in the muscle cells. Its consumption while the animal is underwater does not require additional pumping of blood through capillaries. Because of this there will be a reduced cardiac output, and thus a lower heart rate, during a normal dive, but not to the extent as in a forced dive. By dispersing part of the oxygen store close to where it is needed in this way, the myoglobin is functionally matched to the enforced high blood blow during the brief breathing cycle. Since the blood must all be moved through the lungs for respiratory gas exchange, it would seem just as easy to also have it pass through the muscle capillaries, where it will reload the myoglobin.

The original purpose of the myoglobin, which is present in selected muscles in most mammals, was to act as an oxygen store during an extended period of contraction. The muscle under tension has an internal pressure which temporarily throttles the flow of blood. When contraction ceases capillary circulation is restored and the myoglobin is reoxygenated. Millikan demonstrated this convincingly in transmission monitoring of the changing oxidized and reduced myoglobin in cat muscle. This use of myoglobin as a brief temporary oxygen source is greatly expanded in the diving mammals. Both the amount of myoglobin, and the period over which it must supply oxygen are greatly increased. And, because the amount of myoglobin is so much greater, a correspondingly longer subsequent period of blood flow is needed for its reloading. But the analogy in these two cases is clear. Again, diving adaptation has merely enlarged on an already existing mammalian capability.

The original evolutionary invention for the purpose of oxygen transport throughout the body was hemoglobin-type pigments. The limited carrying capacity of blood that resulted has always been a severe restriction to the emergence of ever higher energy life styles. Such a large protein molecule to carry the relatively small oxygen molecule would seem a poor choice. The respective molecular weights are 16 000 and 32. Even in such a concentrated solution as the 50% hemoglobin of blood, the oxygen is moved at only one part in a thousand. Thus much blood must be pumped to supply the animal's total oxygen needs. But this apparent size mismatch has shown no tendency towards evolutionary change. It can probably now be considered an evolutionary invariant. One can only speculate about the expanded diving biology that would have resulted with a 10 to 100 times greater oxygen capacity in the blood. Because of such evolutionary restrictions, marine mammals can manage only a modest aerobic underwater existence.

The longest repeated breath-holding we have seen is that of gray seals. They sometimes stay underwater for periods of 10 min or more, separated by a couple of minutes of surface breathing (M. Fedak, pers. comm.). Although such an aerobic performance seems impressive, it is really a reflection of the low oxygen demand during sleeping or resting, drawing on a total oxygen store which is much larger than that in terrestrial animals. These seals, as well as other marine mammals, would be relatively poor divers when active if, like humans, they were always forced to rely only on their aerobic resources.

We have found that when a seal starts to swim, this breath-hold time between surface breathing periods decreases. With the relatively modest speed of 3 knots, it is rarely more than 2 min. This shows how the several-fold increase in metabolic rate during swimming draws the seal's oxygen reserves down that much faster, dictating a correspondingly shorter time before the animal must breathe again.

The longest natural divers, such as the sperm whale (1–2 h) and the Weddell seal (70 min) depend on anaerobic reserves. Probably all marine mammals can do the same if forced to. Through this means they have access to an additional supply of metabolic energy, much larger than the aerobic one, after the animal's oxygen has all been consumed. In order to understand the physiology behind this impressively longer diving, we should know something about the phylogenic origin of the anaerobic biochemical system present in all vertebrates from fishes onward through evolution.

ENERGY PATHWAYS

Some general features of metabolic biochemistry involved in all vertebrates are now clear. Such animals have two major biochemical pathways which generate energy from their food, one with and one without molecular oxygen. Any air-breathing animal can call on both of these aerobic and anaerobic reserves during a dive. The former is represented by the oxygen in the lungs as well as that combined in the circulating blood when the whale or porpoise goes beneath the surface. In the other pathway, sugars can be reduced to lactic acid without oxygen. This anaerobic glycolytic scheme has a low overall efficiency, but much of the loss is temporary and can be recovered when oxygen becomes available again on the animal's return to the surface. This paying off of the oxygen debt incurred during a dive is analogous to the recovery necessary after any intense muscular exertion. The size of the anaerobic reserve is determined by the amount of sugars, mostly in the form of muscle glycogen, that can be burned to lactic acid. This glycogen is a few percent of most vertebrate muscles.

Reptiles and fishes have very feeble aerobic systems when compared with mammals and birds. But their anaerobic system is fully as competent. So their ability for greater instantaneous exertion, such as that during fast movement, determined by the muscle mechanics which are similar in all vertebrates, is equal to that in warm-blooded species. But they must stop suddenly when all the glycogen has been converted. This fatigue point can be quite dramatic. Such an animal can suddenly barely move. When one is chased by an alligator the first 100 m are the most dangerous. The subsequent aerobic recovery of a tired reptile is quite slow. The course of this recovery after exertion can be followed by the gradually decreasing heart rate as the lactic acid is converted back to sugar. We have determined this recovery time, as an indirect measure of an animal's anaerobic capacity, by using telemetry to monitor heart rates in a variety of free-ranging vertebrates. Man is pretty well recovered in 15 min, whereas a fish such as a flounder or cod needs 15 h.

Most vertebrate tissue, such as skeletal muscle and viscera, can tolerate long periods without oxygen. A tourniquet, used to control bleeding after injury, can be applied to a human arm or leg for 30–60 min with little apparent permanent damage. The oxygen in the entrapped blood is gone in a few seconds. The source of energy in the meantime is by glycolysis, with the production of lactic acid. After circulation is restored, this lactic acid floods out in the blood and is metabolized aerobically in other tissues, particularly in the liver. In a sense, the rest of the body quickly restores the normal aerobic situation after a local anaerobic event.

Much the same situation occurs when intense local muscle exertion consumes ATP faster than it can be produced aerobically. Local muscles are quickly depleted of oxygen in chinning or push-ups and begin to hurt from the accumulating lactic acid. But there is only moderately increased breathing, showing little overall fatigue. The fraction of the muscle mass involved is small, so the lactic acid and oxygen levels in the circulating blood remain near normal. From such observations we conclude that there is likely to be little harm to blood-starved tissue during the routine anaerobic stress of diving.

Some vertebrate tissues are known to be quite sensitive to oxygen lack. The human brain may be the extreme case. Consciousness is lost in seconds when its circulation is cut off. This fainting comes well before oxygen in the blood has all been consumed. And in the inevitable anaerobicity which soon follows, there can be irreversible tissue damage in a matter of minutes. An analogous time scale of anaerobic injury occurs in the heart when a coronary artery is blocked. These two most vital tissues in man are ill-adapted for surviving protracted periods without oxygen.

From our open-sea breath-holding experiments we could conclude that the bottlenose dolphin, and presumably other cetaceans, deal with this problem in various ways. Cetaceans have a brain that can, temporarily at least, run without oxygen. This is hardly unexpected. One would not reasonably expect these marvelous animals to show fainting and drowning the first time they didn't quite get to the surface.

DEEP DIVING AND ANAEROBIC RESERVES

The physiological events in deep diving are analogous to those in man when running in a middle distance race. Both aerobic and anaerobic metabolic competence are important as in the performance of an athlete. The amount of aerobic effort in diving is set by the size of the oxygen stores the animal can carry with it when it leaves the surface. This oxygen-dependent energy scheme is used up first by some animals such as cetaceans. Then the porpoise or whale must switch to anaerobic glycolysis, which we have seen all muscles have as part of their mammalian and vertebrate heritage.

The extended deep diving one observes in sperm whales and Weddell seals is possible only because these species can make use of the extensive anaerobic reserves present in all vertebrate muscle. Man is prevented from this by an obligatory requirement of oxygen to his brain. When this supply is cut off, consciousness is lost in a few seconds, and, if he were in water, drowning would follow. But some, and perhaps all, pinnipeds and cetaceans appear temporarily to run their entire body anaerobically. Quantitatively understanding the impressive dive performance of these marine mammals requires that we look further at the ability of tissues to run without oxygen. The temporary metabolic energy made available by this process can greatly exceed that due to the oxygen stores of blood and muscle.

Since the earliest beginnings, animals in nature have been chasing and eating one another. The resulting prey–predator nature of life has thus always placed a high premium on quick, reliable, high-speed locomotion. We have already noted that the resulting biochemical strategy in muscles has rested on an anaerobic glycolytic pathway, which is self-contained in the cells. It is ready to rapidly supply the ATP needed for muscle contraction. Such glycolysis was already present in the earliest fishes. It has remained largely unchanged through all of the subsequent vertebrate evolution.

But glycolysis is only 5% efficient in terms of recovering biologically useful energy from food. In order not have to eat 20 times as much, animals gain access to the rest (36 of 38 ATPs) by aerobic pathways, depending on molecular oxygen as a hydrogen acceptor. Thus most mammals and birds depend most of the time on an oxygen transport system to ventilate their energy metabolism. The quick energy from glycolysis is so inefficient that it is used only in crisis situations.

We can think of the cellular need for oxygen coming first during evolution. This provided the driving force for the development of the transport system. But the means for moving large amounts of oxygen into the mitochondria was an imposing evolutionary barrier. And the high-energy life style that has finally emerged in mammals and birds has had to wait on a sufficiently competent respiratory and cardiovascular system to accomplish that expanded transport. The need for oxygen is still frequently greater than the supply, even with these complex and massive transport systems, so evolution has not completely caught up, except possibly in birds. Athletic training, particularly in endurance races, is largely the process of enlarging the oxygen transport system in order to improve extended aerobic performance.

Early vertebrates, such as fishes, were largely anaerobic animals. They were capable of an intense anaerobic muscle exertion, fully equal to the work capability of a mammal, but only for a brief time. They then had to follow this effort with a long slow recovery. In these cold-blooded animals the ability of the muscles to do work may exceed the oxygen supply by 100-fold. The result is that a brief period of exertion in a crocodile mandates a long slow aerobic recovery before another such effort is again possible.

Mammals have retained this anaerobic capability for rapid energy mobilization. And they have added to it a greatly increased aerobic capacity through an expanded lung and circulatory system. This has allowed a much greater sustained aerobic effort, such as that seen in running a marathon. And the increased oxygen transport also permits a correspondingly faster pay-back of any oxygen debt due to a sudden burst of effort. What the crocodile may take an hour to do in its recovery, a mammal can accomplish in minutes.

The size of the glycolytic energy pool is the important parameter in extended diving performance. It is determined by the amount of sugar, mostly in the form of glycogen, which is available for conversion to lactic acid.

The numbers can be estimated as follows: Consider a muscle with 2.5% glycogen. Each gram of carbohydrate yields 4000 calories aerobically. Anaerobic efficiency is only 5% of this, or 200 calories. The 2.5% of glycogen in muscle can thus generate 5 calories anaerobically. This is equivalent to the energy released from 1 ml of oxygen. A seal that is 15% blood will have 0.150 ml of blood per gram of tissue, which contains 0.4 ml of oxygen, yielding 2 calories of energy. Thus we can see that, even in the expanded oxygen storage of a seal, the anaerobic capacity is many times the aerobic one.

IN SEARCH OF THE ELUSIVE DIVING REFLEX

Man, like all terrestrial animals, lives in a sea of oxygen. When he needs more he has only to inhale. Dealing with an oxygen lack, such as in diving, has probably not played an important part in his survival. As a result he has not been greatly penalized by his need for a constant oxygen supply to his brain. Without this supply he faints, and would drown if in water.

We have already noted that most physiological research flows from medical schools, and is essentially human oriented in its intellectual bias. So it is not surprising that man's obligatory oxygen to the brain has been generalized to all other mammals. This is probably true for terrestrial species. It does not seem to be so for marine mammals.

For half a century the accepted wisdom on the physiology of diving animals required them to somehow always maintain a flow of oxygen to the brain. The reflexes which accomplish this were first detailed by Scholander from observations on ducks and seals held underwater. But our attempts to observe this on unrestrained animals were negative. A key element in the diving reflex, a greatly reduced heart rate, was rarely, if ever, seen when an animal was allowed to dive voluntarily. So one of us set out (with Sam Ridgeway of the Navy Undersea Center) to test this accepted dogma. Does the brain in true divers have a steady need for oxygen? We used unrestrained porpoises, diving voluntarily in the open sea, as experimental animals.

In a conventional physiology laboratory, with a dog or human running on a treadmill, the physiologist can measure many parameters involved in exercise. Tubes can even be sutured in place so that samples of arterial and venous blood are available on demand. We were much more restricted in the possible data that could be taken from our free-swimming trained subjects. We chose lung gas composition, as something that was both possible to monitor and would give information on the disappearance of the oxygen stores during a dive.

We trained a bottlenose dolphin, Tuffy, to exhale on command, under a funnel. In this manner we could collect the expired breath for analysis. An underwater buzzer was the signal for him to perform. He received three fish as a reward if he put his nose against the buzzer and exhaled. Our inverted funnel was placed where it would be above the blowhole and catch the gas. Of course, Tuffy did not know what was going on but at times we almost imagined he did. His training told him only what he had to do to get the fish. We also used another buzzer as a signal for Tuffy to begin holding his breath. By varying the time until the collection buzzer we could get lung samples at different lengths of breath-holding. Tuffy would readily hold his breath for as long as 8 min. After a few minutes he would become increasingly impatient for the collection signal. He would look up at us and gnash his teeth, apparently anxious to give us his sample so he would be free to breathe again. He had no way of knowing how long a breath-hold we would ask for, so he presumably had to plan for a maximum.

In the resulting measurements we were surprised at how quickly the oxygen concentration in his lungs dropped. After 4 min, halfway through a dive time that Tuffy could do with ease, the oxygen in the lungs dropped to below 3%. So during the next 4 min he was nearly, if not completely, anaerobic.

After such a long breath-hold, Tuffy would not "perform" again until he completed several minutes of rapid breathing. He was obviously recovering from the oxygen debt he had incurred. We could follow the course of his recovery by calling for a breath sample with our collection buzzer at any time. The oxygen consumption was temporarily higher, indicating the repayment of an oxygen debt from the dive.

Equally important was the slower discharge of the carbon dioxide that had accumulated. Because of its chemical buffering in the blood, this gas takes longer to unload than does the resupply of oxygen.

We were forced to conclude that, during the latter parts of a dive, Tuffy regularly survived several minutes of nearly oxygen-free breath-holding. He quickly recovered during an ensuing period of rapid breathing, and was ready to go under again in a few minutes. He did not lose consciousness, as man would under such a low oxygen tension. He appeared to run his brain with the same kind of temporary anaerobiosis as all vertebrates use in their muscles. If there was any redistribution of blood circulation in order to conserve oxygen for the brain, it would have been only initially effective in the dive.

We could now understand why there had been confusing evidence on the expected diving reflex. Marine mammals make only a partial effort, if any at all, to conserve the initial store of oxygen. Their heart rate, and therefore their circulation, drops relatively slowly and the oxygen stores are quickly consumed by blood flowing to most of the body. When the oxygen is gone, a general anaerobiosis develops in all the tissues, including the brain. We can only speculate whether similar cardiovascular details happen in a sperm whale when it is deep in the ocean for so long. It seems highly likely that such a whale has incurred a large oxygen debt. It will not go down again for 10–20 min, during which time it is presumably paying off this debt.

This ability of an animal to temporarily run all its tissues without oxygen was probably a crucial evolutionary turning point, allowing mammals to live in the water and to dive deep for their food. Then they could stay below as long as their total anaerobic capacity would allow. All species have about the same initial concentration of glycogen, and therefore equivalent anaerobic capabilities. With its inherently higher metabolism, a smaller species will run through this reserve more quickly. But its higher aerobic capacity allows it to more quickly pay back the oxygen debt, represented by the lactic acid in the tissues.

Anaerobic repayment after a dive represents a temporary vulnerability similar to that seen in any animal that has just put forth a maximum effort and is out of breath. We have found that all aquatic animals avoid this whenever possible by diving aerobically for shorter periods. This includes diving birds, seals, and manatees, as well as porpoises. There appears to be a phylogenetically broad aversion towards being caught with reserves down. Even fish, which are predominantly anaerobic animals, avoid exertion and the resulting vulnerability of enforced activity.

HUMAN DIVING

When one is working in the sea with porpoises and seals one becomes painfully aware of how ill-adapted one is to an aquatic life. If man could conserve most of his limited oxygen supply for his brain, it would take much longer before the critical oxygen level at which he faints was reached. In this way, humans could in theory benefit from a proper cardiovascular diving response more than the true marine mammals. The resulting anaerobic stress to man's muscles could be repaid in the same manner as in an athlete after a race. But at present there seems to be little indication of such a diving adaptation in humans.

There is, however, a considerable redistribution of blood circulation in humans under some stress conditions, such as during intense exercise. Blood then preferentially goes to the muscles, while the circulation to the other organs, such as the kidneys and skin, is almost completely shut down. But during exertion the muscles in man are the largest oxygen user, and it is this exercise consumption which would deplete the oxygen stores and thereby allow only a brief submersion before the oxygen was gone. In addition, the needs of thermal physiology also frequently involve altering local blood flow, as when man is too hot or cold. But the patterns used here appear of little relevance to that which could be of use in breath-holding. We can only speculate on why man has not developed his circulatory capability. One would gather that his terrestrial evolutionary history has placed little value on extended underwater performance.

THE PASSIVE DEFENSE RESPONSE

We have recently come to the conclusion that there is possibly something similar to a diving reflex which is able to extend man's survival time when he is accidentally trapped under water or snow. This is a passive defense response, an innate fear or terror response which involves a dramatic slowing of the heart and ventilation rates. It is a parasympathetic response which is automatically activated during crisis or life-threatening situations. We first saw this fear bradycardia in ground-nesting ptarmigan hens. When the incubating hen was threatened by an approaching man or dog, she froze, that is, she became motionless in a certain body position and stayed in this position until the intruder disappeared or a "critical distance" between the hen and the intruder was reached. This behavioral response was accompanied by a sudden drop in heart rate. In some instances it dropped 10-fold in one beat. The ventilation was slower and shallower and the response could be sustained for 20 minutes. The ptarmigan hen can escape but her eggs cannot. An escape attempt would reveal her eggs. A combination of freezing and cryptic coloration seems to protect both the hen and her eggs in most instances.

The passive defense response, accompanied by bradycardia, has been seen in animals as varied as fish, turtles, alligators, birds, opossums, mice, rabbits and deer (Wardle and Kanwisher 1974; Smith and DeCarvalho, 1984; Smith and Allison, 1974; Gabrielsen et al., 1984; Gabrielsen and Smith, 1984; Rosenmann and Morrison, 1974; Smith and Worth, 1980; Espmark and Langvatn, 1979). Even a diving bird, such as an eider duck, shows this response when provoked whilst on the nest. It is easy to imagine that such a bradycardia response may have been the starting point to that which had been seen in forced diving studies, particularly those on birds. So we set out to check some of this laboratory data under the more convincing conditions of unrestrained animals.

We observed heart rates by telemetry in naturally diving birds such as cormorants and ducks. These were hand-reared as pets at home. We were puzzled to see little change in heart rate when such birds voluntarily went beneath the water. It was only when we grabbed them and forced them under water that there was sharp drop in heart rate. We were forced to conclude that there is a great similarity in the fear or freezing response, and that seen in forced diving studies. No matter what causes it, the end result of stretching the oxygen supply could be the same. But whether animals ever use this in normal unrestrained diving remains in doubt.

Now, it seems that such a passive defense response may sometimes operate in man under accidental drowning conditions. There are several cases of drivers going off bridges and surviving after being trapped underwater in a car for periods of longer than 30 minutes. Similar survival times are also known for people buried in an avalanche. In addition, heart slowing, or even stopping, is suspected when a person suddenly faces a situation of stark terror. This is literally being scared to death. This dramatic cardiac response could also be a factor in the sudden crib death of young infants. Such leads suggest an interesting future in studies of the passive defense response. But caution is in order, particularly when physiological theory is pushed far beyond current experimental results.

In our own marine mammal research we have essentially been trying to discover how nature has designed a high-energy animal which can live in the open sea. Physiologists may, and frequently do, have their own ideas on what such an animal should be like. The result, in current research on diving physiology, of this temptation toward trying to outguess nature is a growing disparity between laboratory theories and experiments as compared with observations made on unrestrained seals and whales diving in the open sea. The persistence of the concept of a diving reflex is partially explained by this tendency.

REFERENCES

Blix, A. S. and Folkow, B. 1984. *Cardiovascular responses to diving in mammals and birds, Handbook of Physiology*, American Physiological Society, Washington, D.C. (In press).

Espmark, Y. and Langvatn, R. 1979. Cardiac responses in alarmed red deer calves, *Behav. Processes*, 4, 179–186.

Gabrielsen, G. W., Kanwisher, J. and Steen, J. B. 1984. Defense responses in willow ptarmigan hens (*Lagopus lagopus lagopus*), *Science* (submitted).

Gabrielsen, G. W. 1984. Free diving and habituation of the initial "dive reflex" in forced diving in ducks, *Acta Physiol. Scand.* (submitted).

Gabrielsen, G. W. and Smith, E. N. 1984. Altered physiology during feigned death in the American opossum, *Nature* (submitted).

Kanwisher, J., Gabrielsen, G. and Kanwisher, N. 1981. Free and forced diving in birds, *Science*, 211, 717–719.

Kanwisher, J. and Ridgeway, S. 1983. The physiological ecology of whales and porpoises, *Scientific American*, June, 102–111.

Robinson, D. 1939. The muscle hemoglobin of seals as an oxygen store in diving, *Science*, 90, 276–277.

Rosenmann, M. and Morrison, P. 1974. Physiological characteristics of the alarm reaction in deer mouse, 47, (4), 230–241.

Scholander, P. F. 1940. Experimental investigations on the respiratory function in diving mammals and birds, *Hvalrådets skrifter*, 22, 1–131.

Scholander, P. F. 1963. The master switch of life, *Scientific American*, **209**, 92–106.

Smith, E. N., Allison, R. D. and Crowden, W. E. 1974. Bradycardia in free ranging American alligators, *Copeia*, 770–772.

Smith, E. N. and DeCarvalho, Jr., M. C, 1984. Heart rate responses to threat and diving in the ornate box turtle (*Terpene ornata*), *Physiol. Zool.* (submitted).

Smith, E. N. and Worth, D. J. 1980. Atropine effect on fear bradycardia of the Eastern Cottontail rabbit, *Sylvilagus floridanus*, in *A handbook on biotelemetry and radio tracking*, C. J. Amlander, Jr. and D. W. MacDonald (eds) Pergamon Press, Oxford, pp. 549–555.

Wardle, C. S. and Kanwisher, J. W. 1974. The significance of heart rate in free swimming cod, *Gadus morhua*; some observations with ultrasonic tags, *Mar. Behav. Physiol.*, **2**, 311–324.

9

Resuscitation

V. A. Negovsky, Professor Academician, Laureate of State Prizes USSR, Head of laboratory of Reanimatology, of the USSR Academy of Medical Sciences, Moscow, USSR

Studies of resuscitation of drowned animals are described and implications of the results for resuscitating people drowned in the Arctic are considered. Treatment of patients after resuscitation is briefly outlined.

The subject matter of my short communication is resuscitation from drowning in the Arctic seas. Unfortunately, drowning in the Arctic seas, as well as in other seas and oceans, is a frequent occurrence. It is natural that in saving people one should use all the means (available under certain conditions) that the modern science of reanimatology has at its disposal. (The term "reanimatology" derives from a Latin word "reanimare" which means to resuscitate.) Although this science came into being only some 30–40 years ago, it has rendered the resuscitation of organisms much more effective, in comparison with what the long empirical struggle for the life of a dying man has yielded. And yet this science has comprised the whole experience accumulated over the many centuries of resuscitating dying people. Modern reanimatology began its development in pre-revolutionary Russia, with subsequent intensive augmentation in the USSR on the basis of general pathology (pathological physiology). It envisages the task of investigating the regularities of dying and resuscitating all the tissues and organs of the human body and organism as a whole; on this basis, it has elaborated scientifically substantiated methods of restoring the still extinguishing or just extinguished vital functions of the organism (clinical death, cardiac arrest).

A more profound study of the process of an organism's dying and resuscitation showed quite vividly in the course of time that in resuscitation particular importance should be attached to the brain; that the restoration of cardiac activity and respiration as well as all the other organs and functions is indispensable because it maintains alive this most vulnerable organ of the human organism and helps it overcome the hypoxia sustained, that is to say, to revive. Experimental and clinical practice has already quite vividly shown that in a number of cases, when the incipient stage of resuscitation, along with the restoration of cardiac activity and respiration, is not supplemented with measures aimed at restoring the functions of the brain, this may entail tolerable restoration of cardiac activity and respiration, but incomplete restoration of the brain, especially its higher portions (that is to say, decorticated with patients to a certain extent). Still more disastrous results are registered in delayed resuscitation, when only the heart is revived, and the brain dies. In this case resuscitation becomes senseless. It goes without saying that the earlier and more completely we restore cardiac activity and respiration, the less damage to the brain will be inflicted by the process of dying and the initial stages of resuscitation, and the more completely its functions will be restored.

When speaking of the first step in resuscitation, we mean the restoration of the cardiac functions and respiration, and the use of such premedical measures as artificial "mouth-to-mouth" respiration and external massage of the heart. These methods have justified themselves well, and still yield positive results, although in many cases they are insufficient.

As early as in the 1920s the Soviet investigators Bryukhonenko and Chechulin pointed out that artificial blood circulation is a powerful way of restoring cardiac activity. Since the foundation of my laboratory in 1936, this method, in more developed and profound modifications, was also used to restore the extinguished heart under both normothermia (Bozh'jev, 1971) and hypothermia (Soboleva *et al.*, 1970). Thus, it was shown that animals survived 10- and 15-min circulation arrest caused by artificially induced fibrillation of the cardiac ventricles. Then the animals were resuscitated with the aid of a "heart–lung" apparatus. Following 10-min circulation arrest, cardiac activity was restored by a discharge of an electrical impulse defibrillator 5–6 min after resuscitation initiation, and, following 15-min circulation arrest, 10–12 min after resuscitation initiation. The restoration of cardiac activity led to transition from complete artificial circulation to assisted circulation, until maintained stabilization of hemodynamics and pulmonary ventilation. The total duration of artificial circulation was 20–30 min. Artificial pulmonary ventilation was not restored to either group, during resuscitation or after switching off the "heart–lung" apparatus. Following treatment and suturing the wounds, the animals received no special maintenance therapy.

The animals resuscitated with the aid of the "heart–lung" apparatus after 10- and 15-min circulation arrest were in a state of "flaccid" coma for 6–9 hours. Over 1–3 days they gradually recovered hearing, vision and muscle. They were able to drink and eat 1–2 days afterwards. Three to four weeks later these animals did not differ outwardly from other (healthy) animals.

Artificial respiration is useless under such conditions (the pulmonary circle does not participate in blood circulation), and is not obligatory, since oxygenation is ensured on account of the oxygenated blood used. Similar investigations have been carried out in our laboratory under conditions of drowning.

It is known that the peculiar extinguishment of the physiological functions related to drowning poses great difficulties for resuscitation. The analysis of clinical observations, as well as of experimental data, indicates that favorable results are obtained in cold water drowning (i.e. under hypothermia) or near drowning (i.e. with preserved respiration or cardiac activity) (Redding *et al.*, 1961; Kvitingen and Naess, 1963; Hegendorfer *et al.*, 1970; Pearn *et al.*, 1976; Levin, 1980).

In experimental complete drowning it proved possible to restore the organism's functions, after being underwater for no more than 3–6 min, by using conventional methods (Lebedeva, 1966; Fedorov, 1967; Genaud *et al.*, 1965). By using artificial circulation, Gerya (1966) managed to restore the vital functions of a dog following 9–18 min drowning.

Zaplatkina (1978, 1983), an associate of my laboratory, has conducted experiments to establish quantitative assessment of an animal's neurological restoration following resuscitation after lengthy periods of clinical death caused by drowning. In addition, the efficacy of different resuscitation techniques after clinical death has been compared in two series of experiments. In both series death was caused by drowning in 1% sodium chloride solution heated to the temperature of the animal's body at the moment of drowning (35–37 °C). The duration of clinical death was 19–23 min.

In the first series of experiments, resuscitation was performed with the use of complex measures, including external cardiac massage and artificial pulmonary ventilation with 100% oxygen under constant positive pressure at expiration. To combat endogenous intoxication, the recipient's blood was replaced by a donor's by 200% after cardiac activity restoration.

In the second series of experiments following the same periods of clinical death, the dogs were resuscitated with the aid of the donor variant of artificial blood circulation. As has been revealed by the previous experiments, the method of artificial blood circulation is more effective than cardiac massage, but the donor variant of artificial blood circulation proved even more effective. In this method the blood of the animal under resuscitation, while passing through the donor's organism, is not only saturated with oxygen, but is also purified to a great extent from endogenous toxins which are sure to accumulate in any dead organism. The essence of the method consists in the fact that the donor's living organism is used instead of an oxygenator. The blood of the recipient (the dog under resuscitation) is pumped out by an artificial blood circulation apparatus into the donor. The donor's arterial blood is pumped into the recipient's arterial system. The experiments in both series were carried out on dogs under urethane–chloralose narcosis (urethane, 10 mg/kg of bodyweight; chloralose, 60 mg/kg of body weight).

The process of initial restoration of neurological functions was assessed by the appearance of a corneal reflex and electric activity of the cerebral cortex, and that of late restoration was assessed by the ECOG high frequency component. Neurological deficit was determined in points according to the table of supravital evaluation of CNS functional restoration (Safar *et al.*, 1976). Histological investigations of the brain and internal organs of the survivors were carried out a month after resuscitation.

In the dogs of the first series, the brain's functions were not restored. Partial restoration of the functions of other organs and systems proceeded slowly and with complications. Cardiac activity was restored in 7 out of 11 dogs. Their sinus rhythm was restored 4–10 min after resuscitation initiation. This was paralleled by varying disturbances in atrioventricular and intraventricular conduction. All the dogs developed a progressive decline of arterial pressure. Despite multiple administration of epinephrine and ephedrine, arterial pressure could not be stabilized. Five animals of this group were resuscitated and 6 died during the first five hours of the post-resuscitation period.

In the second series — in resuscitation by the donor variant of artificial blood circulation — 10 dogs out of 11 survived. Bioelectric activity of the heart appeared in the animals 20–30 s after artificial blood circulation initiation. The sinus rhythm was restored during the first minute of resuscitation. Complete visible restoration of the neurological status in the survivors occurred on the second to third day. The animals ate, drank, moved and reacted to pain, sound and light stimuli. Electrophysiological and histological investigations of the brain cortex and the degree of neurological deficit determined according to Safar *et al.* implied complete restoration of the functional capacity of the brain cortex cells in the survivors.

The aforesaid prompts the view that the donor variant of artificial blood circulation has great advantages over other resuscitational measures. This method reveals potential possibilities of the CNS with respect to functional restoration and provides complete and adequate resuscitation of dogs following 19–23 min clinical death caused by drowning in salt water.

There naturally arises the question as to what extent these positive results of resuscitation can be obtained after drowning in the cold waters of the Arctic seas. The above literature data on successful resuscitation of people drowned in winter prompts the idea that under those conditions the results *shall* be no worse, and may be, better. This assumption can be proved by the positive role of hypothermia preventing the death of an organism's tissues upon blood circulation arrest. The aforesaid can be supplemented with the work of Soboleva *et al.* (1967), who showed that, if dying is proceeding under hypothermia, resuscitation is sometimes possible as late as 2 hours after clinical death with complete restoration of the brain's higher functions. However, more detailed investigations devoted to cold water drowning are in store for us.

It may be assumed that the results of experimental investigations cited will be another incentive for a broader application of the method of artificial blood circulation for resuscitation in clinics. Naturally, donor blood circulation is rather hard to realize in clinical practice. Our experimental investigations have shown that the use of hemoperfusion (hemosorption) and hemofiltration (Negovsky *et al.*, 1979; Negovsky and Konovalov, 1983) also yields positive results, being but slightly inferior to the donor variant of artificial blood circulation.

In my report I have no opportunity to dwell upon all the aspects of the therapy of patients who have sustained a severe terminal state and who are suffering from the so-called post-resuscitational disease. These envisage further more perfect oxygenotherapy, combat against endogenous intoxication (hemosorption = hemoperfusion), removal of excess liquid from the organism (hemofiltration), control over the normalization of metabolic processes, restoration of sharply depleted resources of carbohydrates in the liver and so on, demanded by modern reanimatology. One's attention should be focused on the idea of not "losing" the brain, of actively helping the brain to restore its functions. This also includes methods of dilating the brain vessels, nourishing the brain (which is particularly important), carefully stimulating it at certain periods of resuscitation, and many others.

This concept can be expounded as follows: it is necessary to help the brain restore its functions as soon as possible, and the restored (revived) brain will most adequately help restore the functions of the internal organs.

Reanimatology has already greatly contributed to understanding the regularities of dying and resuscitation, and has made resuscitation of tens of thousands of people a current reality. But many more questions are yet to be solved. The mystery of death is not yet revealed. We do not know how to delay proteolytic processes, especially in the brain during dying and resuscitation, or the nature of these processes and ways of surmounting them. We do not know the optimum correlation of excitation and inhibition processes in the reviving brain. We sometimes do not know the failure of which link in the chain of the organism's vital activity provokes a disastrous catastrophe or death, and many other things.

A further detailed study of the processes of an organism's dying and revival will expand man's powers over this most complicated phenomenon, and will make it possible to prevent many unnecessary deaths.

REFERENCES

Bozh'jev, A. A. 1971. The use of nodes of the ISL-2 and AIK-RP 64 "heart–lung" apparatus for artificial blood circulation in resuscitation, *Bull. eksp. biol. i med.*, 72(8), 127–128. (In Russian)

Fedorov, M. I. 1967. The medicolegal and clinical significance of postasphyxic states, *Kazan*, (In Russian)

Genaud, P. E., Legendre, R. and Saury, A. 1965. External massage drowning, *Lancet*, 1, 1383–1384.

Gerya, Yu. F. 1966. The use of artificial blood circulation for the resuscitation of dogs drowned in salt water, *Fiziol. zh. AN USSR*, **XII**(2), 225–231. (In Russian)

Hegendorfer, U., Stoeckel, H. and Ditzel, W. 1970. Pathophysiologie und Therapie bei Ertrinkung — Sunfallen, *Z. prakt. Anasth.*, **Bd5**, 260–272.

Kvitingen, T. D. and Naess, A. 1963. Recovery from drowning in fresh water, *Brit. Med. J.*, 2, 1315–1317.

Lebedeva, L. V. 1966. Pathogenesis and treatment of terminal states caused by drowning. *Osnovy reanimatologii, Moscow*, 353–363.

Levin, D. Z. 1980. Near drowning, *Critical Care Med.*, 8(10), 590–595.

Negovsky, V. A. 1975. *Current Problems of Reanimatology*, Mir Publishers, Moscow.

Negovsky, V. A. and Konovalov, G. A. 1983. Resuscitation aspects of ultrafiltration and haemodialysis, *Resuscitation*, 10, 167–172.

Negovsky, V. A., Zaks, I. O. and Shapiro, V. M. 1979. Experimental extracorporeal haemosorption in the post-resuscitation period in dogs, *Resuscitation*, 7, 145–149.

Pearn, J., Nixoh, J. and Wilkey, J. 1976. Fresh-water drowning and near-drowning accidents involving children: a five-year total population study, *Med. J. Aust.*, 2, 942–946.

Redding, J. S., Richard, A. *et al.* 1961. Resuscitation from drowning, *J. A. M. A.*, 178, 12.

Safar, P., Stezoski, W. and Nemoto, E. 1976. Alemioration of brain following 12 minutes cardiac arrest in dog, *Arch. Neurol.*, 33(2), 91–95.

Soboleva, V. I., Mushegjan, S. A. and Super, N. A. 1967. The use of artificial blood circulation in the resuscitation of animals following prolonged clinical death under deep hypothermia, *Bull. eksp. biol. i med.*, 64(8), 14–17.

Soboleva, V. I., Tolova, S. V., Sidora, A. K. *et al.* 1970. Peculiarities of the restoration period in animals after clinical death and resuscitation by artificial blood circulation, in *The restoration period after resuscitation*. Proceedings of the Symposium, Moscow, pp. 262–265. (In Russian)

Zaplatkina, A. I. 1978. Restoration of vital functions in dogs during resuscitation by different methods following lengthy periods of clinical death provoked by drowning. *Transcripts of the 2nd All-Union Symposium "Acute ischemia of the organs and early post-ischemic disturbances"*, Moscow, pp. 404–405. (In Russian)

Zaplatkina, A. I. 1983. On the possibility of adequate restoration of dogs following prolonged clinical death provoked by drowning, *Anesteziologiya i reanimatologiya*, No. 3, 35–38. (In Russian)

Part II

Diving Operational Management

10

Arctic Diving

P. Nuytten, President, Can Dive Services, North Vancouver, British Columbia, Canada

An account is given of the author's experiences in commercial diving operations in the Canadian Arctic. Problems of equipment are highlighted, and some innovative devices described.

The dictionary defines "difficult" as follows: DIFFICULT/dif-i-kəlt/(hard to do, make, or carry out — hard to deal with, manage, or overcome). Perhaps no other single word describes arctic underwater operations quite as aptly. Carrying out routine underwater work in near-freezing or ice-covered waters *is* difficult. Special "one-off" scientific expeditions or cold tank laboratory studies are one thing — the ability to perform a variety of complex tasks on a routine, efficient, everyday basis is quite another.

Our firm, Can-Dive Services Ltd., has been engaged in working under these environmental extremes for nearly 20 years. We have performed dives using conventional "bounce" techniques, saturation diving, atmospheric diving suits, and submersibles to a total exceeding 6000 individual exposures, under polar and sub-polar conditions. The learning curve was a slow one — not because we are particularly slow-witted, but because arctic underwater work was almost totally without precedence. Virtually everything we did was done for the first time, at least, to our knowledge.

It is interesting to look back over the past couple of decades. Many of the solutions seem obvious now, but were anything but obvious when we first encountered the problems. Our early exposure to arctic conditions was in the form of participation in a number of scientific expeditions to the Canadian high Arctic.

We worked closely with the well-known Canadian physician, Dr Joseph MacInnis. On these expeditions we carried out the first arctic mixed-gas dives — the construction of the first polar subsea habitat (Sub-Igloo); the first under-ice saturation dives; the first under-ice environmental studies; and so on. Looking back, it seems that we spent a large portion of the early 1970s designing and building new equipment for cold water use, modifying existing off-the-shelf gear and conducting round after round of interminable testing under field conditions in the Arctic.

Can-Dive still participates in various research projects in the Arctic; oil-under-ice studies photography and monitoring of pressure ridge growth and that sort of thing. One of the most interesting recent trips involved underwater archaeology on the world's most northern shipwreck — the Breadalbane.

The Breadalbane was lost in 1853 while assisting in the search for survivors of the John Franklin expedition, an expedition to discover a northwest passage through the Canadian Arctic. Breadalbane lies in 100 m of water offshore Beechey Island, about 50 miles from the north magnetic pole. Camp was set up on the sea ice approximately a mile offshore and directly over the wreck site. Two holes were cut through 6 feet of ice, one for a remote-controlled vehicle that acted as a camera platform and the other to allow the entry and exit of the ADS "WASP". On 4 May 1983, I made the first dive onto the wreck and became the first human to touch her decks in 130 years. It was a great thrill. The surface temperature at the wreck site was a steady $-8.8\ ^\circ$C ($16\ ^\circ$F) and the water temperature was approximately $29\ ^\circ$F ($-1.6\ ^\circ$C). The water was crystal clear, with visibility over 300 feet — I could actually make out the bright ice-hole above from the deck of the Breadalbane.

The WASP ADS was not heated and after about 2 hours on the bottom, I began to get very chilled. By the time I surfaced, there was ice building up on the inside of the vision dome (from my breathing) and my wools were starting to freeze to the inside of the WASP unit. The WASP functioned perfectly even under such extreme conditions. A second WASP unit was maintained on surface standby through the dives. Both WASP units and their handling systems were flown to the village of Resolute NWT and then loaded aboard a tractor caravan and transported 70 miles

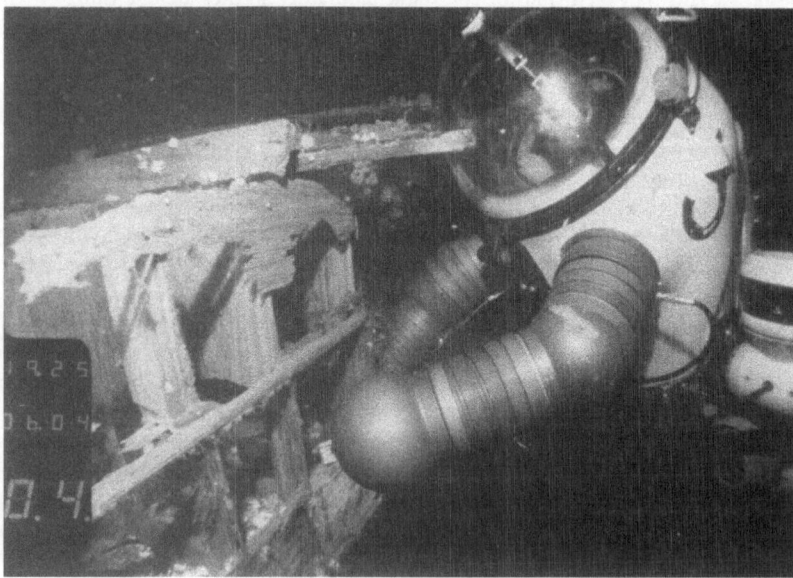

Figure 1 Diver manoeuvers the ADS WASP next to the instrument cabinet of the Breadalbane, sunk in 1853. Breadalbane is the most northerly shipwreck yet located and lies in 320 ft (100 m) of arctic water.

across the sea ice to our campsite. From the time the equipment arrived at Beechey Island until we were set up and ready to dive occupied less than 24 hours.

A photo of the WASP being raised through the ice hole of the first dive was featured on the cover of the July 1983 issue of *National Geographic Magazine*. None of the other work that we have done in the Arctic has received such international attention. There seems to be something about shipwrecks and diving through the ice at the top of the world that appeals to everybody!

The various scientific expeditions were a good way to acquire data on hardware and human performance but they could not compare with the value of routine everyday work in cold water. Can-Dive began working to support off-shore petroleum exploration, off Canada's east coast, in the early 1970s. The surface conditions, in the summer, were excellent but the water temperature was near freezing all year round. Working offshore Newfoundland and Labrador was a good proving ground for much of the equipment and techniques that we have developed. Concurrent with the East Coast Canada operations, Can-Dive conducted winter construction and pipeline work in the Northwest Territories and the Great Lakes areas of Canada. All of this experience was invaluable to us, because it simulated arctic conditions without the logistical difficulties associated with transporting and maintaining a team in the Arctic.

Figure 2 Dough Osborne, Can-Dive Services Ltd., guides the WASP up through the ice. The pilot, Phil Nuytten, had been the first on the deck of the Breadalbane since her sinking in 1853. Water temperature was 29 °F (−1.6 °C).

In 1975, we were awarded a major support contract with Dome Petroleum, working out of an Eskimo village called Tuktoyaktuk, near the Beaufort Sea. During that same year we made the first arctic atmospheric dives for Panarctic Oil Ltd., using the JIM ADS and diving through 16 feet of ice into nearly 1000 feet (304 m) of arctic water.

We currently average about 1000 dive exposures per season in the Beaufort Sea region of the Canadian Arctic. Can-Dive operates six saturation spreads in the Beaufort Sea alone. Much of the equipment is heavily modified to allow routine performance in freezing waters. We continue to operate off Canada's East Coast, carrying out work with saturation systems, remotely operated vehicles, manipulator bells and the usual matrix of oil exploration support equipment.

The work in the Beaufort Sea and the Great Lakes is usually shallow and seldom exceeds 100 m (328 feet) in depth. Offshore Newfoundland we have carried out some very deep saturation exposures. In 1983, for example, we made a series of saturation dives off Eastern Canada, the longest being 37 days at 750 feet (228 m).

Can-Dive is in the midst of development of a number of innovative devices. The problems are the same as anyone else would encounter with one major difference: our hardware must work, and work well, in freezing water. One such device is called the NEWTSUIT. It is an ultra-lightweight atmospheric diving suit that will allow exposures down to a depth of 750 feet; the manipulative dexterity of the limbs will approach those of a conventional saturation diver and the total design weight, complete with life-support system, buoyancy control and all the odds and ends, is targeted at 200 lbs (90 kg). Like other ADS units, the key to the NEWTSUIT is in the patented pressure joints; unlike other suits designed to operate in more temperate climes, the NEWTSUIT must work well in the Arctic. It must work where the cabin temperature is 75 °F (23 °C) and the outside water temperature is 29 °F (−1.6 °C). The joints are fluid-supported and have both an internal and external seal. One seal is warm, the other is cold. As you can imagine, this presents an interesting engineering problem!

A second device, that we have recently launched, is a manipulator submersible called DEEP ROVER. The primary designer, Graham Hawkes, is the same engineer who developed WASP, MANTIS and BANDIT. The first DEEP ROVER was built in Halifax, Nova Scotia, to go on contract for Petro-Canada Ltd., the Canadian National oil company. It will see service in the sub-polar waters offshore Nova Scotia. Graham also was confronted with the problems of building a state-of-the-art machine that must be useful in very cold waters. A computer study of differential contraction and expansion indicated that things were not as simple as originally anticipated. Some of these sorts of problems moved Graham to remark "You know, Phil, we first discussed this design — this underwater helicopter — when we were both in Hawaii". He shook his head. "It seemed so simple then!"

Design of cold water equipment is not particularly difficult. It does require, however, close attention to detail and at least a basic understanding of how materials will react in a frozen sea. If it is impossible to live with loss of tolerances, or dimensional shifts, then the device must be kept warm, by whatever means are available. Where conditions are optimum, the design can be rigid, but where conditions are extreme and rigid, then the design must be forgiving, compliant, able to move and change, as required.

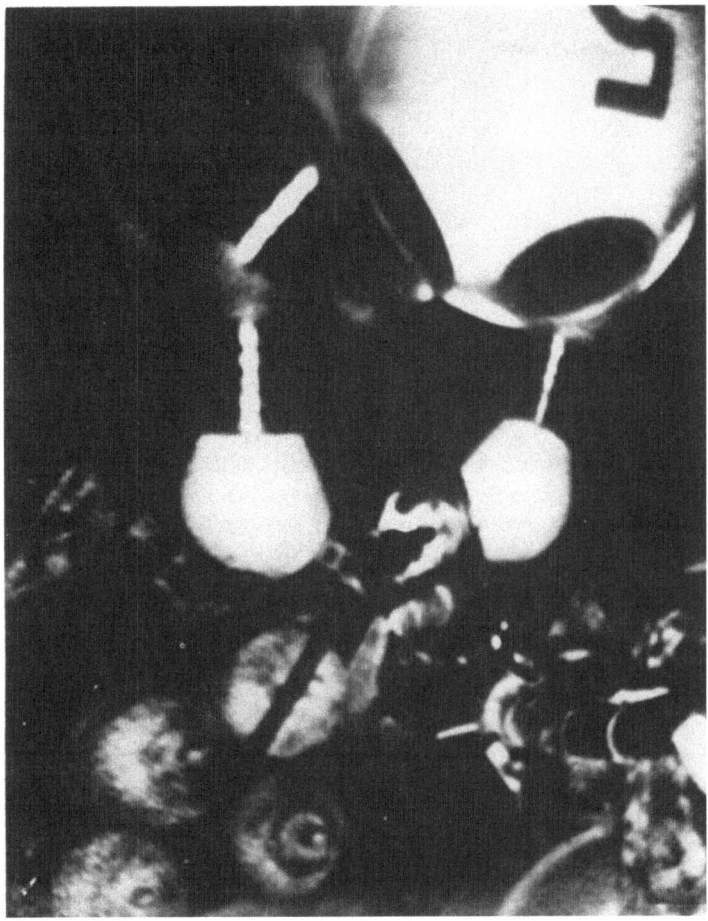

Figure 3   This photo, taken from a video monitor, shows the ADS JIM at work on a BOP stack in the Canadian Arctic. Dives were conducted in 1975 through 16 ft (5 m) of ice, and 905 feet (282 m).

The cold water equipment that we design and build today is, to use the terminology of computer manufacturers, "user friendly" — it is specifically built to be compliant and forgiving. It is ironic that, in the end, it is that most adaptable of beasts — the human — that turns out to be the restricting factor. We were neither designed nor constructed to handle great shifts in our optimum operating temperature, and like the Eskimos of a hundred centuries must use our wits and our skills to allow us not only to function but to perform well under conditions that approach the worst found on this planet. I doubt that the average drilling foreman has any idea of what a complex of experience, effort and ingenuity he calls into play when he says: "Looks like we have a problem — better call the divers out".

We repeatedly found that the most effective solution to a temperature-related

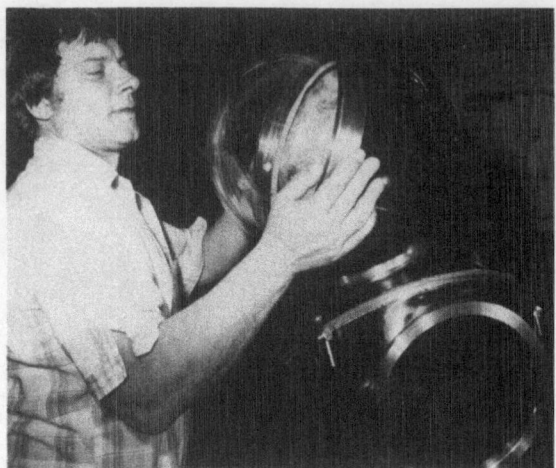

Figure 4 The author clamps the vision dome to the first prototype of the ADS NEWTSUIT. The torso, shown here, is made of high-strength, composite fiber. Design depth is 235 m (750 ft) and unit weight of 100 kg (220 lb).

problem was the simplest. This general rule is, of course, not restricted to arctic diving but is certainly true there also.

In the 1960s and 1970s we experimented with diver garment heating and breathing gas heating: such elaborate solutions as chemicals using the latent heat of crystallization, with electric resistance wires, with heat pumps, thermal oils, tubulated undergarments, thermal regenerators, in fact almost everything except multiple layers of grease. What do we use now? Open circuit hot water. Primitive and brutally inefficient, but simple. We can live with the inefficiency because the hot water system is rugged, reliable and easily available on the drilling vessel. We have foregone elegant, sophisticated solutions and devices because we simply cannot afford the down-time cost of high maintenance, low reliability contraptions.

The same philosophy held true for our breathing mixes. A decade ago we tailored the gas mixes to the exact depth, wrung the last ounce of efficiency out of the tables, saved a little in decompression time but were confronted with a Pandora's box of technical problems. Now we use simple, preset mixes for a given range of depths, accept the penalty of slightly increased decompression and dive day after day with very low bends incidence and very high reliability.

This simplicity that follows complicity seems a very obvious choice now, but it was necessary to actually go through the complexity to really understand the problems and to know what the real choices were. Some readers are probably thinking "So what you're really telling us is that there is no magic — just use the same equipment that everyone else is using and we can operate in the Arctic with no problem?" I honestly wish that were the case. There is no magic, it is true, but there are a hundred details; procedural, hardware, materials, processes and the like that are peculiar to routine cold water operations. There is no real substitute for having

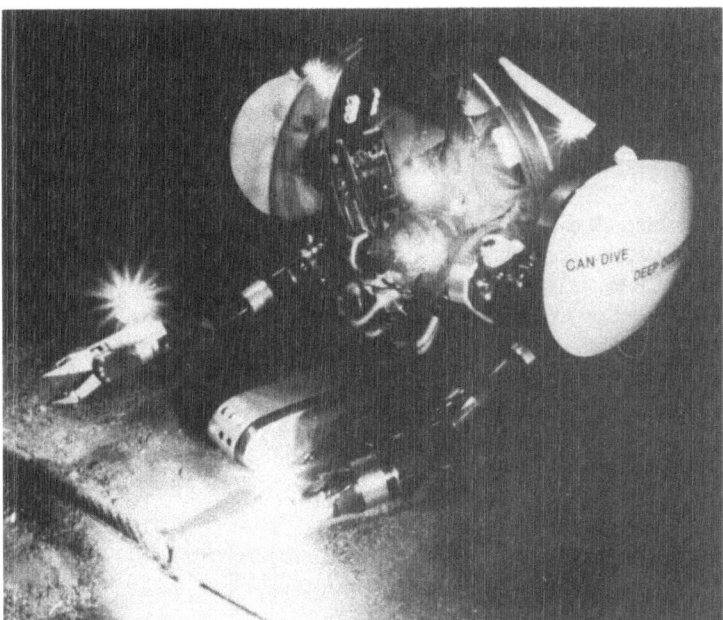

Figure 5 Model of Can-Dive's new submersible "DEEP ROVER". This sub has been compared to an underwater helicopter. Note extensive use of acrylic, nylon and other high strength plastics to improve suitability for polar and sub-polar operations.

made the mistakes. It is a wonderful fact of nature that such a great number of ways to do things differently spring into your mind when your buttocks are freezing off!

Not all of our research and development efforts have been carried out for our own requirements. We have done equipment development work for the Canadian Government, the oil industry, and for other commercial diving companies. In the mid-1970s, for example, we developed and tested a closed-loop diver and breathing gas heating system for Vickers Oceanics in the UK. We delivered and installed the system aboard a Vickers lock-out submersible and were able to extend diver outside excursion times to an average of six hours, where they had only been able to tolerate 1–2 hours using passive systems. Another major research programme was conducted on behalf of the Polar Gas consortium in 1980. This group proposes to build a 36 inch gas transmission line to convey gas from the Arctic Islands to southern markets. The cost estimate exceeds $7 billion. By 1980, Polar Gas had invested some $70 million in engineering and feasibility studies. Can-Dive was contracted to develop a system to allow coded underwater welds to join up the pipeline over a length of 99 miles of submarine crossings between the Arctic Islands and the mainland. The work had to be carried out through the ice and at maximum depths of 1700 feet.

What can I tell you about Arctic underwater operations that you don't already know? First, my opening statement is correct — it *is* difficult to work efficiently and perform well under environmental extremes. Some things become apparent fairly early. Just like the physiological effects of pressure, it is not the lower temperature itself that is the problem, it is the temperature differential. The various pieces of diver life-support hardware are warm on the inside and cold on the outside. This differential temperature across a given fitting, penetrator, seal on joint sets up material stresses, causes loss of tolerance and generally causes changes in the operating characteristics of much of the equipment that was designed for warmer climes.

So what about all these valuable lessons, these short cuts, these "tricks"? Are they secret? Not at all. Space precludes a detailed listing of specific procedures and hardware, but we have learned a lot over the past twenty-odd years and we are willing to share that knowledge. I believe there is a bond of empathy between those of us who must work in freezing and ice-covered waters wherever in the world those waters are located.

11

Deep-diving in Canadian Waters

B. E. Townsend, Diving Officer, Department of Fisheries and Oceans,
Winnipeg, Canada

The Western Region diving program is summarized and training and organizational
strategies used to prepare divers for hostile environments are reviewed. Examples are
given of the needs and problems associated with Arctic research operations.

INTRODUCTION

The Western Region of Canada's Department of Fisheries and Oceans carries out
fisheries research and management activities over one half the land area of Canada.
Arctic activities are conducted in the region's northern sector (Fig. 1) with an overall
area of 3.4×10^6 km^2, a freshwater area of 133 294 km^2 and a coastline of
110 880 km. This northern area provides unique opportunities for underwater
research and discovery.

Canadian Arctic diving operations were initiated in the mid-1950s by naval divers
during surveys of the Royal Canadian Navy's Arctic Patrol vessel, HMCS Labrador
(Wilson, pers. comm.). Research diving began in the late 1960s when the Char
Lake–Resolute Bay Project logged 1174 dives (Rigler, 1972; Welch, pers. comm.)
in support of both freshwater and marine environmental research (Welch and Kalff,
1974, 1975). During this time, the James Allister MacInnis Foundation also
supported marine underwater research (MacInnis, 1972a; Green and Steel, 1975)
and MacInnis and Curtsinger (1973) provided popularized accounts of this project.
In addition to research application, the MacInnis expeditions tested and evaluated
operational strategies, equipment and human performance under arctic conditions
(MacInnis 1972b, 1974; Anderson, 1973, 1974). In recent years, oil and gas
development has stimulated diver supported research projects (Cross, 1982) as have
Arctic mining activities (Fallis, 1982). Papers by Bright (1972) and Jenkins (1976)
also provide valuable information on operational procedures for Arctic research
diving.

This paper summarizes the Western Region diving program and reviews the
training and organizational strategies used to prepare scientific divers for remote,
frequently hostile assignments. Examples are cited within the text to demonstrate
some of the needs and problems associated with Arctic research operations. The

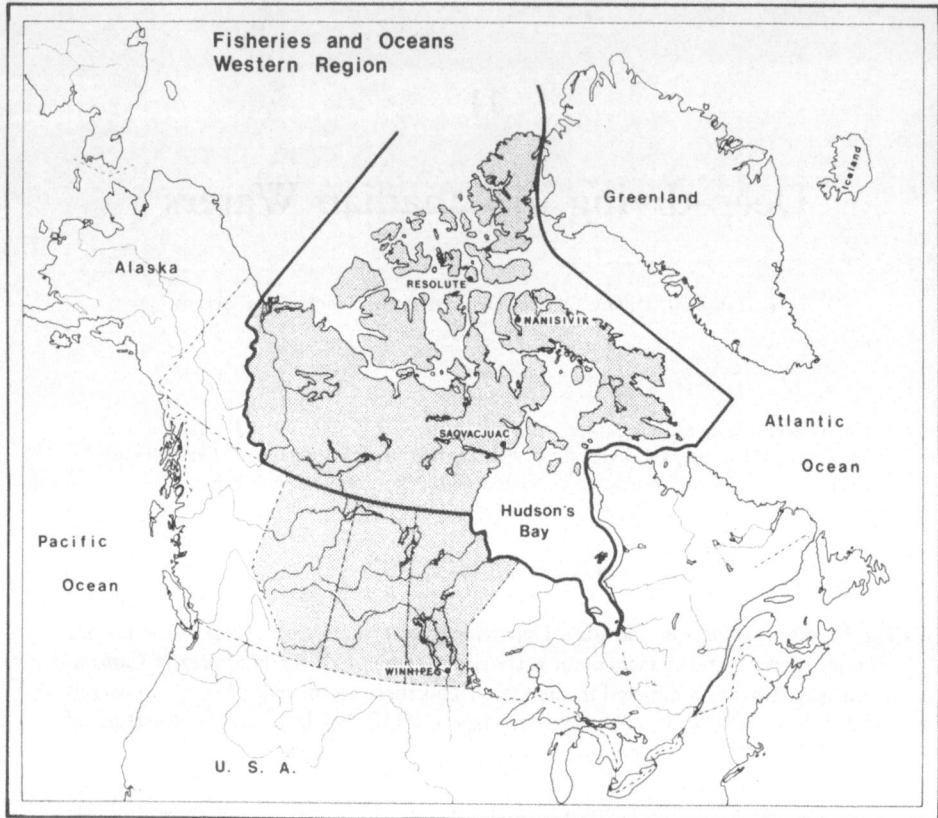

Figure 1   Map of Western Region (shaded) with Arctic area outlined with solid black line.

regional diving program is not restricted to Northern programs (see Fudge, 1984; Townsend and Butler, 1984) and many of the strategies outlined in this paper apply to research operations in general. Arctic projects, however, are most demanding and a great deal of preparation and training is needed to ensure safe effective research operations.

PROGRAM

The Western Region Diving Program is regulated by the scientific community it serves. This self-regulation enhances both the safety and the effectiveness of the program because research personnel control both the operational and experimental components of diving.

Each year, an average of 40 divers participate in the Region's diving program. Dives rarely exceed 20 m and although SCUBA is most common, surface supplied helmet equipment is also used for certain applications. Underwater projects include specimen collection, fish census, quantitative photography and *in situ* experiments involving radiochemicals. Approximately half the divers are full-time staff members and the remainder are university students working during the summer, university graduate students studying with scientists in the region and other individuals doing research contracts. This scenario parallels the recent description of the average Canadian scientific diving operation (Sparks, 1984).

TRAINING

Program efficiency and safe diving attitudes are greatly enhanced by progressive training procedures, Inflexible regulation of activities tends to produce "avoidance responses" in research personnel and this attitude tends to decrease safety (see also Flemming, 1980). Common sense approaches to training ensure the interest and cooperation of the participants.

Most Arctic field projects involve short term, very intense periods of activity and the "part-time" diving scientist must be gently reminded on frequent occasions that limits and skills decrease with age and with an inactive life style. During summer (ice-free) operations, free-swimming divers are involved with projects where long underwater tracks and long surface swims are common. Tides and currents are frequently encountered and physical fitness is a very important factor for diver performance and safety. Training programs allow divers to recognize for themselves their personal limitations and to learn to operate within these boundaries. It is better, of course, to establish these in controlled training sessions than during a remote cold water diving operation.

Each year, every person who proposes to dive must complete a comprehensive diving medical conducted by a physician who understands the rigors of diving. Pulmonary functions are performed and an active ECG is obtained during a treadmill or bicycle stress test. The stress test also provides a V_{O_2} maximum value which is considered to reflect fitness level (Stegemann, 1981). In our experience, the test also has an additional benefit in that it appears to promote friendly competition and is a real training incentive.

After passing the medical, each diver participates in a poll check-out where abilities are assessed, reviewed and recorded by a fully accredited instructor. An 800 m swim in full SCUBA equipment is conducted with a target time of 20 min. This time period reflects a fitness level suitable for free-swim operations. Following the pool session, emergency contingencies for drowning and hypothermia are reviewed and each diver is certified in CPR and first aid.

Advanced training is made readily available and many of our scientific divers have become SCUBA, CPR, and first aid instructors. We also offer special incentive courses such as underwater photography, search and rescue and underwater orienteering. In addition to these in-house training programs, a research diver course is conducted in cooperation with the local universities. During this course, research diving skills are practiced and a wide variety of underwater research topics are covered to demonstrate research applications.

EQUIPMENT

The success of Arctic diving operations is greatly enhanced by the proper selection and use of equipment. Well-serviced gear is essential and effective thermal protection for divers and equipment increases efficiency and safety. Such preparation is particularly important for marine operations where salt and sub-zero water temperatures place extra stress on personnel and equipment. Constant volume neoprene-type dry suits are commonly used in marine operations and, when combined with polypropylene or wool underwear, they provide reasonable thermal protection for divers. Yokohoma diving dress is used during helmet operations and the dry head and dry hand capabilities of this suit maximize thermal protection and increase diver effectiveness. Extra underwear is required when Yokohoma canvas type dress is used because these suits are thin and poorly insulated.

During summer operations, or for short-ice dives, wet suits are often used in freshwater operations. Wet suits allow increased mobility and improved buoyancy control and for free-swimming, short-duration dives, diver effectiveness may actually be enhanced. One-eighth inch (3 mm) vests and bib-type pants extend dive time considerably. A comparison of 1970–71 Char Lake dive times, however, showed that dry suits extended underwater time threefold (Welch, pers. comm.) and demonstrate the effectiveness of dry suits even in Arctic freshwater operations.

Both single and double hose regulators are used in Arctic operations and Bright (1972) compares their performances. In adverse conditions, regulator freeze-up remains a problem but the following procedures are used to lessen the frequency and seriousness of the event. Regulators are kept warm prior to diving and the diver does not breathe through the unit until it is submerged. As a safeguard against malfunction, each diver is outfitted with two regulators. These regulators are fitted on a double studded yoke which links two 1.5 m^3 aluminum cylinders. Some divers also use a spare regulator on a separate pony tank as emergency back-up. Double hose regulators with dry suit whips on the hooka port are used with Dacor $\frac{3}{4}$ inch tank valves fitted with a submersible pressure gauge. Because the pressure port of this valve is exterior to the valve seat, the submersible gauge is turned off when the cylinder is not in use.

To further minimize regulator freeze-up, dry breathing air is supplied. Each compressor is outfitted with purification and sorbent cartridges to assure pure dry air. As a matter of routine, an ambient dewpoint of $-60\ ^{\circ}C$ is required and air purity is within CSA Z180.1 Compressed Air Breathing Standards. Filters are replaced on an elapsed time basis and compressors are overhauled annually. Each year, an independent air analysis tests air purity.

OPERATIONAL CONSIDERATIONS

Physically and mentally fit, well-equipped divers operating under simple but well-planned diving strategies maximize the safety and effectiveness of Arctic operations. Operational remoteness cannot be stressed too strongly.

Before any field operation is initiated, emergency evacuation contingencies are reviewed and tested. Evacuation routes and emergency transport procedures are established. Hyperbaric facilities and appropriate regional medical personnel are contacted to establish pre-emergency liaison. Each diver is issued a laminated emergency procedure card which lists regional contacts and their telephone numbers. Each card also contains a microfiche of that person's entire medical history. Each diving team is issued oxygen resuscitators with a four-hour supply of oxygen.

Arctic transportation and evacuation contingencies are uncertain even in ideal situations. All contingencies are limited by weather conditions. Distances to definitive medical facilities are very large and transportation times could easily exceed 12 hours. Thus as with any operationally remote venture, prevention is the key.

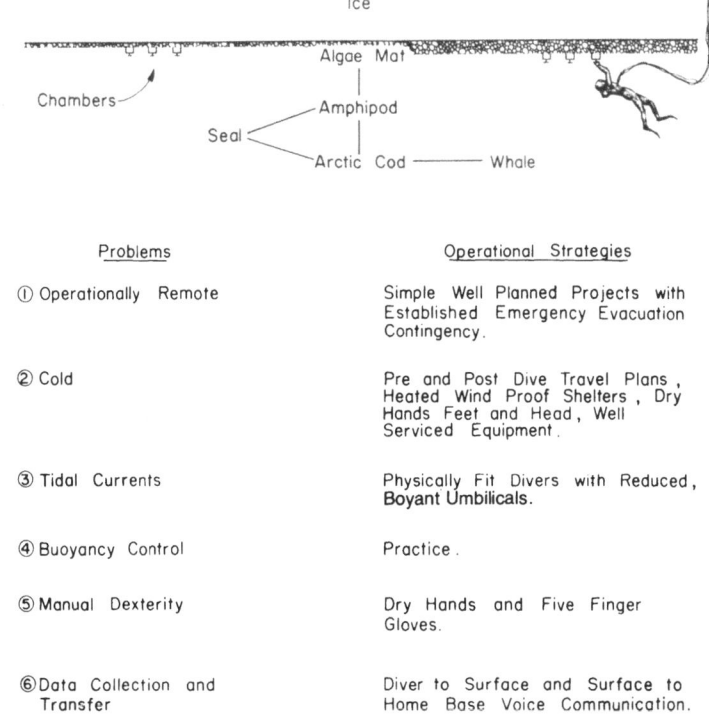

Problems	Operational Strategies
① Operationally Remote	Simple Well Planned Projects with Established Emergency Evacuation Contingency.
② Cold	Pre and Post Dive Travel Plans, Heated Wind Proof Shelters, Dry Hands Feet and Head, Well Serviced Equipment.
③ Tidal Currents	Physically Fit Divers with Reduced, Boyant Umbilicals.
④ Buoyancy Control	Practice.
⑤ Manual Dexterity	Dry Hands and Five Finger Gloves.
⑥ Data Collection and Transfer	Diver to Surface and Surface to Home Base Voice Communication.

Figure 2 Sub-ice algae study with food chain interactions and operational strategies used during the experiment.

Winter project

A good example of an Arctic winter diving project was the study designed to measure sub-ice algae production near Resolute Bay during the early spring. Although the sub-ice algae production represents only 5% of the total annual production (Horner and Schrader, 1982; Newbury, 1983), it is a critically important component of the food web in the early spring. Figure 2 demonstrates possible under-ice food chain interactions as well as operational strategies used during the experiment.

Delicate ice algae communities were isolated within plastic boxes attached to the ice by divers (Fig. 3). These water-filled boxes, incubated *in situ*, permitted primary production rates to be monitored. Divers used a modified push—pull diving system which consisted of bottled air piped through a control panel and carried to and from the diver via pressurized air hoses. Air was supplied on demand and was then exhausted back to the surface. This surface to diver air loop and demand breathing mechanism eliminated in water exhaust bubbles that would have destroyed the integrity of the biological community being investigated.

Diver efficiency and overall operational effectiveness was greatly influenced by pre- and post-dive procedures. A warm, wind-proof staging area was a very important part of the operation, and ice hole shelters were used at permanent dive sites. For more transient operations, large motorized flex track vehicles were used for transportation and shelter (Fig. 4).

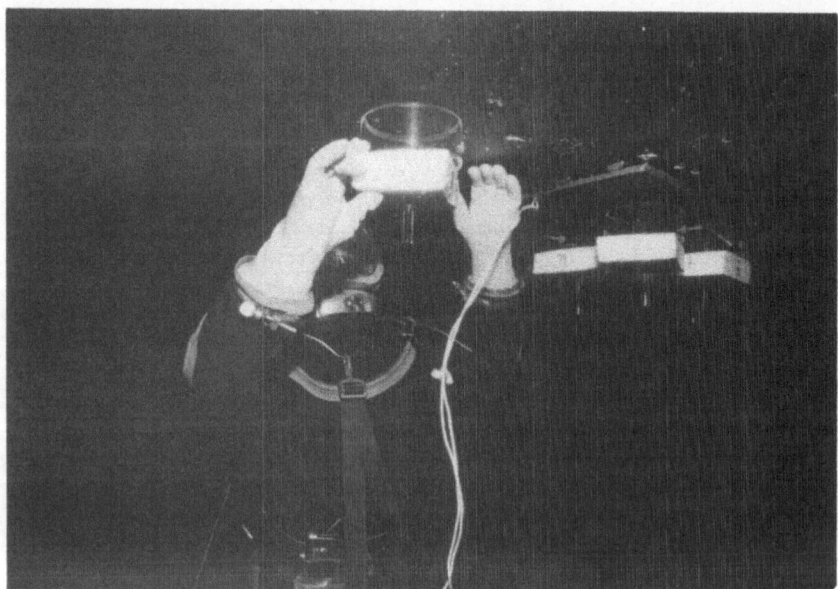

Figure 3 Diver places incubation chamber under ice. Bubbles were introduced by photographer using SCUBA.

Figure 4 Hypothermia is very real, both below the surface and above. Pre- and post-dive travel plans are critical and winter operations must ever be guided by judgements regarding weather conditions.

During the experiments, a variety of motor skills were required and experimental procedures depended heavily upon manual dexterity and buoyancy control. Proper hand apparel was a compromise between thermal protection and manual dexterity. Dry hands greatly improved diver efficiency and insulated five-finger gloves permitted reasonable dexterity. Buoyancy control was also very important since experiments were suspended at the ice–water interface. Divers had to maintain a neutral position without disturbing the experiment. Bulky suits and shallow water volume changes often limited diver effectiveness and practice was needed to master neutral buoyancy procedures. Immobilized, neutrally buyoant divers were also more susceptible to cold, and extra thermal protection had to be worn. An effective base camp to dive site radio link, as well as diver to surface voice communication, greatly enhanced operational safety and effectiveness.

Summer project

Although summer diving projects are considerably more pleasant than winter operations, certain problem areas must be recognized and addressed (see Fig. 5). A good example of a typical summer operation was the trace metal monitoring program which assessed the impact of the Nanisivik lead–zinc mine effluents on the marine biota of Strathcona Sound (Fallis, 1982). Sediment and biota samples were collected from stations within the Sound and divers were able to select the required organisms for metal analyses. The survey was conducted in September and divers had to contend with snow flurries, moving ice, cold water, and currents. Dives were

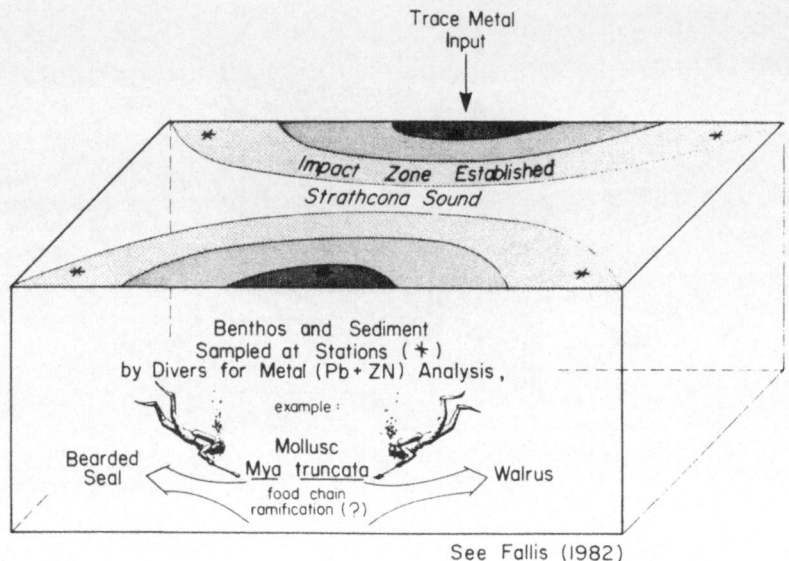

See Fallis (1982)

Problems	Operational Strategies
① Moving Ice	Effect U.W. Communication (Recall).
② Hostile Marine Mammals (Walrus, Bear)	Adequate Surface Support with Recall Capabilities.
③ Hunters (Rifle carrying)	Advise Local Villages of Activities.
④ Surf and Tides	Physically Fit Divers with Adequate Surface Support.
⑤ Repetitive Diving	Well Serviced Gauges and Trained Top Side Personnel with Oxygen.
⑥ Operationally Remote	Simple Well Planned Projects with Established Emergency Evacuation Contingencies.

Figure 5 Mine effluent study with sample location and operational strategies used during the monitoring program.

shallow and of short duration but as many as three dives per day were common. Repetitive dive schedules were continually and meticulously monitored and each regulator was fitted with a maximum depth indicator and a bottom surface interval timer.

In addition to environmental and physiological problems, biological hazards must also be considered. The polar bear, an aggressive predator, can be a serious threat to diving operations. The walrus, always tempermental, is best given a wide berth. In many summer surveys, divers also encounter rifle-carrying seal hunters and locals must be kept informed of diving locations.

CONCLUSIONS

Arctic diving requires a great deal of attention to preparation, training and on-site management. These commitments are considerable but, when the returns from underwater research operations are reviewed, the effort proves very worthwhile. Projects such as the trace metal monitoring program and the studies of the productive sub-ice communities could not have been accomplished without the use of skilled, careful divers. Thus, it is possible to take a diverse group of very independent individuals with strong motivations regarding what they wish to accomplish scientifically and train them not only in basic diving techniques but also in safe, effective diving techniques in a very hostile environment.

ACKNOWLEDGEMENTS

I very much appreciate the support provided by Bruce W. Fallis and H. (Buster) E. Welch during the effluent and sub-ice algae projects, R. D. Hamilton and Ian Davies spent considerable effort editing the text. Donna Glowacki provided friendly and efficient typing services.

REFERENCES

Anderson, B. 1973. *Arctic III Expedition: Diving Equipment and Human Performance During Diving Operations in the High Arctic*, Office of Naval Research and NOAA-MUST Office, Washington.

Anderson, Briger G. 1974. Diving equipment and human performance diving undersea operations in the High Arctic, in *Proc. Symp. The Working Diver*, Marine Technology Society, Columbus, Ohio, pp. 325–340.

Bright, Chester V. 1972. Diving under polar ice, in *Proc. Symp. The Working Diver*, Marine Technology Society, Columbus, Ohio, pp. 145–157.

Cross, W. E. 1982. Under-ice biota at the Pond Inlet ice edge and in adjacent fast ice areas during spring, *Arctic*, 35, 13–27.

Fallis, B. W. 1982. *Trace metals in sediments and biota from Strathcona Sound, NWT; Nanisivik Marine Monitoring Programme, 1974—1979*, Can. Tech. Rep. Fish. Aquat. Sci., 1082.

Flemming, N. C. 1980. Safety procedures and training for the scientific diver, in *Technical and Human Aspects of Diving and Diving Safety*, Vol. 1, (Doc N300/80E), pp. 137–176.

Fudge, R. J. P. 1984. Scuba assisted research on Southern Indian Lake, Manitoba, in *Proceedings of the Third Canadian Ocean Technology Congress, Toronto*.

Green, J. M. and Steele, D. H. 1975. Observations on marine life beneath sea ice, Resolute Bay, NWT, in *Proceedings of Circumpolar Conference on Northern Ecology*, National Research Council of Canada, Section II, pp. 77–86.

Horner, R. A. and Schrader, G. C. 1982. Relative contributions of ice algae, phytoplankton, and benthic microalgae to primary production in nearshore regions of the Beaufort Sea. *Arctic*, 35, 485–503.

Jenkins, W. T. 1976. *A Guide to Polar Diving*, Office of Naval Research, Arlington, Virginia AD-AO30067.

MacInnis, J. B. (ed) 1972a. *Arctic I and II Underwater Expeditions. James Allister MacInnis Foundation Arctic Diving Expeditions*, The Arctic Institute of North America.

MacInnis, J. B. 1972b. Arctic diving and the problems of performance, in *Proc. Symp. The Working Diver*, Marine Technology Society, Columbus, Ohio, pp. 159–172.

MacInnis, J. B. and Curtsinger, W. R. 1973. Diving beneath Arctic ice, *Natl. Geogr. Mag.*, 144(2), 248–267.

MacInnis, J. B. 1974. Arctic diving operations. Results of five expeditions, in *Proc. Symp. The Working Diver*, Marine Technology Society, Columbus, Ohio, pp. 7–28.

Newbury, T. K. 1983. Under landfast ice, *Arctic*, 36, 328–340.

Rigler, F. H. 1972. Director's review, in *Char Lake Project Annual Report 1971—1972*, Canadian Committee International Biological Program.

Sparks, R. 1984. Canadian scientific diving: results of the 1983 Canadian Association of Underwater Sciences National Survey, in *Proc. Third Canadian Ocean Technology Congress*, Toronto.

Stegemann, J. 1981. *Exercise Physiology: Physiologic Basis of Work and Sport*. Skinner, J. S. (trans. and ed), Georg Thieme Verlag, Stuttgart.

Townsend, B. E. and Butler, J. E. 1984. Use of diving in marine and freshwater research: overview of Fisheries and Oceans activities in Western Region, in *Proc. Third Canadian Ocean Technology Congress*, Toronto.

Welch, H. E. and Kalff, J. 1974. Benthic photosynthesis and respiration in Char Lake, *J. Fish. Res. Board Can.*, 31, 609–620.

Welch, H. E. and Kalff, J. 1975. Marine metabolism at Resolute Bay, NWT, in *Proc. Circumpolar Conference on Northern Ecology*, National Research Council of Canada, Section II, 69–75.

12

Diving in Antarctica

*P. A. Berkman**, Marine Biology Research Division, Scripp's Institution of
Oceonography, University of California La Jolla, CA USA

Research diving in Antarctica during the winter darkness is represented by a
continually changing suite of environmental constraints. The different circumstances
for each dive influence the dive equipment and procedures, and ultimately the
experimental designs. Preparing for these contingencies will determine the success of
the diving program.

INTRODUCTION

SCUBA is a powerful tool for working in aquatic environments. In the early 1950s
scientists began to recognize the freedom of this diving technique (Bascom and
Revelle, 1953), and by the end of the 1950s they already were using SCUBA to
explore the frigid world of the Southern Ocean (Fane, 1959; Neushul, 1960, 1961).
As a Staff Research Associate for Scripp's Institution of Oceanography, I participated
in an expedition to study the ecology of benthic foraminifera in McMurdo Sound,
Antarctica, and made 110 dives between January and November 1981. Fifty-five of
these dives were made during the period of darkness, which lasted from April to
August. This paper discusses SCUBA diving during the austral winter.

DIVE SITES

Our winter diving was staged at two sites at the southeastern edge of the Ross Sea in
McMurdo Sound (Fig. 1). Both dive locations were along the coasts of volcanic
islands, but the majority of the diving was conducted in front of McMurdo Station
on Ross Island in an area where diving had occurred since the early 1960s (Peckham,
1964).

*Present address: Graduate School of Oceanography, University of Rhode Island, Kingston, Rhode Island
02882–1197

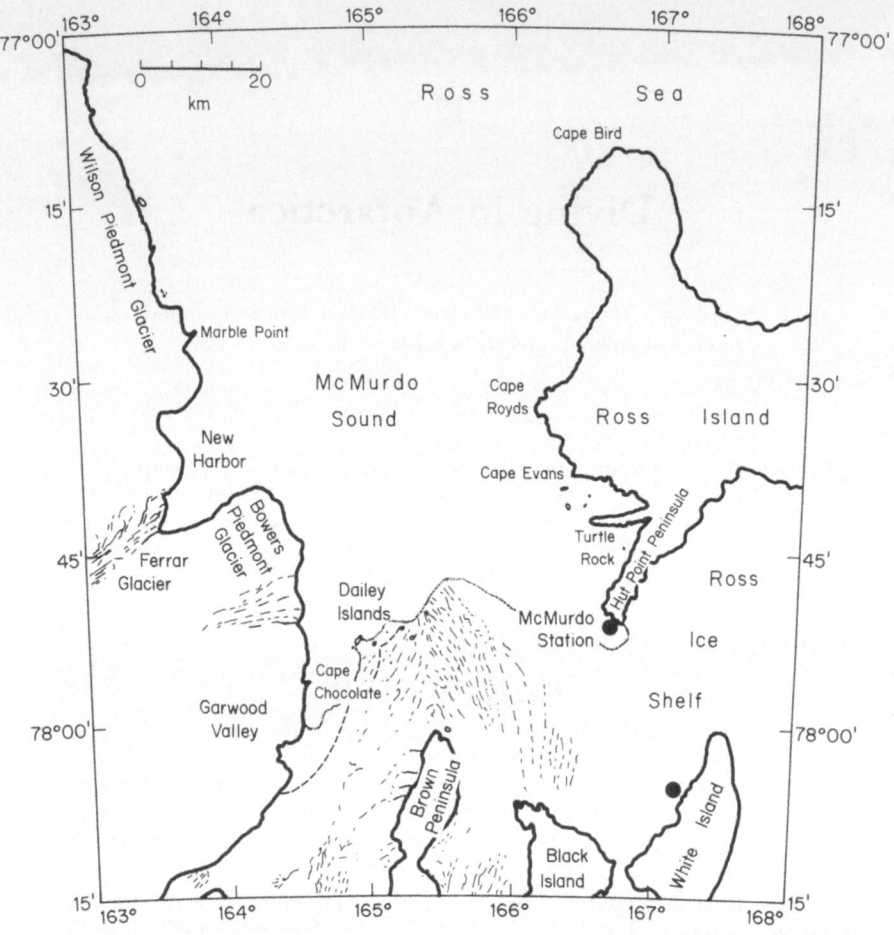

Figure 1 Map showing McMurdo Sound, Antarctica. Dive sites are marked by ●.

We dived through a hole created by the brine ejected from the water distillation plant, and from this hole we would swim down the nearshore slope to depths no greater than 30 m and not further than 75 m from our point of entrance. Unfortunately, during our year the sea ice in front of McMurdo Station was too unstable to position a hut. However, at White Island a hut was located over a 16 m deep hole which was drilled through the McMurdo Ice Shelf (Davis *et al.*, 1984). The difficulty with diving at White Island was that there were no previous diving experiences in the region. Divers at McMurdo Station were exposed to the extreme surface environment, whereas at White Island they were confronted with the unknown (Fig. 2).

Figure 2   Diver surfacing through brine hole at McMurdo Station.

DIVE PREPARATION

Three divers, each of whom had been certified by Scripp's Institution of Oceanography, composed the diving team: a surface tender, a principal diver who performed the experiments, and a safety diver who held the underwater light and checked for potential hazards. Prior to each dive we scrutinized the tasks to be completed and the procedures to be followed. Dive duration, calculated as total water time, was based on the no-decompression limits described in the *US Navy Diving Manual* (1973), and all dives were limited to a maximum of 30 min. Changes in the wind, surface temperature, sea ice, presence of animals, equipment, types of experiments, and the demands on the divers were always involved in our pre-dive discussions. After submerging, the light line served as a connection for line signals between the safety diver and the tender. Diver–diver and diver–tender communication followed the procedures outlined in the *NOAA Diving Manual* (1975), and special line signals were developed to deal with rapid surface temperature changes and marine mammals in the area. Any modifications to these procedures were made prior to the dive and for the purpose of dealing with unique circumstances.

The diving equipment complemented the tasks and contingencies of each specific dive. Thermal protection was required to protect the divers against the sub-zero degree water temperatures. Divers suffered the risk not only of hypothermia from not being well insulated, but also the loss of dexterity and mental agility. Even if the vital organs were protected, decreases in the hand skin temperature below 13 °C would have impaired manual performance (Gaydos and Dusek, 1958). We used the Poseidon UNISUIT with thermal underwear for body warmth and protection, and neoprene three-fingered gloves with wool inliners filled with hot water for our hands. The hood of the UNISUIT was pulled over the upper edge of a conventional dive mask so that only our lips were exposed. Inflation of the UNISUIT was used to maintain neutral buoyancy, and any transport of heavy objects was facilitated by other flotation devices. Except for situations when we were cold before a dive or when there was a serious suit leak, the diving outfit described above provided good thermal protection and enabled us to perform our experiments effectively. Over the UNISUIT we wore twin 72 cubic foot steel tanks with a double-hose regulator. Although these regulators never froze shut, there were several incidents of serious free-flow. On one dive the air from my tank escaped at a rate close to 200 p.s.i. min^{-1}, and based on this experience we learned never to be in the water with less than 800 p.s.i. tank pressure. Repairing and modifying our equipment was a task which we never completed.

McMURDO STATION DIVE SITE

Light

Although we began practicing our night diving techniques in late March while there was still daylight, diving during the first two months of the winter required constant re-evaluation (Fig. 3). Eventually we developed a lighting system which enabled us to travel up to 75 m from the dive hole, to effectively illuminate the area of study and to have a safety system with several levels of back-up. On each dive we used two 100 W Burns—Sawyer lights, one of which was connected to the station electrical system and the other to a portable 6 kW generator. One of these lights was suspended directly beneath the brine hole and could always be seen by the divers. Interestingly, the absence of solar illumination distorted our perception: underwater objects seemed much further away and much larger than they had during the

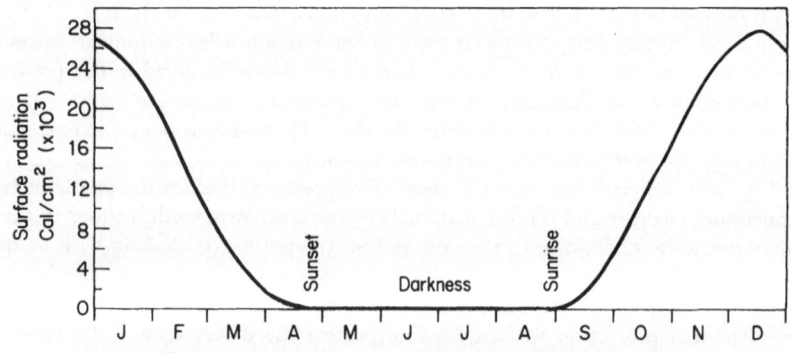

Figure 3   Monthly solar illumination profile at 78° S.

previous summer. The other light was carried by the safety diver and served as a tether for surface attachment and communication. With this light the safety diver illuminated the area of study for the principal diver. Dives were terminated if either of these lights malfunctioned. As an added precaution each diver also wore a battery-powered light on his wrist so that the principal diver could locate the safety diver holding the tether to the surface. The thought of losing the dive hole during the darkness forced us to develop contingencies.

Meteorological conditions

At McMurdo Station the weather was categorized with regard to the surface temperature, wind speed and visibility (Fig. 4). Class 3 represented unrestricted visibility and a wind-chill factor warmer than −60 °C. Class 2 occurred when there was diminished visibility and wind-chill between −60 °C and −75 °C. Class 1 existed when the wind-chill dropped below −75 °C and there was less than 150 m visibility. Surface weather conditions influenced the distribution of marine mammals, the growth of sea ice, the turbidity of the nearshore water column when there was no ice cover, and the safety of the diving team.

We dived throughout the winter in relative comfort when the wind-chill did not go below −40 °C. (However, it must be recognized that we probably exhibited some degree of physical or psychological adjustment to the cold surface temperatures after having been in the Antarctic for four months.) We suited-up for each dive in the warm aquarium building and were only exposed to the surface weather while we put on our dive fins and filled our gloves with hot water. Coming out of the water

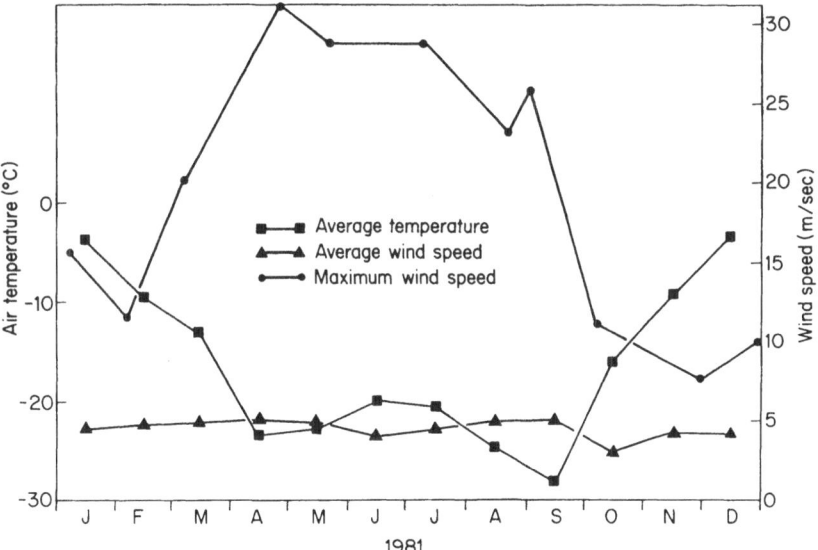

Figure 4   Monthly meteorological conditions at McMurdo Station, Antarctica (77° 51′ S, 166° 40′ E).

was much more severe. Often the immediate necessity to tend to our biological samples required us to be exposed for several minutes after a dive. Our suits stiffened, walking became very difficult and exposed facial tissue sometimes froze.

The most dangerous aspects of the wind-chill factor were the sudden and extreme changes. On one dive we entered the water when the wind-chill was −35 °C and by the time we exited the effective temperature had dropped a further 40 °C. Because the temperatures could change so rapidly, it was important for the tender to be aware of the weather patterns and be prepared to signal the divers to return to the surface if the weather became worse. If at any point during a dive the tender became uncomfortable, the dive was terminated to protect the tender and the divers whose safety he must monitor.

Ice conditions

The Antarctic has only about 15% multi-year ice because of the absence of any continental land barriers (National Academy of Sciences, 1970). The distribution and behavior of this sea ice can be characterized by its growth and disintegration, drift and deformation, and the zone of open water (Sater, 1969). The thickness of the sea ice, the ambient surface temperatures, and the prevailing winds all were related to the stability of the ice cover (Fig. 5).

In late April, the sea ice was almost 1 m thick; a storm with 89 knot winds, in less than three hours, blew away the entire ice cover of McMurdo Sound. Evaluating the stability of the ice cover was always a crucial consideration.

Ice also affects objects beneath the sea surface. Icebergs scour the bottom and represent a serious hazard to any structures with which they come in contact. Another ice phenomenon which deserves attention is anchor ice. As described by Dayton *et al.* (1969), anchor ice is composed of small aggregations of platelet ice which form on the ocean bottom when the water temperature is lower than −1.8 °C. These clusters of ice crystals have a buoyant capacity sufficient to lift benthic objects as heavy as 25 kg, have been observed to depths of 33 m, and may have a significant role in the nearshore ecology (Dayton *et al.*, 1970).

Currents and tides

Before diving in an ice-covered region, previous accounts of the hydrology should be reviewed. This will at least provide some sense of those regions which may pose a dangerous risk to divers. For example, "strong" currents have been reported in the vicinity of McMurdo Station since the early expeditions of Robert Falcon Scott

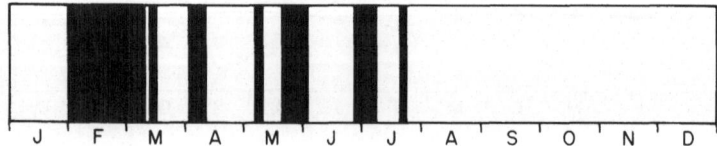

Figure 5 Ice cover in McMurdo Sound, Ross Sea, Antarctica, 1981. Black areas = ice present; white areas = ice absent.

(Hodgson, 1907). Contemporary large-scale hydrographic studies in the Ross Sea indicate that water motion is strongly influenced by bottom topography (Jacobs *et al.*, 1970). Currents also can be observed from the surface by suspending a light under water and determining whether it deviates from the vertical, or when in the water by evaluating the position of benthic objects. If the position of benthic objects indicates that a persistently strong current exists in an area, diving in that area should be avoided and the divers should surface immediately. Divers must be cognizant of the presence of currents while diving under ice because of the risk of being swept away from the dive hole.

In nearshore areas, tides will create a crack between the ice and the shore. During high tide this tidal crack will be pushed closed and during low tide it will be pulled open. Recognizing when the relative tides occur could be important if the divers were to use the tidal crack as an entrance or exit, or in an emergency as an escape route.

Animals

Many large marine mammals migrate into the polar seas during the summer because of the increased food production and the decreased ice cover. Some marine mammals, however, can survive in regions of fast ice throughout the year. During the Antarctic winter we periodically encountered Weddell Seals. These seals have the ability to ream holes in the ice through which to breathe, and were frequently encountered out on the sea ice when the surface temperature was warmer than -40 °C. Generally, the Weddell Seals did not give us any problems, but there was one instance when an aggressive individual chased us out of the water. This incident occurred shortly after we found a small Crabeater Seal with a badly mauled tail flipper hauled out near our dive hole. These seals may have been competing for a breathing hole which just happened to be our dive hole.

A more serious incident occurred in early July when we encountered a Leopard Seal trapped in McMurdo Sound during one of the periods between the ice breaking out and refreezing. Leopard Seals, some of which are larger than 4 m, are the top carnivores in the Antarctic and eat other seals and penguins. In the past, these seals have endangered man (DeLaca *et al.*, 1975). Normally, Leopard Seals are restricted to regions of pack ice and when I encountered this enormous seal with a cheshire-cat grin staring at me under water I was extremely intrigued. It poised itself 2 feet in front of me for about 30 s before my partner shone the 1000 W light in its eyes to chase it away. Suddenly it darted toward my partner who, in a defensive motion, swung his collecting bag. The seal then circled through the water, and finally came to rest on our light line. Realizing that we did not have many options, we proceeded back-to-back toward shore until we were able to exit through our hole. After this dive, regardless of the species and what we knew about its ecology, we treated each large marine animal with the deference that a potential aggressor warrants.

Research

Foraminifera, testate single-celled animals, exist both in the water column and in the sediment. Our study focused on some of the benthic species which were known

Figure 6   The common Antarctic sea anemone, *Urticinopsis antarctica*.

or assumed to have affinities with species in other regions. The 1000 W light provided excellent illumination for performing reproducible experiments and describing the faunal characteristics at the study site. The UNISUIT buoyancy control allowed us to hover over the bottom so that we could conduct our research without floculating the fine-grained sediment. Small coring devices, 5 cm³ syringes which were capped with 1 cm diameter corks, were easily manipulated while wearing the three-fingered neoprene gloves. Throughout the winter we were able to conduct our experiments in a manner which enabled us to learn about the small-scale patchiness, dynamics and life habits of Antarctic benthic foraminifera.

As in all other localities, documentation is an important aspect of diving, not only for data collection, but for critiquing and improving each dive operation. Eilerston (1978) discussed the utility of accurate documentation in terms of protecting the assets, and the type of documentation (photographic, written or instrumental) should reflect the goals of the entire diving program. The uniqueness of the Antarctic marine environment should compel divers to observe natural events with penetrating vigilance (Fig. 6).

WHITE ISLAND DIVE SITE

As discussed earlier, we made several dives underneath the McMurdo Ice Shelf. These dives were made through a hole melted by a 1 m diameter copper coil heated by 100 °C circulating glycol (Davis *et al.*, 1984). The purpose of this hole was to provide easy breathing access for Weddell Seals in the region and to facilitate the collection of information related to the general diving physiology of this species. Unfortunately, the placement of an aluminum tube in the ice to aid in the maintenance of the hole also frightened the seals away.

Diving through the 16 m deep hole in the McMurdo Ice Shelf was treated as every other night dive. However, because there was no information on the local hydrography or ice dynamics, we were especially cautious. Dives were shorter than 20 min, and each diver was tethered and did not stray far from the underside of the dive hole. Because we were diving from a hut, meteorological conditions were less important to this diving operation.

Directly beneath the dive hole, the brash ice seemed to have increased over the winter. Also, small rocks were found on the underside of the Ice Shelf which suggest that anchor ice may grow on the ocean bottom to depths greater than 70 m.

COMMON SENSE

Winter research diving in Antarctica is affected by an ever-changing combination of environmental factors. Dives are affected by solar illumination, wind and surface temperatures, ice conditions, currents and tides, animals and the research considerations. Before each dive the contingencies should be assessed, and the diving protocol should be flexible to accommodate those situations which are not intially envisioned. However, after all the preparation is "said and done", the responses of the divers will determine the safety and success of the diving program.

ACKNOWLEDGEMENTS

I wish to thank Dr T. E. DeLaca for the wonderful opportunity to study in Antarctica. I also would like to express my appreciation to D. S. Marks and G. P. Shreve for their friendship and support throughout our year together. My respect for Antarctic diving is entirely a consequence of the many discussions and experiences shared with R. Davis, T. DeLaca, D. Marks, R. Moe, G. Shreve and W. Stockton. This project was supported by NSF DPP 80–03432.

REFERENCES

Bascom, W. N. and Revelle, R. R. 1953. Free diving: a new exploratory tool, *American Scientist*, **41**(4), 624–627.
Davis, R. W. *et al.* 1984. Maintenance of an observation hole through the McMurdo Ice Shelf for winter oceanography, *Antarctic Journal of the US*, **18**(4), 12–14.
Dayton, P. K. *et al.* 1969. Anchor ice formation in McMurdo Sound, Antarctica, and its biological effects, *Science*, **163**, 273–274.

Dayton, P. K. *et al*. 1970. Benthic faunal zonation as a result of anchor ice at McMurdo Sound, Antarctica, in *Antarctic Ecology*, M. W. Holdgate (ed), pp. 244–258.

DeLaca, T. E. *et al*. 1975. Encounters with leopard seals (*Hydrurga leptonyx*) along the Antarctic Peninsula, *Antarctic Journal of the US*, 10(3), 85–91.

Eilerston, R. E. 1978. Protecting your assets, in *The Working Diver*, Marine Technology Society, Ohio, pp. 295–299.

Fane, F. D. 1959. Skin diving in polar waters, *Polar Record*, 9, 433–435.

Gaydos, H. F. and Dusek, E. R. 1958. Effects of localized hand cooling versus total body cooling on manual performance, *Journal of Applied Physiology*, 12, 377–380.

Hodgson, T. V. 1907. On collecting in Antarctic seas, *National Antarctic Expedition*, III, 1–10.

Jacobs, S. S. *et al*. 1970. Ross Sea oceanography and Antarctic bottom water formation, *Deep-Sea Research*, 17, 935–962.

National Academy of Sciences, 1970. *Polar Research: A Survey*, Committee on Polar Research, National Research Council, Washington, pp. 49–73.

Neushul, M. 1960. Biological collecting in Antarctic waters, *Veliger* 2(1), 15–17.

Neushul, M. 1961. Diving in Antarctic waters, *Polar Record*, 10(67), 353–358.

Peckham, V. 1964. Year-round SCUBA diving in the Antarctic, *Polar Record*, 12, 143–146.

Sater, J. E. 1969. *The Arctic Basin*, Arctic Institute of North America, Washington, DC, pp. 26–42.

United States National Oceanic and Atmospheric Administration. 1975. *The NOAA Diving Manual: Diving for Science and Technology*, US Dept of Commerce, National Oceanic and Atmospheric Administration, Office of Marine Resources, Washington.

United States Naval Ship Systems Command. 1973. *US Navy Diving Manual*, Navy Dept, Washington.

13

Diver Training

K. Sveinsson, Reykjavik, Iceland

The history of diving in Iceland is reviewed briefly. The author's involvement in teaching and training divers is described, together with information on the problems and dangers of working in cold waters.

The purpose of this paper is to tell you something about how diving is taught to men who go on to work as divers aboard fishing ships in Icelandic coastal waters. I believe diving off fishing ships in the open sea is considered something of an Icelandic speciality.

The story of diving in Iceland begins shortly after the turn of the century, when the Danish authorities started stationing salvage vessels off the Icelandic coast because of increasing traffic in Icelandic waters. These ships had divers on board from whom the Icelanders first learnt diving. Some of these Icelanders went on to become well-known divers with the famous Danish salvage companies which operated in various parts of the world, while others stayed in Iceland, chiefly working on the many harbour projects then in construction all over the country.

When Icelanders themselves took over the Coast Guard and salvage operations off the Icelandic coasts in the years following 1920, Icelanders were engaged as divers aboard these ships and undertook diving for fishing ships when needed. In those times, however, fishing ships always had to be towed to shelter or to the nearest port whenever fishing gear had to be cut loose from the ships' propellers or any other diving operation carried out, since it was impossible to dive in the open sea in the old helmet suits, then the only ones used by divers.

Around 1955 a change took place, with the arrival in Iceland of the first frogmen, men who had learnt this new diving technique abroad and brought it back to Iceland. One of these men, who had been trained in Norway, then started service operations for the herring fleet, accompanying the ships to their traditional fishing grounds off the north and east coasts, cutting their propellers free of fishing gear when needed. At about the same time, Icelanders started fishing for herring using the fishing gear called *hringnót* or purse seine, which is particularly susceptible to being caught up in the propeller. When this happened, the man I mentioned was sent for. He would dive from his own boat and free the distressed ship's propeller.

He is, therefore, a pioneer, the first man to practise open-sea diving in Icelandic waters.

In 1955, the Icelandic Coast Guard adopted this technique for use aboard their ships and an officer was sent for a few months to the United States to learn diving with the American Coast Guard. When he came back, he in turn trained his colleagues and since then there have always been one or two divers with experience of open-sea diving operations aboard every Coast Guard vessel, except during the past two years, when open-sea diving operations have been suspended during a dispute about divers' pay.

When the Icelandic Coast Guard started these diving operations, I was an officer aboard one of their ships and thus was one of the first to receive training. I have therefore about 25 years' diving experience in the seas off the coast of Iceland, or, rather, in the whole of the North Atlantic, from Spitsbergen in the north and as far south as the English Channel, then off Greenland in the west and all the way east to Norway. The Icelandic protection vessels cover vast areas following the fishing fleet.

In 1964, the Icelandic Supreme Court ruled that cutting free a ship's propeller in the open sea should be considered a rescue or salvage operation and that the fee should be 2% of the ship's insurance value. In this particular case, the ruling was in favour of a Coast Guard vessel. The man I mentioned earlier, who pioneered open-sea diving, then decided to charge the same for his services as the Coast Guard, instead of his earlier fixed price. Subsequently, the Icelandic insurance companies decided to buy and operate their own rescue ship to accompany the fleet and handle the rescue operations, in order to limit the greatly increased rescue operations expenses. In 1966, a small rescue ship was bought from Norway and I was engaged as its captain and chief diver, which I am to this day.

At the same time I became involved in teaching and training divers, both for work aboard my ship and aboard fishing vessels. In the course of these 18 years, I have taught and trained many men for this kind of work. This instruction has mostly been given aboard and off my own ship, to one or two men at a time. They are always registered as members of my crew and are given duties in accordance with this. Prior to joining the ship, they have been subjected to a thorough physical examination in order to find out whether they have the physical strength needed for diving, something which doctors can measure and check. Whether a man is mentally up to the job, has the clear head and cool nerve needed, can be ascertained only when we start diving in earnest..

Instruction starts with the usual things that can be learnt from books in connection with diving, but I must admit that I have never laid great stress on this, for I consider the practical skills to be what really count in a job like this. I start practical training in shallow water in a sheltered cove or inlet, of which there are many on the coast of Iceland, letting the men get used to the diving equipment and having them swim long distances in order to develop their strength and endurance. I have always considered it most important for my trainees to be very good swimmers and used to think this was an absolute prerequisite for becoming a good diver. But this is not necessarily so. Take my colleague, for example, a man I taught myself, who has been diving professionally for years. He is a very good open-sea diver, but would drown if he got out of his depths without equipment because he cannot swim;

he has never been able to get the hang of it, although he swims like a seal in his diving togs.

After the initial period, the men train with the equipment generally used to pry loose fishing gear from ships' propellers: chiefly big knives, axes, various kinds of shears and some kinds of compressed-air equipment as well. I entangle the propeller of our own ship with wires, ropes and nets, so they can practise under the most realistic circumstances possible.

The next step is to take them along on a real assignment in the open sea. The ship's fishing gear has become entangled in its propeller and assistance has been requested. However, I take trainees only if the sea is calm and there is little wind. In this way, my trainees get experience gradually as I take them along on more and more difficult assignments. I have found this to be the best way to teach open-sea diving where the conditions can be difficult.

The best age for learning diving is between 18 and 25, for then the men are at the peak of their strength and, even more important, their courage and nerve are unimpaired. I have taught men of 30 and more and, almost without exception, they lose their nerve when we start diving off ships in the open sea, in darkness and rough seas. But, even if they are too old for this kind of diving, they have quite often continued to dive for sport.

During the 18 years I have been involved in teaching open-sea diving, I have taught and trained many men, but the drop-out rate is very high. Some give up quickly, others keep at it for a little longer. This just goes to show how physically demanding and dangerous this kind of diving can be. I do not know how long a man can do this kind of job. I am only 50 and still going strong, and intend to keep at it for many more years. Maybe there will be another ICEDIVE Conference in 1994 and then I will be able to tell you. But it really does not surprise me that men quickly give up this kind of job. It is not exactly fun getting thrown overboard from a rolling ship in darkness, snow and frost, which is the usual case on the fishing grounds off Iceland in winter — and the winters in Iceland are pretty long.

Until 1981, no laws or regulations existed on diving or diving instruction in Iceland, but at the end of that year a law covering both took effect. A total of 80 men then received certificates or diplomas granting them the right to practise diving professionally. Of these men, a total of 10 received diplomas giving them the right to teach diving, but so far no diving school has been established. Hopefully this will come in the near future. Only two of the men who received diplomas are professional divers in the sense that this is their chief profession. The others all dive as a sideline, usually in connection with the fishing fleet when it is in port. Diving is practised by people in many professions as a sideline: in one town on the north coast, the local bank manager undertakes diving when needed and in another town on the west coast the local clergyman used to zip up his diver's suit when called upon.

During the last three winters, the navigation schools in Reykjavik and Vestmannaeyjar have held courses in frogman-diving for future officers in the Icelandic fishing fleet. Approximately 10—15 men have attended these courses each year. They have received some 40 hours of instruction, both theoretical and practical. These courses do not make divers of the men, only prepare them for diving operations, but with good supplementary training these men can become able divers who can dive in the open sea under relatively unfavourable conditions. The goal is to have aboard as many ships as possible men who can dive in the open sea when conditions are fairly good.

The temperature in the sea around Iceland varies greatly, from 4 °C to 8 °C down to −1 °C to −2 °C, off the west and northwest coast, so diving there is undeniably Ice Diving. In this area there is capelan fishing during the earlier part of winter and on into the New Year. It has been my job most of these past years to accompany the fleet and render it assistance when needed. I would say that 95% of the assistance rendered consists in cutting loose fishing gear which has become entangled in the ships' propellers. The capelan fishing mainly takes place in waters 45–50 nautical miles off the coast for 2–3 months of the fishing season, whereas for the last 4–5 weeks the ships fish fairly close to the coast. When the ships are fishing farthest away from land it is most important from the financial point of view to be able to free propellers on the spot instead of having to tow the ships to shelter in order to disentangle the gear. It is in cases like this that I and my men, those who dive with me off my ship, come to the aid of these distressed ships. And I can tell you that it has often been damned cold diving in these chilly waters. For about 15 years I did all my diving in a wet suit, in fair weather or foul, warm or cold.

When diving in waters of −1 °C to −2 °C in a wet suit, I was able to work for 20–30 min at a stretch, until I became numb with cold. My signal for "enough" was when I started having a bad pain in the neck. I then had to come up into the rubber dinghy that we use when diving and return to my ship for a hot shower in order to regain body warmth. It was also very good to drink some hot soup or coffee. Hot tea was not as effective. After about an hour's re-heating one was able to continue diving, but then one got cold much quicker. And, if the work at hand was not finished in the 5 or 10 min one was able to withstand the cold, the task had to wait a few hours while one returned to normal once again, took a hot bath, had something to eat and slept.

From 1966 to 1969, the Icelandic herring fleet fished in the waters off Iceland and all the way north to Spitsbergen or Svalbard(i), as I prefer to call it. The fleet went all the way up to 80° N in its search for herring. All these years I accompanied the fleet on my ship to these northern parts. Sometimes we were up to four months at sea without entering port, and our supplies and other necessities were brought to us by the large tankers which transported the herring from the fishing ships to shore. During these years hundreds of diving operations were executed by myself and my men in these waters. It was coldest to the north and northwest of Jan Mayen, usually below zero. At the time, we nearly always had four trained divers on board, so that it was possible to alternate teams of two men, one team taking over from the first while these men got warm again.

From my ship we always dive in teams of two, because the greatest danger when diving in open sea, trying to get fishing gear disentangled from ships' propellers, is becoming entangled in the gear oneself. There should always be two men together, so that they can help each other if one of them becomes entangled in the gear while working to free the propeller. This happens quite often, but for a diver who is well-trained and has experience in open-sea diving, this is no great problem. Experience has taught him how to react in such an event and he has the necessary skill and knowledge.

Luck has been with us in our open-sea diving operations off Iceland; we have had neither fatalities nor serious accidents. However, three fatal accidents have occurred in the last ten years, in shallow waters close to shore. In two instances, those involved were young men with little or no diving experience. Their drowning remains unexplained, a mystery to this day, for their equipment, which I inspected after the accidents, was in perfect order. The third to lose his life was a seasoned diver, with about 15 years' diving experience, who had worked with me for a number of years on diving operations, but later gone ashore and since dived only as a sideline. He was working on the propeller of a big fishing ship in Reykjavik harbour when the propeller was set in motion all of a sudden, with the inevitable tragic results. It never became clear why the propeller was set in motion. But this goes to show how dangerous diving around ships can be, anywhere, if the most stringent safety precautions are not observed.

Accidents due to divers' sickness or other deep-sea diving ailments have not occurred, which is just as well, for until now there has been no de-pressurizing chamber in Iceland. The one closest to us was in Holy Loch in Scotland and, if an accident of that sort had occurred, one would have had to fly the patient to Scotland for treatment. But in late 1983 there was a change for the better, with the arrival of two de-pressurizing chambers, both of which are located in Reykjavik. One of these takes two men and a doctor; the other is for one man only and easily transportable. The larger one belongs to the two truly professional divers now operating in Iceland; the other to Slysavarnafélag Islands (the Society for the Prevention of Accidents).

Before ending this paper, I would like to recount two stories about diving. One says something about the dangers which are inherent to open-sea diving and the other shows how necessary it is for most ships to have diving equipment on board and a man who knows how to use it.

In early winter 1983, I was aboard my ship off the southwest coast of Iceland when a request for help came in from a trawler located about 50 nautical miles southwest of Iceland, its warp had got caught in its propeller. We were there in about four hours. By that time we had a northeasterly force 8 gale and pretty high seas, so that it was impossible to dive and cut the warp loose. The trawler was therefore towed towards shore, into shelter, where we started diving operations in the middle of the night, with a stiff gale blowing but a fairly calm sea. The propeller was very badly entangled and it took two of us seven hours of diving to free it. When we were about halfway done with the job, I had to come up to change an oxygen bottle. In the meantime, the other diver carried on with his work. When I came back down again, a few minutes later, I did not see the other diver at once due to the darkness and the fact that our lights cover only a fairly small area. Then I found him, hanging on a hook beneath the screw; the end of a broken chain from the warp had caught on one of the two taps on his oxygen bottles. He was unable to unhook it from the tap and also unable to cut it with his knife. An experienced diver, he decided to sit it out, for he knew I would be returning shortly. I quickly managed to pry him loose and we continued our work as if nothing had happened. But if this had happened to an inexperienced diver, with little previous training, and had he been working alone, I am afraid the story would not have such a happy ending. To my mind, this story illustrates that one can never be too careful when diving and that those who engage in this business must have very good instruction and training.

Now for the second story. In the middle of August 1981, the Norwegian ship *Nina Profiles*, a big oil prospecting ship, contacted me at my home in Reykjavik. The ship was working off the east coast of Greenland at 67° 00′ N and 28° 00′ W. The propeller was full of heavy electrical cable which they could not budge and was therefore, firmly stuck. There was a helicopter on board, which could be sent for me, if I was willing to provide assistance; I immediately said I was. Together with my equipment, I was taken to the ship, which was located a short way from the edge of the pack ice off Greenland. The weather was good, a light southwest wind was blowing and there were waves of about 0.5 m. It took me about two hours to work the cable loose from the propeller. This is a kind of cable used for some types of sea bottom soundings. When this diving operation was over, a heavy fog had set in, something quite usual near the pack ice, so that they were unable to fly me back immediately. The ship was therefore turned around and the compass set for Iceland in order to try to get out of the fog. Meanwhile, I had ample time to inspect this great, modern ship, which was equipped with the most advanced positioning equipment and millions of dollars' worth of various kinds of computer equipment used in oil prospecting, as well as the earlier mentioned helicopter. One thing was lacking aboard, however, which did not cost much in comparison with all the rest: diving equipment and men who knew how to use it correctly. The expedition leader who took me around the ship told me that this would be the last trip he took without such equipment and men; he would see to it that the position was put right as quickly as possible. After about three hours' sailing, the fog lifted and I was flown back to Iceland. One more exciting diving operation was over.

14

Cold Water Rescue

L. A. Laitinen, Department of Clinical Physiology, Central Military Hospital, Helsinki, and *Seppo Sipinen*, Finland

Factors affecting survival times in cold water are considered, together with strategies for preventing hypothermia. The organization of sea rescue operations in Finaldn is described.

Finland has 60 000 lakes and a coastline of 1500 km. In most parts of the country the temperature may be below 0 °C for six months annually.

It is known that water temperature is a major factor in cold water survival. Table 1 shows some of the mean survival periods in water of different temperatures. Additionally, many factors influence survival times. For example, the movement of water can increase the heat loss to 200-fold compared with still water. The cooling effect of wind is important. This is presented at various ambient temperatures in Table 2. With a wind speed of 4.5 m/s and with an ambient temperature of +5 °C in air, the cooling effect on the skin is −2 °C. If the wind speed increases three-fold, the cooling effect can increase six-fold. Additional factors influencing the survival time are the victim's clothing, body size, behavior in water, age and sex as well as physical condition.

The best treatment of hypothermia is a preventive one. If a special survival suit is not available, the best protection is given by wool underwear, underneath a watertight overall. Because 30% of the heat escapes through the head and neck, they have to be especially protected. There are two additional ways to prevent hypothermia when the victim is already in cold water. One is the so-called heat escape lessening posture, which mimics the fetal posture; in this way the victim can reduce the heat loss by 50%. If there is more than one victim , it is wise to assume a huddling posture, with the victims staying as close together as possible, holding their arms around one another's necks.

Table 1
Mean survival periods in water of different temperatures

Water temperature (°C)	Time taken for tiredness and unconsciousness to result	Time taken/for death to result
0	less than 15 min	less than 15–45 min
0– 5	15–30 min	30–90 min
5–10	30–60 min	1–3 h
10–15	1– 2 h	1– 6 h
15–21	2– 7 h	2–40 h
21–27	3–12 h	3h–undefined

Table 2
Other authorities participating in sea rescue operations

Aviation authorities	air rescue centers and section centers flying equipment
Police and customs authorities	boating equipment
The National Board of Navigation	pilot stations, pilot boats ships maintaining navigability icebreakers other vessels
Fire authorities	boats for extinguishing fire, preventing oil pollution, and for patient transport
The armed forces	navy signal centers vessels of navy and coast artillery sea surveillance stations and regional centers flying equipment of air force
The postal and Telecommunications	coastal radio stations
The Roads and Waterways Administration	ferries
Ahvenanmaa county council	archipelago ferries vessels for bad weather conditions and transport
The Finnish Sea Rescue Society	rescue ships rescue boats private boats

However, the greatest protection against hypothermia is given by special emergency suits. They are divided into three categories: immersion, survival, and rescue suits. In Finland, it is strongly requested that all naval and commercial ships provide easily available emergency suits for their crews. It is natural that in a country like Finland there is a very advanced water rescue service. Because of the climate, cold weather conditions have been explicitly considered in all rescue activities in our country.

ORGANIZATION OF SEA RESCUE SERVICES

The law and statute on sea rescue services are quite recent in Finland. The law was enforced on 1 September and the statute on 15 September 1982. According to the law and the statute, the Frontier Guard is responsible for the sea rescue services. The Sea Rescue Service Office, located in the headquarters of the Frontier Guard, directs

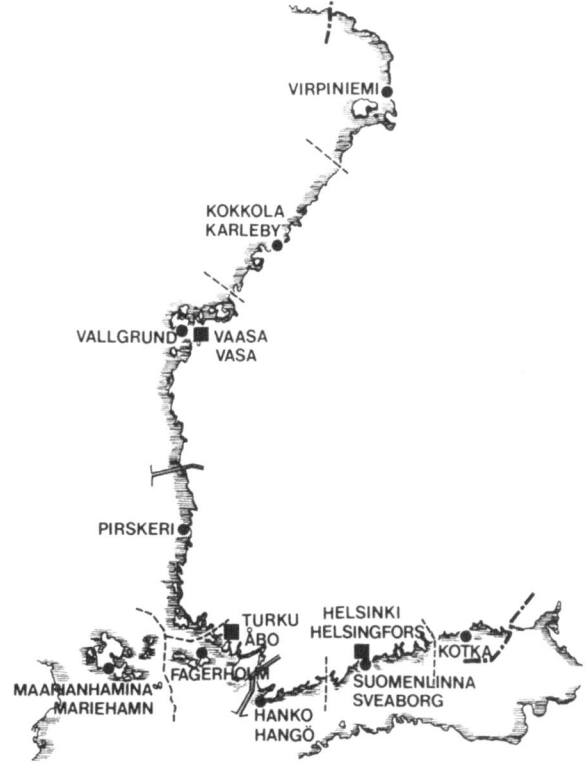

Figure 1 The naval guards with their sea rescue centers are located in Helsinki, Turku and Vaasa. Each of them has three regional section centers which are labelled with black dots (●). From north to south they are: Virpiniemi, Kokkola, Valgrund, Piiskeri, Maarianhamina, Fagerholm, Hanko, Suomenlinna and Kotka.

the collaboration of sea rescue planning and activities as well as supervises the execution of these services.

Finland has territorial agreements on collaboration in rescuing human victims with the Soviet Union and Poland, relating to the Baltic Sea and the territorial waters of the parties concerned. Moreover, the Aviation Boards of Sweden and Finland have an official record on collaboration in air rescue operations.

The responsibility for the execution and direction of sea rescue operations lies with the naval guards subordinated to the Frontier Guard. There are three naval guards, which serve the areas of the Gulf of Finland, the Archipelago Sea and the Gulf of Bothnia. The headquarters of these naval guards with their sea rescue centers are located in Helsinki, Turku and Vaasa. The centers direct the sea rescue services and are assisted by three regional section centers, whose locations and areas of responsibility are shown in Figure 1. For the execution of sea rescue operations, the naval guards use boat patrols, guard ships, guard planes, and light and medium-weight helicopters.

USE OF FLYING EQUIPMENT

After the recent accidents at sea, the application of flying equipment in rescue operations in particular has been improved in Finland. The guard aeroplane squadron, with bases in Helsinki, Turku, Vaasa, Kajaani, Rovaniemi and Ivalo, is responsible for the air operations of the rescue services run by the Frontier Guard. The flying equipment consists of seven Agusta Bell helicopters, Piper Navajo aeroplanes, two Beaver aeroplanes and three MI-8 helicopters (Fig. 2). The Guard Aeroplane Squadron has about 70 employees, 50 of whom are pilots or mechanics.

The most efficient is the squadron stationed in Turku. It is able to maintain continuous helicopter guard duty with a 15 min take-off time during office hours and a 1.5 hour take-off time during other periods. The surface rescuers and mechanics of the squadron have received first aid training with the objective to achieve the competence requirements of the ambulance services training given by the National Board of Vocational Education. The availability of first aid group doctors and nurses for helicopter rescue operations is being negotiated with the Turku University Central Hospital. The helicopters have basic equipment consisting of resuscitation appliances, bandages, splints and medicines as well as blood transfusion outfits. Furthermore, the doctor can carry a complementary medical package.

Only in 1980 were the search and patient flights and the related training in sea rescue operations introduced by the Naval Guards. In 1981, 5000 hours were used for training. In 1982, the search and patient flights amounted to 93 hours with the MI-8, 49 hours with an aeroplane and with hired civil aeroplanes, totalling 502 flight hours. Of this amount, 10% were search, patient and rescue flights. The search and archipelago flights conducted were 153 in total, of which 66 were patient flights with a total of 71 patients on board. The latest actual cold water rescue was performed in connection with the sinking of the ship *Malmi* in November 1979.

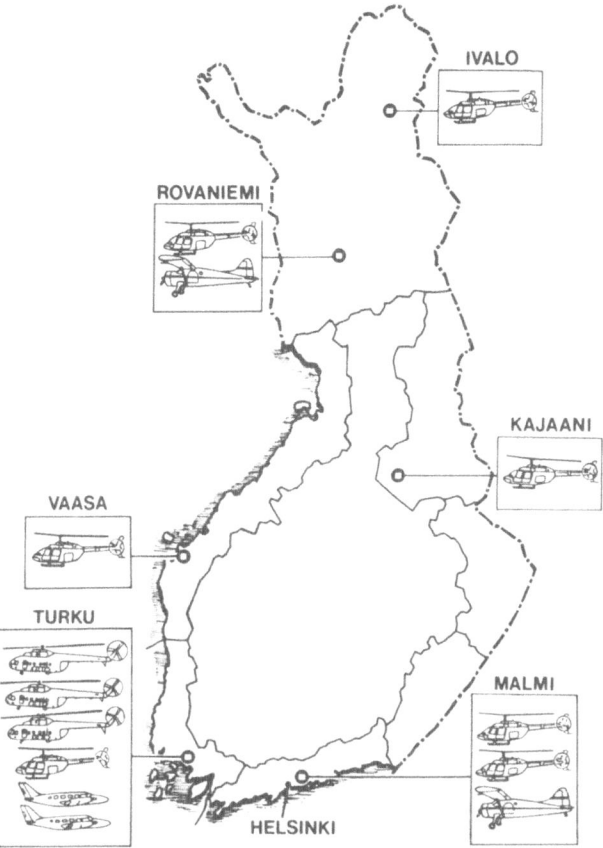

Figure 2 The guard aeroplane squadron, with bases in Helsinki, Turku, Vaasa, Kajaani, Rovaniemi and Ivalo, is responsible for the air operations of the rescue services run by the Frontier Guard. The flying equipment of each squadron is shown.

OTHER AUTHORITIES PARTICIPATING IN SEA RESCUE OPERATIONS

In Finland, the aviation, customs, and fire authorities and the police, the Board of Navigation, the armed forces, the Postal and Telecommunications, the Roads and Waterways Administration, the Ahvenanmaa county council and the Finnish Sea Rescue Association participate in sea rescue operations under the leadership of the Frontier Guard (Table 2).

ACCIDENTAL DROWNINGS

The latest reliable statistics on drownings are the figures given by the Central Statistical Office for 1980. During that year, 401 persons drowned. Of these, 119 were boatsmen or people in boat traffic. Of the 119, 5 drowned from ships or vessels with ship classification. The unclear cases totalled 170, including also possible murders and manslaughter. These figures include 19 cases of falling into water and 22 cases of drowning in ice. Of the 401 cases of drowning, 231 were accidents. Only preliminary statistics are available for 1981, 1982 and 1983. In general, it should be observed that men predominated, 10% were women and 1–2% children.

What are the practical rescue measures when a water accident happens in Finland? When there is an emergency in the lake districts, the alarm is received, through the regional alarm center, by the nearest fire brigade or by a health care unit, which is in readiness. In the sea districts, the alarm is received directly by the naval guard which in its district takes care of the rescue services. Additionally, as the regional alarm center considers necessary, units from the Ministry of Defence, Customs, Fire brigades and the police may participate in the rescue operations.

15

Deep Diving in Mountain Lakes

Underwater maintenance of dams in Switzerland is considered. An efficient rescue system for divers has been devised. Safe decompression procedures have been developed for different altitudes; two Hyperbaric Treatment and Research Centres are described.

Part I: Practical Diving Activities in Switzerland

Roland Pralong, Forces Motrices Neuchâteloises, S.A. Hydrovision Department, Corcelles, Switzerland

As a small country in the hub of Europe, Switzerland is not considered as an important market by the major companies involved in underwater operation. Nevertheless, minimal basic activities are regularly under way and most of the time display some interesting specialities which are the subject of the present report.

Switzerland's geography is distinguished by three different zones: in the north, the Jura mountains; central, the Plateau, and in the south, the Alps. Switzerland produces about 50 000 GWh of electricity per year, covering 68% of her needs. An important part of this electricity is produced by means of 125 major dams (Fig. 1).

Forces Motrices Neuchâteloises and its Hydrovision Department are specialized in all underwater inspection and maintenance works as well as the dredging of these dams, down to a depth of 200 m. The main peculiarity of these operations is that they take place at an altitude between 1000 and 3000 m, sometimes with an outside temperature of -25 °C (-13 °F).

Diving is managed according to tables specially computed for altitude operations. These tables have been established by the DKLZ (Druckkammerlabor in Zürich) and Mr Schenk. All decompressions handled according to these altitude tables have been absolutely normal and no trouble has been experienced. Consequently we are in a position entirely to trust these documents and can warmly recommend them without hesitation.

In order to give you a better idea of the tasks we have to perform, we shall look at

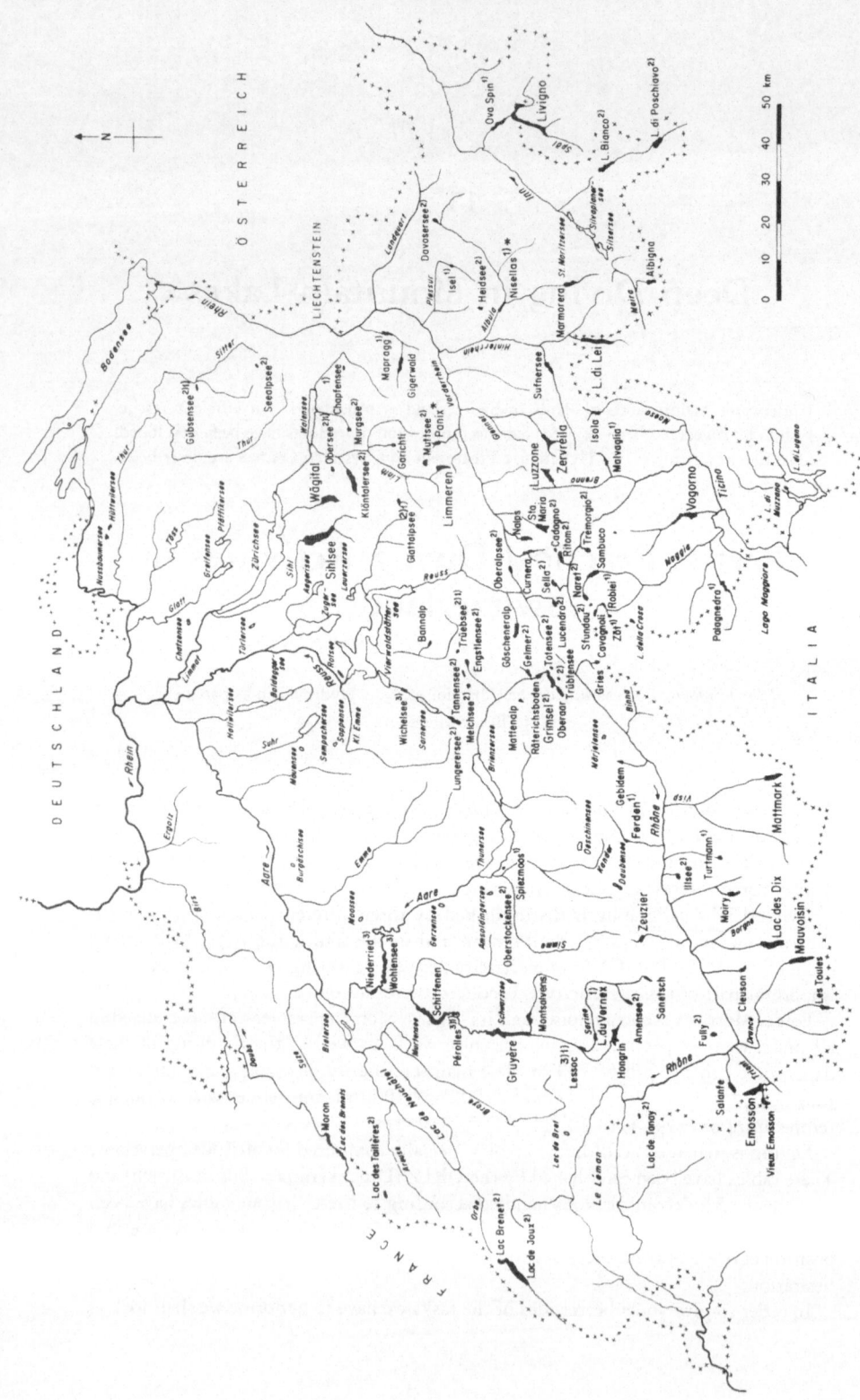

Figure 1 Major dams in Switzerland.

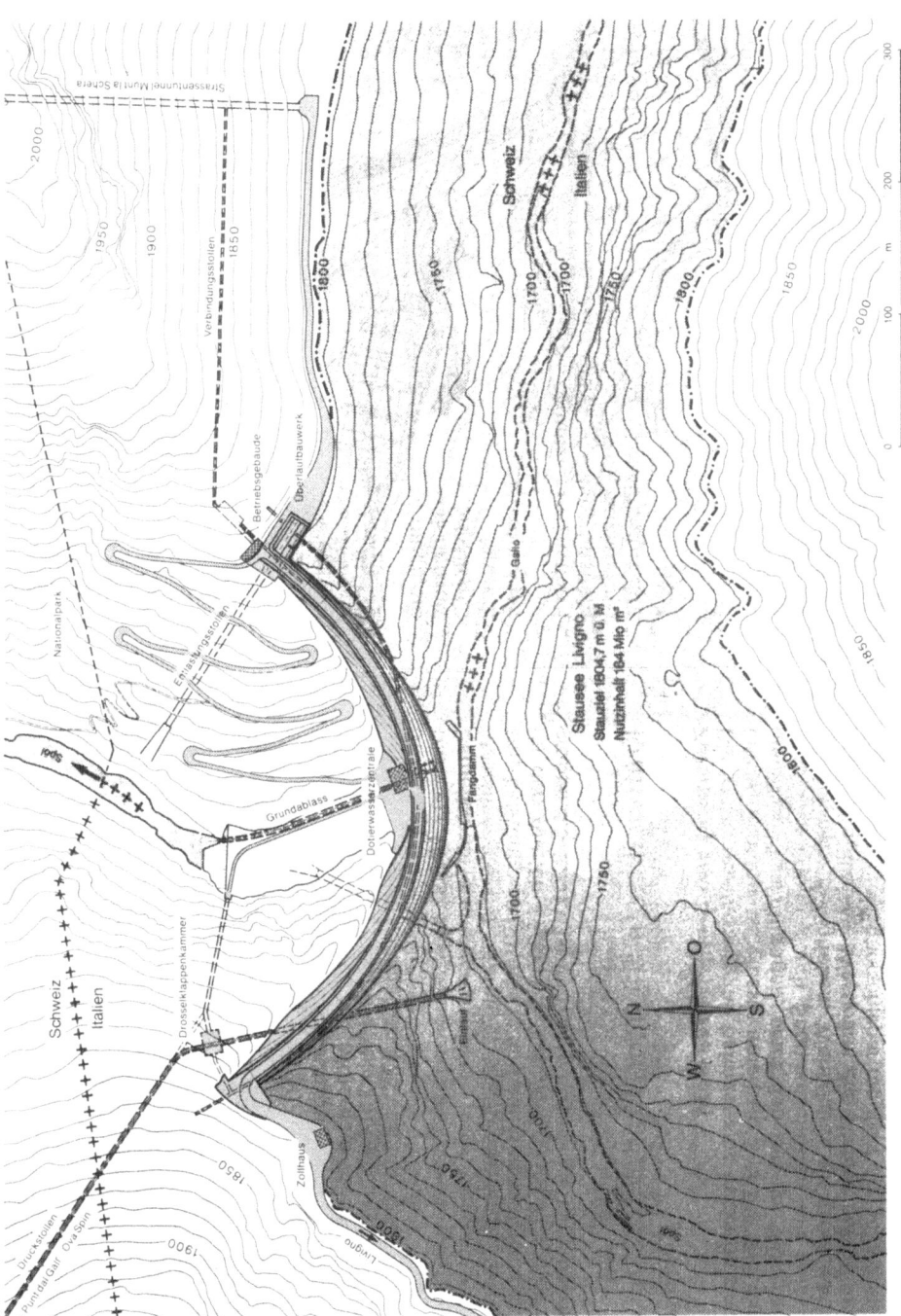

Figure 2 130 m height dam.

the typical structure of a dam in more detail. Figure 2 shows the main structural parts: the concrete dam, the spillway, the intake and the bottom outlet. Underwater inspection and/or maintenance work is performed mainly on the two last items, that is the intake and the outlet.

As an example of a recent underwater job completed by our Hydrovision Department, we cite the maintenance of a dam intake, the surface of which is 200 m^2. During a underwater TV inspection, the divers had noticed that two of the grids (6.0 m $\times$ 1.0 m) had been torn off. It was decided to effect an immediate repair to prevent foreign materials entering the pipe and damaging the turbines down at the plant. On another job, TV cameras used for underwater inspection showed the intake being blocked to one-third of its height by sedimented materials from the lake.

Figure 3   Dam at altitude of 1800 m.

Our divers work down to a depth of 50 m with compressed air open-circuit systems. A minor incident or a major accident may happen at any time; that is why we always bring a single-seater recompression chamber (10 bars). This chamber, built in our workshops, can be linked to the multi-seater chamber of the DKLZ in Zürich. Nevertheless, because of the local geography and the often difficult access to the sites with a land vehicle, transporting an injured person to a hospital equipped with a multi-seater chamber may take hours. For this reason, a specialized and efficient rescue system has been organized.

Part II: Principles for Diving Operations in Switzerland

B. Schenk, Head of Operation and Engineering/Techn. Research, Hyperbaric Laboratory, University Hospital, Zürich, Switzerland

TOPOGRAPHIC SITUATION OF SWITZERLAND

Highest point	4674 m above sea-level (Dufourspitze)
Lowest point	193 m above sea-level (Lago Maggiore)
Total area	41 293 km^2
Area covered with water approx.	1527 km^2 (about 3.7% of total area)

Lakes

highest	approx.	3000 m above sea-level
lowest		193 m above sea-level (Lago Maggiore)
deepest		372 m (Lago Maggiore)

Rivers

longest	approx.	375 km
total length of all watercourses	approx.	2500 km

In addition to well-trained divers equipped with the necessary materials the following main components are of great importance with respect to the specific topographic situation in Switzerland to provide the essential basis for diving operations under such conditions:

(a) safe decompression tables for different altitudes;
(b) well functioning rescue organisation in case of diving accidents;
(c) hyperbaric treatment and research centres, and adequate treatment methods.

Standard air decompression tables for different altitudes

With reference to the diving depth and short bottom times required, practically all diving activities in Switzerland are based on air-diving. Therefore, the standard air decompression tables, calculated for different altitudes and in use since 1973, are reproduced in Appendix 1.

Methods of calculation

In 1970, decompression profiles for several dives at lower atmospheric pressure were calculated at the Hyperbaric Laboratory, Department of Internal Medicine, University Hospital Zürich. Previous diving experiments at sea-level were taken into consideration, in particular those including tolerated supersaturation factors of $^{14}N_2$ half-times from 5 to 635 min. These factors are lower than those in the tables of the US Navy (1963) and of the Royal Navy, and considerably lower than those in the French GERS tables.

According to the calculation, based on our good experience with our supersaturation factors, 106 simulated dives were performed by 50 different subjects at a final ambient pressure of 0.7 bar (3000 m above sea-level). The data collected from these experiments performed near the border-line of sufficient decompression provided the basis for the calculation of our standard air decompression tables.

For the calculation, we assumed that the initial P_{N_2} in all tissues of a diver is 0.8 bar at the beginning of the first dive. We determined "no decompression times" for each altitude and integrated a system for calculating repetitive dives. Therefore, each dive has been marked by a letter which represents a definite $P_{N_{2011}}$. This system can also be used to calculate the safety flight level for flying after diving.

Actual Dives at different Altitudes

Between 1971 and 1975 a total of 278 actual single-controlled test dives by 143 different subjects were performed in open water at altitudes of 900 and 1700 m above sea-level. Of these 278 dives, 184 were repetitive dives. All these dives were performed by male divers and in most cases by divers of the Swiss Army and Water-Police Organisation. In the meantime, the tables have been in use by practically all Swiss scuba divers and a great number of divers in other countries.

No symptoms of insufficient decompression — such as disorders of the central nervous system, inner ear, skin bends, muscle strain, fatigue or joint bents — has been reported to date.

A Deco-Computer "Deco Brain" carried by the diver and based on the above-mentioned decompression concept has been on the market since 1983.

Rescue organisations for diving emergency

Owing to the topographic situation in Switzerland, a lot of our mountain lakes are accessible only over mountain passes at higher altitudes than the lakes. Therefore, the rescue organisation must be equipped with specific first aid material, including transportable monoplace recompression chambers, aircraft, road and cross-country vehicles and, last but not least, with a communication network for organising and commanding the necessary crew.

In cooperation with the Treatment Centres, there are two important groups involved which operate according to the following scheme:

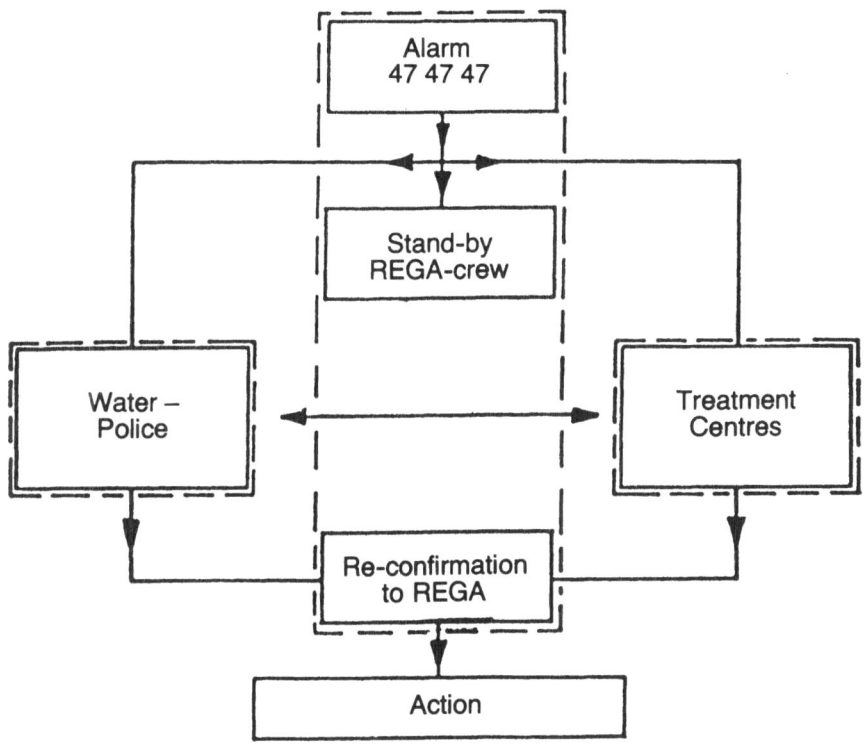

The Swiss Air Rescue Service, REGA

In case of diving accidents, the REGA is the alarm centre for receiving the emergency calls. Through their specific communication and radio network, with links to the Water-Police System, the rescue action is planned, coordinated and controlled.

In Switzerland, the REGA provides more than 90% of the air rescue service. The REGA is a charitable, humanitarian organisation, founded in 1952. REGA activities are based on the principles of the Red Cross and aid is offered to anyone in need.

The REGA currently has 15 bases, located so that any operational area in Switzerland can be reached within 15 min of flight. These bases are in a state of readiness 24 h a day. During the day, the crews are on call in the bases; at night they must be available at any time and ready to take off within 30 min.

In addition, the REGA maintains an ambulance jet base in Zürich-Kloten, from where all repatriation flights abroad are effected.

The REGA has ten Alouette III, SA 319 B, all equipped with rescue winches. Additionally, the bases of the large agglomerations Zürich, Basle and Berne have three twin-turbine Bölkow BO-105 CBS at their disposal. To date, the Canadair Challenger and the two Learjets have called on more than 500 airports in over 70 countries.

The REGA alarm centre is staffed 24 h a day. It handles emergency calls from all over Switzerland and abroad through the telephone number (Zürich) 47 47 47. For the immediate application of the means at its disposal, the REGA uses the only private 2m-band radio network covering all of Switzerland. More than 300 walkie-talkies are in operation.

From 1960 to date, the REGA has carried out more than 40 000 operations, in the course of which more than 35 000 persons were saved or transported.

In case of a diving accident, a specially designed mobil monoplace recompression chamber (Dräger type, without flange, 8 bar) is available for the pressurised transportation of patients in aircrafts. In addition to this possibility, the REGA field of operation also includes:

Mountain rescue missions in summer in rock and ice, search operations for missing persons. Aid for seriously ill persons in remote locations.

Mountain rescue missions in winter. Assistance for seriously injured skiers, aid in case of avalanche accidents through air transport of emergency doctors and avalanche dog teams.

Aid for victims of traffic and occupational accidents.

Transport of organs, blood, medicaments and serum.

Relocation flights for emergency patients to medical centres. Transport of high-risk newborn babies, patients with spinal cord injuries and serious burns.

The REGA carries out the worldwide repatriation of seriously ill or injured patients in its own ambulance aircraft or on commercial carriers.

Aid for mountain farmers, transport of animals, fodder and material.

Catastrophe aid.

The Swiss Air Rescue Foundation is financed chiefly through donations. Roughly two-thirds of the annual financial requirements, which amount to 30 million Swiss Francs, are covered by such voluntary contributions.

The Water-Police Rescue Organisation

Each Canton or district in Switzerland has its own water-police organisation which, in case of diving accidents, is responsible for the rescue part in its own area or waters. For national actions they cooperate together through a central headquarters.

The Water-Police Rescue Organisation, with its own divers and the necessary

Table 1
Mobile Monoplace Recompression
Chambers available in Switzerland

Chamber 1	Zürich	REGA
2	Zürich	Water-Police of town
3	Oberrieden	Water-Police of canton
4	Schlieren	private diving firm
5	Brugg	Army, Diving Dept.
6	Lugano	SLRG
7	Ascona	SLRG

rescue equipment, first-aid material, mobile recompression chambers and transport vehicles, has links with the Diving Department of the Army, the Road Ambulance Service and private rescue organisations, if needed. As a result of this cooperation on ground, six additional mobile monoplace recompression chambers (Spiro-Flange type, 5 bar) (Table 1) and crews, located in Switzerland, are available in case of emergency.

Hyperbaric treatment centres for diving accidents

There are two hyperbaric treatment centres integrated in the national organisation; the main centre is located in Zürich, the other in Lausanne.

The Centre for Hyperbaric Treatment and Diving Research, Zürich

The centre is located at University Hospital Zürich, Switzerland, and is accessible from the airport Zürich-Kloten, and the helicopter landing ground at the University Hospital. It has been in operation since November 1975. The hyperbaric chamber system has been designed and built to act as:

Deep diving simulation unit down to 1000 msw.

Hyperbaric facility for treatments in connection with diving accidents, with the possibility of flangemounting transportable recompression chambers or taking over patients under pressure in transport chambers without flange system, such as the REGA one.

Oxygen-filled chamber system for various hyperbaric oxygen applications.

The chamber system was developed and designed by Ing. B. Schenk, University Hospital Zürich. The pressure vessels were manufactured by Gebr. Sulzer, Winterthur, and 33 Swiss and foreign firms manufactured the auxiliary systems.
The use of the chamber and its range of applications are as follows:

Medical research and studies for deep and safe diving. Testing of various compression and decompression calculations.

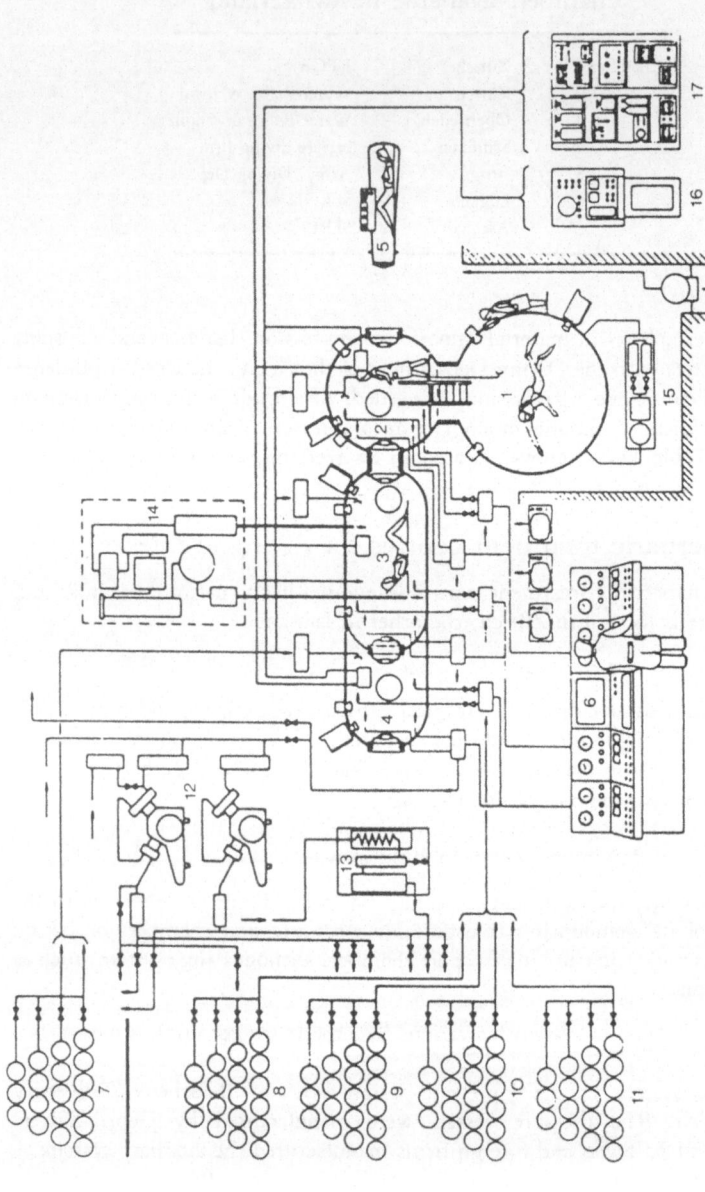

Figure 1 Schematic diagram of the monitoring and supply peripherals of the diving simulator at the Universitätsspital Zürich 1, wet chamber; 2, transfer lock; 3, decompression chamber; 4, lock; 5, one-man conveyor; 6, control and monitoring centre; 7, breathing gases; 8, oxygen; 9, gas mixtures; 10, air; 11, helium; 12, compressors; 13, helium recovery unit; 14, pressure chamber air conditioning and treatment unit; 15, water treatment unit; 16, medical surveillance; 17, gas analysis systems.

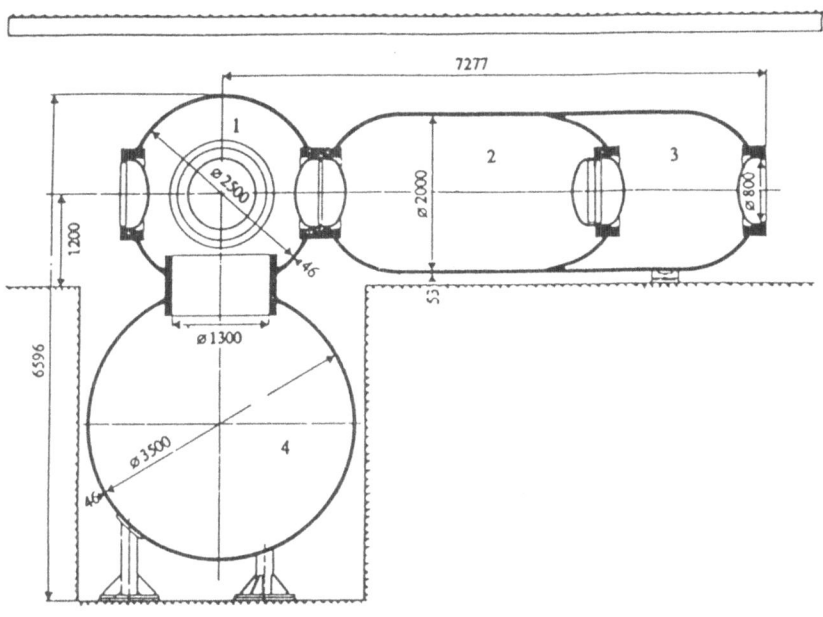

Pressure Range *Pressure Range*   in all chambers 0.5–100 bar

Dimensions	Chamber 1	Chamber 2	Chamber 3	Wet Chamber 4
Diameter (ID)	2.5 m	2.0 m	2.0 m	3.5 m
Length (internal)	—	3.2 m	1.6 m	—
Doors (ID)	0.8 m	0.8 m	0.8 m	1.3 m
Volume	8.0 m³	10.0 m³	6.5 m³	22.0 m³

Life-Support System (LSS)—Operating Range

Temperature	115–45 °C	15–45 °C	15–45 °C	4–45 °C
Humidity	30–100%	30–100%	30–100%	—

Chamber gas	pure oxygen or air or heliox or others
Mask breathing gas	pure oxygen or air or heliox or others
Max. rate of pressure change	up to 10 bar min⁻¹ possible
Personnel locks	chamber 1 and chamber 3
Material locks	into chamber 1, 2 and 3
Transport chamber connections	available at chamber 1

Figure 2   Diagram of hyperbaric chamber complex with some chamber data.

Research and development into operational equipment to extend the human sensing and manipulative capabilities under hyperbaric conditions.

Human factors' studies in connection with problems arising from long periods in confined spaces under hyperbaric conditions.

Testing of technical equipment under controlled simulated conditions.

Recompression treatments and hyperbaric therapies in connection with diving accidents, decompression sickness, gas embolism or hyperbaric oxygen applications.

Training in hyperbaric operation of diving personnel, doctors and medical personnel, supervisors and staff for hyperbaric facilities. Selection of diving personnel.

The Hyperbaric Treatment Centre, Lausanne

The centre is located at Hôpital Cantonal Lausanne, and is accessible from the helicopter landing ground near the hôpital Cantonal Lausanne. It has been in operation since 1964 (National Exhibition in Lausanne). The hyperbaric chamber system has been designed and built to act as:

Hyperbaric facility for treatment in connection with diving accidents, *without* the possibility to flagemount transportable recompression chambers.

Hyperbaric facility for various hyperbaric oxygen applications.

Development, design and manufacture was carried out by Roberto Galeazzi, Italy. The chamber is used for recompression treatments and hyperbaric therapies in connection with diving accidents, decompression sickness, gas embolism or hyperbaric oxygen applications. Chamber data are as follows:

Working pressure	4.5 bar
Compartments	2
Total length	3.1 m
Diameter	1.5 m
Chamber gas	Air
Mask breathing gas	Air or oxygen
Personnal lock	1
Material lock	1
Saturation capability	partly possible
Transport chamber connection	not available

Tables of treatments used in case of air-diving accidents

The methods of treatment used are adapted periodically to new needs. Since 1982, the following Treatment Tables, developed at the Hyperbaric Laboratory, University Hospital Zürich, have been in use for air-diving accidents (see also Appendix 2).

Table 1	In case of cerebral and/or spinal symptons after air-dives of 12 m or less
Table 2	In cases of cerebral and/or spinal symptoms after air-dives between depths of 13 and 50 m.
Table 3	In case of cerebral and/or spinal symptoms after air-dives deeper than 50 m.
Table 4	Hyperbaric oxygen therapy in case of cerebral and/or spinal symptoms after a diving accident occurred more than 48 h ago, or in case of other hyperbaric oxygen applications, such as additional therapy after a treatment according to Table 1, 2, 3 or 5.
Table 5	In case of "minor symptoms" (bends, muscle pains, joint pains) after air-dives.

APPENDIX 1   STANDARD AIR DECOMPRESSION TABLES*

1.   0– 700 m above sea-level (760–700 Torr)
2. 701–1500 m above sea-level (699–635 Torr)
3. 1501–2000 m above sea-level (634–596 Torr)
4. 2001–2500 m above sea-level (595–560 Torr)
5. 2501–3200 m above sea-level (559–510 Torr)
6. Altitude adaptation
    feet sea-water $= m \times 3198 \ (\pm 5\%)$

*© Deep Diving Research Laboratory, University Zürich, Switzerland. These tables have been published in Böni, M., Schibli, R. A., Nussberger, P. and Bühlmann, A. 1976. Air decompression tables for different altitudes, *Undersea Biomed. Res.*, 3(3), 189–204.

Table 1

DECOMPRESSION - TABLE 0 - 700 m ABOVE SEA-LEVEL

Depth m	Bottom-time min	15	12	9	6	3	Repet. group
15	90					5	J
	120					10	K
	150					15	L
20	50					4	H
	60					7	J
	75					18	K
	90					23	K
	105				3	31	K
25	30					5	G
	40					8	H
	50					12	J
	60					20	K
	75				3	30	K
	90				10	38	L
30	25					5	G
	30					9	H
	35					12	J
	40				5	14	J
	50				5	25	K
	60			6	8	32	K
	75			7	15	42	L
	90			8	25	52	L
35	20					5	G
	25					9	H
	30					12	J
	35				5	17	J
	40				7	30	J
	50			3	10	35	K
	60			7	15	45	K
	75			10	15	62	L
40	15				2	5	G
	20				2	10	H
	25				3	12	J
	30			2	5	20	J
	35			2	10	30	K
	40			5	15	35	K
	50		2	7	20	40	K
	60		2	10	25	50	L

Depth m	Bottom-time min	18	15	12	9	6	3	Repet. group
45	10						4	F
	15					2	6	G
	20					3	11	H
	25					5	20	J
	30				3	10	30	K
	35				5	10	35	K
	40			2	5	15	45	K
	50		2	5	10	20	55	L
	60		2	5	20	25	60	-
50	10						5	F
	15					3	8	H
	20					5	17	J
	25				3	10	27	J
	30				5	10	35	K
	40		2	5	12	15	50	K
	50		2	10	15	20	60	L
	60		3	12	20	30	70	-
55	10				1	2	5	F
	15				1	4	11	F
	20			1	4	8	24	G
	25			2	7	10	32	J
	30		1	3	9	13	38	K
	40	1	5	8	17	18	58	K
60	10				2	3	5	H
	15				3	5	15	J
	20			2	6	10	29	J
	25		2	2	10	10	35	K
	30		2	4	12	15	40	L
65	10			1	2	3	6	H
	15			1	3	8	18	J
	20		1	2	6	13	33	J
	25		2	4	10	13	40	K
	30	1	2	8	14	18	46	L
70	10			2	3	4	6	H
	15			2	3	10	20	J
	20		2	3	5	15	35	J
	25	2	2	5	10	15	43	K
	30	2	3	10	15	20	50	L

Max. ascent rate 10m/min

No decompression limits (0 - 700 m)									
m	9	12	15	18	20	25	30	35	40
min		200	75	50	30	25	20	15	10
Each dive must include a decompression stop of 3 min at 3 m									

REPETITIVE SYSTEM 0 - 700 m ABOVE SEA LEVEL

SURFACE INTERVAL TABLE (min)

Repetitive group at the end of the surface interval											
L	**K**	**J**	**H**	**G**	**F**	**E**	**D**	**C**	**B**	**A**	**"O"**
L	14	30	47	68	94	127	149	174	206	255	440
	K	16	34	55	80	113	135	160	194	240	426
		J	18	39	65	98	119	145	180	225	409
			H	22	47	80	101	127	160	206	394
				G	26	59	80	106	139	186	372
					F	34	55	80	113	160	346
						E	22	47	80	127	312
							D	26	59	106	292
								C	34	80	266
									B	47	233
										A	186

Repetitive group at the beginning of surface interval

REPETITIVE TIMETABLE (min)

Depth m	6	9	12	15	20	25	30	35	40	45	50	55	60	65	70
L			300	160	96	69	55	45	39	34	30	27	25	23	21
K		>400	250	127	80	59	47	39	34	29	26	24	22	20	18
J		400	150	101	67	50	40	33	29	25	23	21	19	17	16
H		200	113	80	55	42	34	27	25	22	19	17	16	15	14
G	>200	135	85	63	44	34	27	22	20	18	16	14	13	12	11
F	200	108	61	47	34	26	22	17	16	14	13	12	11	10	9
E	113	61	42	34	25	19	16	13	12	11	9	8	8	7	7
D	85	48	34	27	20	16	13	10	10	9	8	7	7	6	6
C	61	37	26	22	16	13	10	8	8	7	6	6	5	5	5
B	43	26	19	16	12	9	8	6	6	5	5	4	4	4	4
A	26	17	12	10	8	6	6	4	4	3	3	3	3	3	3

Repetitive - group

Table 2

DECOMPRESSION - TABLE 701 - 1500 m ABOVE SEA LEVEL

Depth m	Bottom-time min	16	13	10	7	4	2	Repet. group
12	120						5	H
15	40						5	F
	75						8	H
	90						13	H
	105						16	J
	120						19	J
20	20						4	E
	25						6	F
	30					2	8	F
	40					3	9	G
	50					5	11	H
	60					5	19	H
	75					8	25	J
	90					12	25	J
	105					15	30	-
	120					20	37	-
25	15					1	5	D
	25					4	8	F
	30				3	4	9	G
	35				4	5	13	G
	40				5	5	17	H
	50				5	7	23	H
	60				7	12	25	J
	75				10	15	30	J
	90				15	17	38	-
30	10					1	3	D
	15					3	6	E
	20					5	9	F
	25				3	5	10	G
	30				6	6	12	G
	35				7	7	18	H
	40			3	10	10	23	H
	50			5	10	15	25	J
	60			6	15	15	31	J

Depth m	Bottom-time min	16	13	10	7	4	2	Repet. group
35	10					2	3	D
	15				3	5	8	F
	20			2	5	5	11	G
	25			3	7	7	13	H
	30			4	8	8	16	H
	35			5	8	8	22	H
	40		2	5	10	10	25	J
	50		3	6	12	15	32	J
40	5					1	3	C
	10				2	2	6	E
	15			3	3	5	9	F
	20			4	5	6	12	G
	25		3	5	7	8	16	H
	30		3	6	8	10	20	J
	35		4	8	8	11	25	J
	40		6	8	9	17	32	J
45	5					2	3	C
	10				3	4	8	F
	15		2	3	5	5	12	G
	20		3	4	6	8	15	H
	25	1	4	6	6	10	19	H
	30	2	5	7	10	16	28	J
50	5				1	3	4	D
	10			3	4	5	10	F
	15	1	3	5	6	6	15	G
	20	3	3	6	6	10	18	H
	25	3	4	8	10	10	25	J
	30	6	6	8	13	18	35	J
55	10		2	4	5	6	12	F
	15	1	3	5	7	7	17	G
	20	4	5	6	7	10	24	H
60	10	1	3	4	6	6	13	G
	15	2	3	4	8	8	18	G
	20	5	6	6	8	10	30	H

Max. ascent rate 10m/min

	No decompression limits (701 - 1500 m)							
m	9	12	15	18	20	25	30	35
min	720	90	30	20	15	10	5	4
	Each dive must include a decompression stop of 3 min at 2 m							

REPETITIVE SYSTEM   701 - 1500 m ABOVE SEA LEVEL

SURFACE INTERVAL TABLE (min)

| Repetitive group at the end of the surface interval | | | | | | | | | |
J	H	G	F	E	D	C	B	A	"O"
J	17	35	59	87	105	125	150	183	265
	H	19	42	71	89	109	134	166	250
		G	23	52	69	90	115	147	230
			F	29	46	67	92	124	207
				E	18	39	64	96	178
					D	21	46	78	161
						C	25	58	141
							B	33	115
								A	81

Repetitive group at the beginning of surface interval

REPETITIVE TIMETABLE (min)

Depth m		6	8	10	12	15	20	25	30	35	40	45	50	55	60
Repetitive - group	J			320	171	120	73	53	43	35	30	26	23	21	19
	H		>225	190	125	93	60	44	35	30	25	22	20	18	16
	G	>330	225	130	92	71	47	35	29	24	21	18	16	15	13
	F	330	132	90	67	53	36	27	23	19	16	14	12	11	10
	E	142	83	59	46	37	26	20	16	14	12	11	9	8	7
	D	103	64	47	37	30	21	16	13	11	10	9	7	7	6
	C	73	48	36	28	23	16	12	10	9	8	7	6	6	5
	B	51	34	26	21	17	12	9	8	7	6	5	5	4	4
	A	32	21	16	13	11	8	6	5	5	4	4	3	2	2

Table 3

DECOMPRESSION - TABLE 1501 - 2000 m ABOVE SEA LEVEL

Depth m	Bottom-time min	16	13	10	7	4	2	Repet. group
12	60						4	H
	90						7	H
	120						12	H
15	30						4	E
	40						6	E
	50						7	F
	60						8	F
	75					1	12	G
	90					3	16	H
	105					3	20	H
20	15						5	C
	20						7	D
	25					2	7	E
	30					2	10	F
	40					3	12	G
	50					5	18	H
	60					6	24	H
25	10						5	C
	15					2	6	D
	20					4	8	E
	25					6	10	F
	30				3	6	12	F
	40				5	7	20	G
	50				7	10	25	H
	60				8	15	25	H
30	5						4	C
	10					1	5	D
	15					4	7	E
	20				2	4	10	F
	25				4	6	11	G

Depth m	Bottom-time min	16	13	10	7	4	2	Repet. group
30	30				6	8	16	G
	40			3	10	12	25	H
	50			3	12	15	30	H
	60			6	15	20	35	H
35	5						4	C
	10					2	5	E
	15				4	5	10	F
	20			1	6	6	14	F
	25			3	7	8	17	G
	30			4	8	10	20	G
	40		2	7	10	15	30	H
	50		3	8	12	20	40	H
40	5					2	4	C
	10				2	3	8	D
	15			3	5	5	11	F
	20			4	6	7	14	G
	25		3	5	8	8	20	H
	30		3	6	10	11	25	H
	40		6	8	11	18	40	H
45	5					3	4	C
	10			1	3	4	9	E
	15		2	4	5	5	14	G
	20		3	5	7	8	19	G
	25	1	4	6	8	10	25	H
	30	2	5	7	10	17	35	H
50	5				2	3	6	D
	10		1	3	4	5	12	F
	15	1	3	6	6	7	18	G
	20	3	4	6	8	10	23	H
	25	3	5	8	10	12	28	H

Max. ascent rate 10 m/min

	No decompression limits (1501 - 2000 m)						
m	9	12	15	18	20	25	30
min	360	50	25	15	10	6	4
	Each dive must include a decompression stop of 3 min at 2 m						

REPETITIVE SYSTEM 1501 - 2000 m ABOVE SEA LEVEL

SURFACE INTERVAL TABLE (min)

Repetitive group at the end of the surface interval								
H	G	F	E	D	C	B	A	"O"
H	17	37	60	74	90	108	130	191
	G	20	44	58	75	93	116	179
		F	24	39	55	73	96	160
			E	15	31	49	72	136
				D	17	35	58	113
					C	19	42	106
						B	23	88
							A	65

Repetitive group at the beginning of surface interval

REPETITIVE TIMETABLE (min)

Depth m		6	8	10	12	15	20	25	30	35	40	45	50
Repetitive - group	H	-	-	361	168	103	63	46	37	30	25	23	20
	G	-	>224	184	119	77	50	37	30	25	21	19	17
	F	>437	224	118	83	57	39	29	23	19	17	15	13
	E	437	119	75	57	41	28	21	17	14	12	11	9
	D	194	89	58	45	33	23	17	14	12	10	9	8
	C	122	65	45	36	25	18	14	11	9	8	7	6
	B	78	45	32	25	19	13	10	8	7	6	5	5
	A	46	28	21	16	12	9	7	6	5	4	3	3

Table 4

DECOMPRESSION - TABLE 2001 - 2500 m ABOVE SEA LEVEL

Depth m	Bottom-time min	16	13	10	7	4	2	Repet. group
12	50						4	E
	60						6	F
	75						8	F
	90					2	10	F
	120					5	15	G
15	20						4	C
	25						5	D
	30						6	D
	40						9	E
	50						12	F
	60						14	F
	75					2	18	G
	90					5	22	G
20	10						4	B
	15						7	C
	20						10	D
	25					2	10	E
	30					3	12	E
	40					5	16	F
	50					6	20	F
	60					8	26	G
25	5						4	B
	10						6	C
	15					3	8	D
	20					5	10	E
	25				1	7	10	F
	30				3	7	18	F
	40				5	10	30	F
	50				8	15	30	G
	60				8	20	35	G

Depth m	Bottom-time min	16	13	10	7	4	2	Repet. group
30	5					1	5	B
	10					3	7	C
	15					5	9	E
	20				3	7	10	F
	25				6	9	18	F
	30				8	10	25	F
	40			3	12	15	33	G
	50			5	15	20	38	G
35	5					2	4	B
	10				1	3	6	D
	15				4	5	13	E
	20			2	6	7	15	F
	25			3	8	9	24	F
	30			5	8	10	30	G
	40		3	7	10	18	45	G
40	5					3	5	B
	10				2	4	9	D
	15			3	5	6	13	F
	20			4	7	8	17	F
	25		2	5	9	10	25	G
	30		3	6	11	15	32	G
45	5				1	3	7	C
	10			1	3	6	10	E
	15		2	4	5	8	15	F
	20		3	5	8	10	25	G
	25	1	4	6	10	12	30	G
50	5				2	4	8	C
	10		1	3	4	6	14	E
	15	1	3	6	7	9	21	F
	20	3	4	6	9	12	28	G

Max. ascent rate   10 m/min

	No decompression limits (2001 - 2500 m)				
m	9	12	15	18	20
min	240	40	15	7	5
	Each dive must include a decompression stop of 3 min at 2 m				

REPETITIVE SYSTEM   2001 - 2500 m ABOVE SEA LEVEL

SURFACE INTERVAL TABLE (min)

Repetitive group at the end of the surface interval								
H	G	F	E	D	C	B	A	"O"
H	17	36	58	72	87	104	124	180
	G	19	42	55	70	88	108	163
		F	23	36	51	69	89	146
			E	14	29	46	66	121
				D	15	33	53	109
					C	18	38	94
						B	21	27
							A	57

Repetitive group at the beginning of surface interval

REPETITIVE TIMETABLE (min)

Depth m		6	8	10	12	15	20	25	30	35	40	45	50
Repetitive - group	H	-	-	>206	180	106	65	47	37	31	26	23	20
	G	-	>280	206	124	80	51	38	30	25	22	19	17
	F	-	280	127	87	59	39	29	23	20	17	15	13
	E	>258	132	80	58	42	28	22	17	15	13	11	10
	D	258	97	63	46	34	23	18	14	12	10	9	8
	C	145	70	47	36	26	18	14	11	10	8	7	7
	B	89	48	34	26	19	13	10	9	7	6	6	5
	A	51	30	22	17	13	9	7	6	5	4	4	3

Table 5

DECOMPRESSION - TABLE 2501 - 3200 m ABOVE SEA LEVEL

Depth m	Bottom-time min	16	13	10	7	4	2	Repet. group
12	40						5	D
	50						8	D
	60						10	E
	75						14	F
	90					2	18	F
15	15						4	B
	20						6	C
	25						8	D
	30						10	D
	40						14	E
	50						18	E
	60						20	F
	75					2	26	F
	90					6	32	F
20	10						5	B
	15						8	C
	20					2	10	D
	25					2	12	D
	30					3	14	E
	40				1	7	18	E
	50				1	10	24	F
	60				2	12	30	F
25	5						5	B
	10						8	C
	15					3	10	D
	20				1	5	13	D
	25				3	7	18	E
	30				4	10	23	F
	40				6	12	35	F
	50				8	15	40	F

Depth m	Bottom-time min	16	13	10	7	4	2	Repet. group
30	5					2	6	B
	10					3	9	C
	15				1	5	12	D
	20				3	8	13	E
	25			2	5	12	20	F
	30			3	9	12	30	F
	40			4	12	16	40	F
	50			6	15	22	50	F
35	5				1	3	6	B
	10				1	4	9	D
	15			1	2	6	14	E
	20			2	6	9	18	F
	25			3	9	13	25	F
	30		1	5	10	15	33	F
	40		3	7	12	22	50	F
40	5				1	3	8	B
	10				2	5	10	D
	15			3	5	7	15	E
	20		1	4	8	10	22	F
	25		3	5	11	14	30	F
	30		3	8	14	18	35	F
45	5				2	3	9	C
	10			1	3	6	12	D
	15		2	4	6	10	18	E
	20	1	2	6	10	14	32	F
	25	1	4	8	12	17	38	F
50	5			1	2	3	10	C
	10		1	3	5	8	14	E
	15	1	3	6	8	12	22	F
	20	3	4	7	11	16	34	F

Max. ascent rate   10 m/min

No decompression limits (2501 - 3200 m)				
m	9	12	15	18
min	150	30	10	5
Each dive must include a decompression stop of 3 min at 2 m				

REPETITIVE SYSTEM   2501 - 3200 m ABOVE SEA LEVEL

SURFACE INTERVAL TABLE (min)

	Repetitive group at the end of the surface interval					
F E	D	C	B	A	"O"
F 22	34	47	63	80	127
E	13	26	42	59	106
D	14	29	47	94	
C	16	34	80		
B	18	65			
A	47				

Repetitive group at the beginning of surface interval

REPETITIVE TIMETABLE (min)

Depth m		6	8	10	12	15	20	25	30	35	40	45	50
Repetitive - group	F	-	>160	145	94	63	41	30	24	20	17	15	13
	E	-	160	89	63	44	29	22	18	15	13	11	10
	D	-	113	69	50	35	24	18	15	12	11	9	8
	C	206	80	51	38	27	19	14	12	10	9	8	7
	B	113	55	36	27	20	14	11	9	7	7	6	5
	A	63	35	23	18	13	9	7	6	5	4	4	3

Table 6
Altitude adaptation More than 12 hrs at altitude

m	701–1500	1501–2000	2001–2500	2501–3200
No-Decompression limits	701–1500	701–1500	1501–2000	2001–2500
Decompression table and repetitive system	701–1500	701–1500	1501–2000	2001–2500
P_{N_2}(tissue ATS)	0.66	0.61	0.57	0.53

When staying more than 12 hrs at an altitude exceeding 1500 m you may use the next lower decompression table.

APPENDIX 2   TABLES OF TREATMENTS USED IN AIR-DIVING ACCIDENTS*

Table 1

Recompression in several minutes
with 1.9 bars (19 m)

	Bresathing gas	
	Air	100% O_2
bars	min.	min.
1.9	15	
1.5	30	
1.2	60 or	30
0.9	60 or	60
0.7	90 or	60
0.5	120 or	60
0.3	120 or	30
0		

8 h. 15 min. with air
4 h. 45 min. with 100% O_2 at 12 m

*Periods of breathing 100% O_2
should be interrupted each hour
by 5 min. of fresh air.

*Published in A. A. Bühlmann, *Dekompression—Dekompressionskrankheit*, Springer Verlag, 1983.

Table 2

Recompression in several minutes with 5.0 bar (50 m)

A			B			
After 30 min, practically sym.pton-free			After 30 min, improvement, but still some symptoms			
	Breathing gas			Breathing gas		
	Air	100% O_2		Air	100% O_2	Air
bars	min.	min.	bars	min.	min.	min.
5.0	30		5.0	90		
4.5	10		4.5	20		
4.0	15		4.0	30		
3.5	15		3.5	30		
3.0	15		3.0	45		
2.5	30		2.5	60		
2.1	60		2.1	120		
1.8	90		1.8	180		
1.5	120		1.6	240		
1.2	120 or	60*	1.4	240 or	60	+ 30
1.0	120 or	60	1.2	300 or	60	+ 30
0.8	120 or	60	1.0	300 or	60	+ 60
0.6	180 or	30	0.8	300 or	60	+ 60
0.4	180 | or	30	0.6	300 or	—	+ 240
0.2	120 or	15	0.4	240 or	—	+ 240
0			0.3	240 or	—	240
			0.2	120 or	—	120
			0			

20 h	25 min. with air	47 h	35 min. with air
10 h	40 min. with 100% O_2 at 12 m	34 h	35 min with 100% O_2 at 14 m

*In 2A, periods of breathing 100% O_2 should be interrupted each hour by 5 minutes of fresh air.

Table 3

Recompression with O_2–He with 9.0 bars (90 m).

Breathing gas at full pressure: 8–10% O_2, 82–84% He[*]

	Breathing gas				Breathing gas		
bars	min.	% O_2	bars	min.	% O_2	min.	% O_2
9.0	180	8[*]	3.0	90	20		
8.5	10	8[*]	2.8	120	20		
8.0	10	8[*]	2.5	120	20		
7.5	15	8[*]	2.2	120	20		
7.1	15	8[*]	1.9	120	20		
6.7	15	8[*]	1.6	120	20		
6.4	20	8[*]					
6.1	20	8[*]	1.4	60	100 +	30	20[*]
5.8	20	8[*]	1.1	60	100 +	30	20[*]
5.5	20	8[*]	0.8	60	100 +	30	20[*]
5.2	20	8[*]	0.6	60	100 +	60	20[*]
5.1	20	8[*]	0.3	—	—	120	20[*]
			0				
5.0	20	15[*]					
4.8	30	15[*]		32 h	25 min.		
4.6	60	15[*]					
4.2	60	15[*]					
3.8	60	15[*]					
3.5	60	15[*]					
3.3	90	15[*]					

[*]It is not necessary to flush out the pressure chamber with O_2 and He at the beginning of recompression.

O_2 concentration in the pressure chamber must be regularly measured.

Deviations of 2% are tolerable.

Room temperature should reach 28–30 °C.

For dives 50 m deep, Table 3 can be substituted for Table 2.

Recompression with 5.0 bars, after a dive at 5.0 bars for 120 to 180 min., pressure reduction down to 4.8 bars (48 m); for further decompression see appropriate table.

Table 4

More than 48 h have passed since the appearance of neurological symptoms, and gas bubbles have again re-formed extensively. Recompression at 5 bars has not caused a speedy reoccurrence of symptoms. In this case, recovery of the damaged nerve tissue can be accelerated by hyperbar oxygen therapy.

Recompression in several minutes with 1.5 bars (15 m)

bars	Breathing gas 100% O_2 min.
1.5	60[*]
1.0	60
0.8	30
0.6	30
0.4	30
0.2	30
0	

4 h

[*]Periods of breathing 100% O_2 should be interrupted each hour by 5 min. of fresh air. This treatment may be carried out twice in 24 h after an interval of 4 h.

Table 5

Illness after deep dives or exposure to compressed air. Complaints at the end of decompression or later.

Recompression in several minutes with 1.0 bar (10 m)

After 10 min. clear				After 10 min., no or only slight improvement			
					Additional recompression with 1.5 bars		
	Breathing gas					Breathing gas	
	Air		100% O_2			Air	100% O_2
bars	min.		min.	bars	min.		min
1.0	10			2.5	10		
0.9	60	or	60*	1.9	15		
0.7	90	or	60	1.5	30		
0.5	120	or	60	1.32	60	or	30*
0.3	120	or	30	0.9	60	or	60
0				0.7	90	or	60
				0.5	120	or	60
				0.3	120	or	30
				0			

6 h. 40 min. with air	8 h. 35 min. with air
3 h. 40 min. with 100%	5 h. 5 min. with 100%
O_2 at 9 m	O_2 at 12 m

*Periods of breathing 100% o_2 should be interrupted each hour by 5 min. of fresh air.

16

Developments in Offshore Medicine

A. M. House, Centre for Offshore and Remote Medicine (MEDICOR), Memorial University of Newfoundland, Canada

The development of offshore resources is particularly important for Canada at a time when conventional supplies of hydrocarbons are decreasing. This development carries with it the need to attend constantly to health and safety. Jurisdictional and legal considerations are reviewed, together with the current provisions for health and safety in offshore drilling units. A pilot project utilising a satellite link between an offshore installation and an onshore health centre is described.

At this time Canada's two areas of offshore hydrocarbon exploration are the Beaufort Sea in the north and the Atlantic area in the east. Most activity in the east is taking place in the Grand Banks area of Newfoundland, particularly in Hibernia and in the waters of the Scotian Shelf off the coast of Nova Scotia. Although significant discoveries of oil and gas have been made, there has not yet been any production. The Hibernia field is very likely to be the first in production, within three to five years. On the Scotian Shelf, commercial quantities of gas have been found but there is little indication to date of significant oil reserves. Limited drilling in the Labrador Sea has not yet been fruitful but is continuing.

The North Atlantic to the east of Newfoundland in the area of current petroleum related activity is known to have some of the most severe climatic conditions in the world. The Beaufort Sea, while colder, does not have a significant iceberg problem and some drilling has been possible through artificial islands. Although the north has its potential problems, it is the Atlantic area which has already demonstrated its savage weather and ice conditions in the Ocean Ranger disaster. The conditions which exist in what has often been referred to as "iceberg alley" are a constant reminder in eastern Canada of the need for constant attention to health and safety in the offshore industry.

Because conventional supplies of hydrocarbons in Canada are decreasing and the tar sands resources in the west are marginally economic at this time, there is tremendous pressure to develop offshore resources as soon as possible. The health and safety of workers are of major concern. Industry and government are continuously dealing with this matter. In phase II of the activities of the Royal Commission on the

Ocean Ranger Marine Disaster, a detailed study of health and safety is being carried out. As part of the Commission's terms of reference, an occupational health study has just been completed by the Centre for Offshore and Remote Medicine at Memorial University of Newfoundland. The report of this year-long study should shortly be available to the public.

Because of certain jurisdictional and legal considerations, the provision of health care in the Canadian offshore is a complex matter. For example, at the time of writing the ownership of the Hibernia area is in dispute, and, although the Supreme Court of Canada is expected to rule on the legal question shortly, it is unlikely that the overall relationship between the federal and Newfoundland governments with respect to exploration and production will be easily clarified. Ownership of the Scotian Shelf and of the Beaufort Sea has been settled in favor of the federal government. With the settling of jurisdictional disputes, there remains another major problem with respect to health care. Health matters in Canada are constitutionally a provincial matter. Therefore, although the Federal government may achieve full regulatory control, it can be seen that the organization and implementation of health services will have to be through provincial organizations and agencies. There is an additional problem: health care providers in Canada (apart from the armed forces) must be registered and licensed in a province. There is no national Canadian licence to practise medicine or nursing.

Despite these major problems, it is possible, through lease agreements with operating companies, to achieve a satisfactory arrangement which should ensure a high quality of health care and safety arrangements. At this time, there are approximately ten mobile offshore drilling units (MODUs) operating off the east coast of Canada. With rare exceptions, medical care on rigs is provided by a rig medic who is either a Registered Nurse with additional experience or training or an ex-armed forces medic with appropriate additional training. A small number of rigs employ a paramedic, described as an individual trained mainly to cope with medical emergencies.

Most companies have medical departments with a medical director (usually located in Calgary) who appoints a regional company physician (or group) to provide supervision of the rig medic, to be available for emergency calls and to perform pre-employment medicals as well as carrying certain other medical duties. There is a consensus that the operator's medical director should have responsibility for all health care on the rig but, although this is the case with some companies, it is by no means the common practice with all operators. On some MODUs the operator and one or more contractors may have different physicians providing care. This frequently applies to diving contractors, where the diving supervisor may relate directly to a separate shore-based physician. This practice has grown because neither the rig medic nor the company contractual physician may have significant knowledge of the medical problems of diving. This situation is gradually changing. The education and training of medics and company physicians in the basics of diving medicine is the aim of company medical directors.

Diving medical expertise is available in St John's and in Halifax. The medical supervision for diving problems in the Beaufort Sea is provided through Calgary. To date, there has been relatively limited diving activity in the Beaufort. The

Table 1
Distances and travel time between shore-based facilities and various sites of offshore exploration

Journey	Approx. distance (km)	Approx. helicopter flying time (h)	Approx. time by supply vessel (h)
St John's to:			
Continental Shelf margin	611	4.2	31.7
Hibernia	321	2.2	16.7
Ungava Peninsula	1609	11.1	92.8
Botwood (supply base)	257	1.7	13.33
Cartwright	740	5.1	38.33
Hopedale	1046	7.2	54.16
Marystown (shipyard)	193	1.3	10.00
Goose Bay		2.5	
Halifax to Venture	338	2.3	17.5
Canso to Venture	187	1.3	9.7
Halifax to St. John's	917	6.3	47.50

Onshore hyperbaric facilities currently existing in St. John's are inadequate but it is expected that these will be upgraded within the next two years. Facilities in Halifax have just been upgraded to a reasonably satisfactory level. There is no capacity for transfer of divers under pressure from offshore sites in Canada.

geographical location of the exploratory activity on the Canadian east coast makes the provision of medical services and diving medical services particularly difficult to deliver (see Table 1).

Onshore hyperbaric facilities currently existing in St John's are inadequate but it is expected that these will be upgraded within the next two years. Facilities in Halifax have just been upgraded to a reasonably satisfactory level. There is no capacity for transfer of divers under pressure from offshore sites in Canada.

Apart from the special consideration for diving medicine, land-based health resources are considered to be adequate to meet the requirements of the developing offshore industry.

As part of an effort to improve health care and safety facilities in the Canadian offshore, the Centre for Offshore and Remote Medicine (MEDICOR) and Memorial's Telemedicine Centre are now participating in a telemedicine pilot project, which is utilizing a satellite link between an offshore installation and an onshore health centre in an attempt to provide reliable communications and to facilitate the meeting of health needs in the offshore. The project is a joint effort of Memorial University, Canada's Department of Communications (DOC) and the Newfoundland Telephone Company. DOC has made available the Anik B satellite channels and earth terminals, including a prototype steerable terminal located on the semisubmersible Sedco 706, now under contract to Mobil Inc./Husky-Bow Valley. Two satellite telephone channels are available. One is used primarily to provide an ordinary dial-up telephone service between the rig and the "switched-network". This provides a world-wide telephone service. The other channel, which is a "4-wire" circuit, provides a dedicated link between the rig's hospital room and the Health Sciences Centre at Memorial University in St John's. Physicians experienced in emergency cases and a full range of medical specialists (consultants), including experts in diving medicine, are therefore available at all times. It is also possible to arrange teleconferences with experts located anywhere in the world, should this need arise.

The high quality link is used for voice and to transmit pictures by slow scan television system (SSTV). Other medical data, including electrocardiograms, can also be transmitted. The system is currently being developed to support the medical care of divers who may be in saturation.

When necessary, the satellite channels can be used simultaneously for medical care, thus allowing discussion between the physician on shore and the rig medic and/or the diving supervisor at the same time that medical data are being transmitted. If necessary, the slow scan system can be linked to the onboard television system of the rig and allow pictures to be transmitted to the shore base. The connections can be used for education and training of medical or diving personnel. These education programs can be supplemented by the transmission of pictures on the two-way slow scan apparatus.

At the time of writing, there have been some technical problems with the satellite terminals, particularly with the steering mechanism of the prototype terminal on the Sedco 709. Despite this, the system has functioned reasonably well and is gradually being improved. The system has been used with considerable benefit in the management of a number of illnesses and injuries. Slow scan pictures have greatly aided in the management of eye injuries and diseases.

Although the priority use of the satellite channels during the pilot project is to support health care, industrial use is encouraged, and, in addition, personal telephone calls are allowed. The system also has the capacity to receive live television transmissions when not required for medical or industrial purposes.

It should be noted that most MODUs have the Inmarsat satellite service in addition to HF and VHF radio. However, because of its high cost, the use of this commercial system is limited by most companies to meet emergency and other important needs. If necessary, slow scan pictures can be transmitted by the Inmarsat system. It is anticipated that within a year the type of satellite system being developed by the Memorial group will be commercially available at reasonable cost for use in the Canadian east coast offshore.

Satellite communications are not usually available for standby vessels and many other ships. This deficiency will probably be rectified if the current Canadian mobile satellite (M-sat) program proceeds as planned. This proposed narrow band system is expected to provide inexpensive satellite communication levels by way of small earth terminals which will be hand-held vertically and are planned to be used on ships, offshore oil rigs, and motorized land vehicles. The launch of this large satellite able to accommodate up to 3000 telephone channels is planned for 1987 and the service should be commercially available within a year or so thereafter.

Because trophospheric scatter systems are not practical on MODUs at this time, it is clear that only satellite systems can provide reliable communications links to remote sites such as the offshore. It is very unlikely that real time live television transmissions will be possible from MODUs in the near future, although systems capable of this will be available in the future at reasonable cost. In the meantime, inexpensive narrow band or telephony systems can be used.

In Canada in the past year or so, there have been a number of other developments which should help to improve health and safety in Canadian waters:

An industry task force should shortly be making public its report after a year-long study on safety.

The Canada Oil and Gas Lands Administration, which is the Canadian Government's "single window" on the offshore, has recently established a Medical Advisory Group made up of medical representatives of government, industry, and educational institutions.

A number of educational institutions have developed and are offering courses for physicians, rig medics and other workers.

An industry Medical Advisory Committee has been established within the Offshore Operations Division (OOD) of the Canadian Petroleum Association.

For a number of reasons, the offshore hydrocarbon activity is developing slowly in Canada. The development of a high quality health care system has been correspondingly slow but recent developments indicate that such a system is now being put into operation.

REFERENCES

Bartlett, R. 1983. *Problems in Providing Medical Support for Remote Diving Operations*, Canadian Offshore Technology Conference II, Toronto, p. 4.

Canadian Government, *M-Sat Phase B Project Plan*, March 1982.

House, A. M. 1981. Telecommunications in Health and Education, *Canadian Medical Association Journal*, 124, 667.

House, A. M., Roberts, J. and Canning, E. M. 1981. Telemedicine in Northern Health Care, in *Proceedings of the Fifth International Symposium on Circumpolar Health*, Copenhagen, pp. 88–90.

Williamson, D. D. 1982. Telecommunication in the energy field, *Journal of Canadian Petroleum*, May/June, p. 687.

17

Monitoring of Breathing Gases

P. Wiesner, Werk Druckkammertechnik, Drägerwerk AG,
Lübeck-Travemunde, Federal Republic of Germany

Modern underwater simulators can be used to study many physiological and technical
problems connected with arctic diving. The German underwater simulator GUS1 is
described together with the gas analysis system. A special analytical technique —
transcutaneous analysis — for measuring partial oxygen pressure in the blood of divers
is considered.

Diving in arctic waters is connected with numerous physiological and technical
problems. Owing to modern underwater simulators, many of these problems can be
studied on the land and under safe conditions.

In December 1983 the new German underwater simulator GUSI was inaugurated
by the successful completion of a 300 m saturation dive, which was performed under
the responsibility of Drägerwerk. With this simulator a unique system for testing of
diving equipment, of underwater working and of underwater welding procedures is
available. In manned experiments an underwater depth of down to 600 m can be
simulated.

The main working chamber of the system (Fig. 1) is a horizontal cylinder with an
internal diameter of 3.50 m, a length of 12.70 m and a volumetric capacity of
110 m^3. This chamber can be filled completely or partly with water. Both gas and
water can be kept at temperatures down to 0 °C to simulate arctic conditions.

An important component for dives in such a simulator, or for real offshore dives,
is the gas analysis system. With the development of the GUSI simulator on the one
hand, and the construction of offshore deep diving systems like the one for the rig
Safe Karinia on the other (Fig. 2), Drägerwerk gathered further experience in
designing, building and using gas analysis systems. In this paper we want to report
on some basic problems in the analysis of breathing gases and on the transcutaneous
measuring of the partial blood oxygen pressure of divers.

What gases are to be analysed? One group is formed by the gases for breathing:
oxygen, helium and nitrogen. Another group contains the exhaled carbon dioxide
together with impurities which form during underwater welding: carbon monoxide,
hydrogen, methane and higher hydrocarbons, nitrous fumes and ozone. Hydrogen

Figure 1 The big working chamber A 1 of the GUSI-Simulator (GKSS Geesthacht).

and methane must be analysed for safety reasons; the others are toxic. All gases are listed, together with the associated usual analysis technique in Table 1.

One question is whether one should analyse the gases inside the chamber, i.e. under pressure, or outside the chamber. The most elegant way, in fact, would be the analysis under pressure in the chamber because most of the analysis principles would give the physiologically important partial pressure directly as result. Unfortunately, there exist only a few instruments which can reliably be used under pressure in the

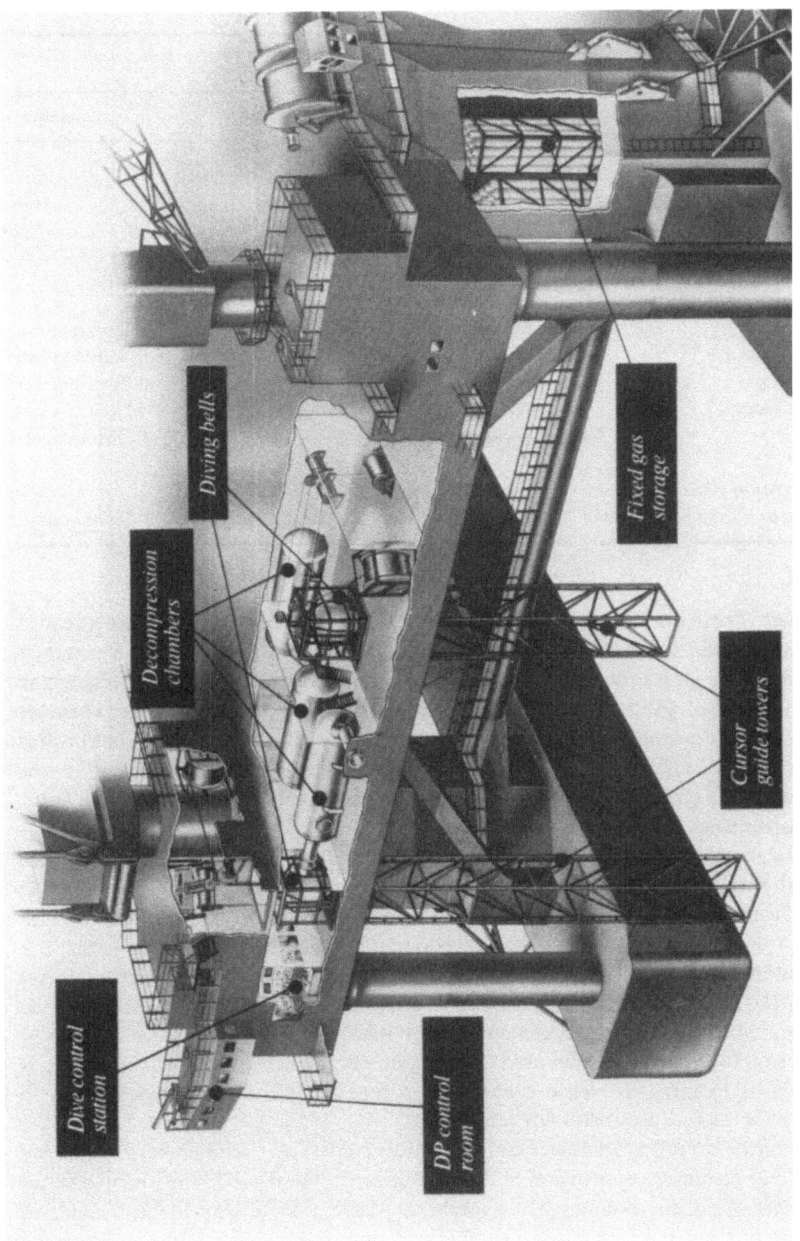

Figure 2 Scheme of diving system on rig *Safe Karinia* (Consafe Offshore).

Table 1
List of gases and the associated analytical methods

Gas	Analytical method		
	Continuous	Periodical	Sampling
Oxygen	Paramagnetic Electrochemical		
Nitrogen		Gas chromatography	
Helium	Thermal conductivity	Gas chromatography	
Toxic impurities			
Carbon dioxide	Infrared absorption		Detector tubes
Carbon monoxide	Infrared absorption		Detector tubes
Hydrocarbons	Flame ionization		Detector tubes
Nitrous fumes		Chemoluminescence	Detector tubes
Ozone	Chemoluminescence		Detector tubes
Other impurities			
Hydrogen		Gas chromatography	Detector tubes

chamber. These are simple portable oxygen analysers, based on the electrochemical principle, and detector tubes, which have been used successfully in pressure chambers for a long time. The few stationary instruments which work under pressure have practical disadvantages: they are too expensive to be installed in every chamber compartment; during longer stays under pressure the instruments have to be checked and calibrated; taking calibration gases into the chamber is unpractical and for the calibration the chamber pressure has to be taken into account; and last, but not least, the instruments have to be certified for fire safety.

In larger diving systems the analysis today is done generally with expanded gas. The advantages are that the instruments can be installed, operated and controlled at one place and, even during the dive, it is possible to check the instruments with calibration gases and to repair them if necessary. The drawback, of course, is that the concentrations are indicated in volume percentage, whereas for the divers' comfort the partial pressures of the gas components are important and not the percentage. So for keeping, for example, oxygen and carbon dioxide within limit values, tables or conversion forumlas are to be used. The latest development (i.e. in the GUSI gas analysis) is to carry out these conversions automatically with a computer and to display the partial pressure values (Fig. 3).

Saturation diving systems and big simulators consist of several pressure chambers because of technical, economical and safety reasons. These are living chambers and their ante chambers, diving bells, hyperbaric rescue chambers (or lifeboats) and, in the case of simulators, working chambers. If one wanted to equip every chamber compartment with one (or, for safety reasons, even with two) instruments of each type necessary, the costs of such a system would be absurd, not to mention the

Figure 3   Computer terminal for the display of chamber atmosphere data in chamber control
board (GUSI Simulator, GKSS).

service problems. So it is reasonable to use a reduced set of instruments, which can
be connected to all chambers sequentially by some kind of switching system.

Depending on the application, there are two different ways for controlling the
switching system. For big diving simulators like GUSI, we developed a computer
controlled switching matrix with magnetic valves, which allows a very versatile
combination between all analysers and chambers. In the case of the GUSI, the
switching matrix is controlled by a process computer. Depending on the demand by
the chamber operator, the computer connects the right instruments with the
chamber, controls the measuring ranges of the instruments and indicates the gas
concentrations on the displays in the chamber control boards. The data are
displayed, either in volume percentage values or converted to partial pressure values.
Moreover, the computer checks and calibrates the instruments with calibration
gases, which frees the operation crew from a lot of work and makes the system safer.
Details of the GUSI gas analysis have been reported elsewhere (Weisner, 1982).

Figure 4   Central gas analysis panel in offshore rig *Safe Karinia*.

In offshore applications, the universal combination of many instruments with many chambers is less important because there are not so many analysis tasks at any one time. Here we lay more emphasis on operational reliability and reduced maintenance requirements. As on the offshore rig *Safe Karinia*, the analysers are then switched to the chambers via flexible hose connections (Fig. 4). In this way is avoided the risk of leakages, which could lead to false readings in a complex magnetic valve matrix (there they have to be excluded by periodical leak tests).

The question of how many analysers of each type a simulator or diving system should have is still open. Generally, one has to define the maximum number of chambers in the system which may be occupied. In this case, every occupied chamber should have one oxygen and one carbon dioxide analyser and there should be at least one additional oxygen analyser for gas mixing purposes or as backup. Actually, most systems have many fewer CO_2 analyser and the gas atmosphere is then checked at regular intervals for CO_2 only. This implies knowledge about the speed of CO_2 increase in a failure mode and a careful procedure. We have a special case with gas regulation systems (e.g. automatic oxygen replacement). Here, there has to be a second analyser beside the instrument in the feedback loop in order to detect any failure in the regulation system and to give the alarm.

So far not much attention has been paid to the cross-sensitivity of the different gas analysis methods. An analyser shows a cross-sensitivity when the reading is influenced by the presence of any second gas component. There are generally two types of cross-sensitivity: the first (type A) shows up as an offset of all values, the second (type B) causes a deviation in the sensitivity (Fig. 5). When using air—oxygen mixtures, these cross-sensitivities can be neglected with most anlaysers, since the main background gas is nitrogen. Cross-sensitivities become important with the use of helium as part of deep diving gas mixtures.

For oxygen, analysers working on the paramagnetic principle are often used as well as electrochemical ones). They make use of the relatively high magnetic susceptibility of oxygen, which exceeds those of other gases by orders of magnitudes ($O_2 = 100$, $N_2 = -0.358$, $He = -0.059$). Nevertheless, the susceptibility of helium or nitrogen may interfere in the analysis of breathing gases for deep dives, because then the oxygen percentage becomes rather low. For example, let us assume that the zero of an analyser was adjusted with nitrogen. Then measuring a typical breathing gas for 400 m depth with 0.3 bar oxygen partial pressure and helium as rest gives a reading of 0.42 bar, which is too high by 40%. This deviation is due to the high helium percentage of the mixture. Therefore, it is important that, in deep diving applications, the analyser is adjusted with pure helium for the zero point and with a helium—oxygen mixture for the span. The deviation by a nitrogen component (trimix-gas) is neglible for helium-adjusted instruments (see Table 2).

The CO_2 content is usually determined with infrared analysers. Depending on the background gas (He or N_2), the reading is influenced by about 10%. This deviation is negligible for the monitoring of breathing gases for CO_2. The same applies for CO analysers, also mostly of infrared type.

Underwater welding produces contaminations like hydrocarbons and nitrous fumes. Hydrocarbons are detected with flame ionization analysers. These instruments show a strong dependence on the background gas, N_2 or He, which amounts to about 30% of the measured value. Even stronger is the effect in chemiluminescence analysers used for nitrous fumes. In this case, the reading may be wrong by a factor of 2, depending on the background gas. In both gases, the deviation should not be neglected and can be avoided by the use of appropriate calibration mixtures.

Finally, we want to report on a special analytical technique: the transcutaneous measuring of the partial blood oxygen pressure. Mainly for monitoring of neonates,

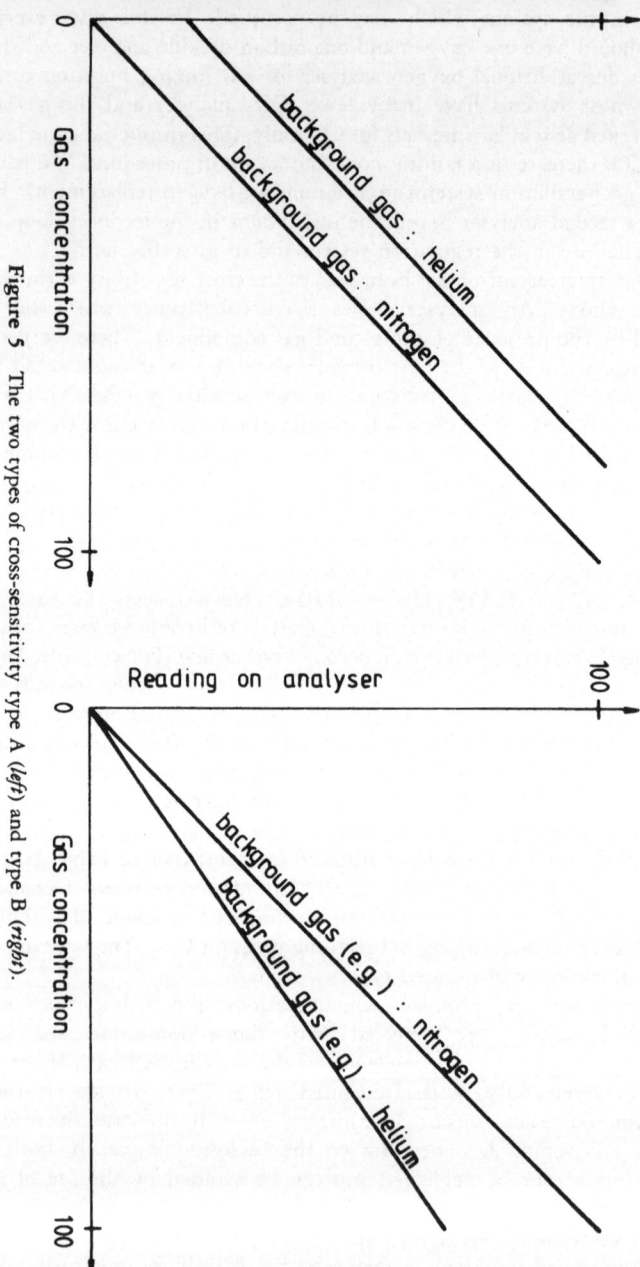

Figure 5   The two types of cross-sensitivity type A (*left*) and type B (*right*).

Table 2
Deviation of analyser

1 Zero adjustment with nitrogen:

$$\text{Reading} = pO_2 + 0.003pHe$$

$$\text{Deviation} = \frac{0.3pHe}{pO_2} \%$$

2 Zero adjustment with helium:

$$\text{Reading} = pO_2 - 0.003pN_2$$

$$\text{Deviation} = \frac{-0.3pN_2}{pO_2} \%$$

pO_2, pHe and pN_2 = partial pressures of O_2, He N_2 respectively.

Dräger manufactures the pO_2 monitor "Oxymeter", with the transcutaneous sensor "Transoxode". Of course, there was the question whether the blood oxygen pressure of divers or patients in pressure chambers could be measured using the same principle. After the first positive results of Huch *et al.* (1977), we can report on detailed experiences of two other institutes.

The "Transoxode" is applied mostly in the left forearm bend (Fig. 6). The

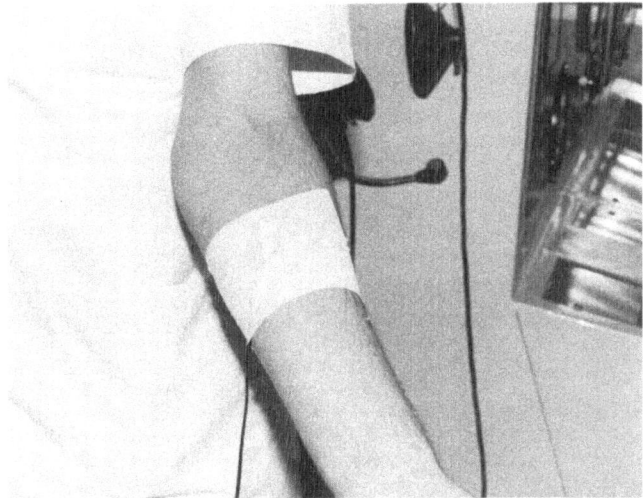

Figure 6 "Transoxode" transcutaneous sensor (affixed with adhesive tape).

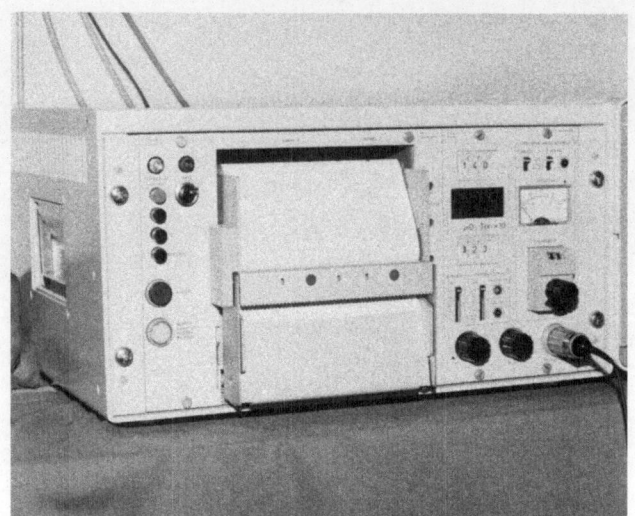

Figure 7 The "Oxymeter" monitor.

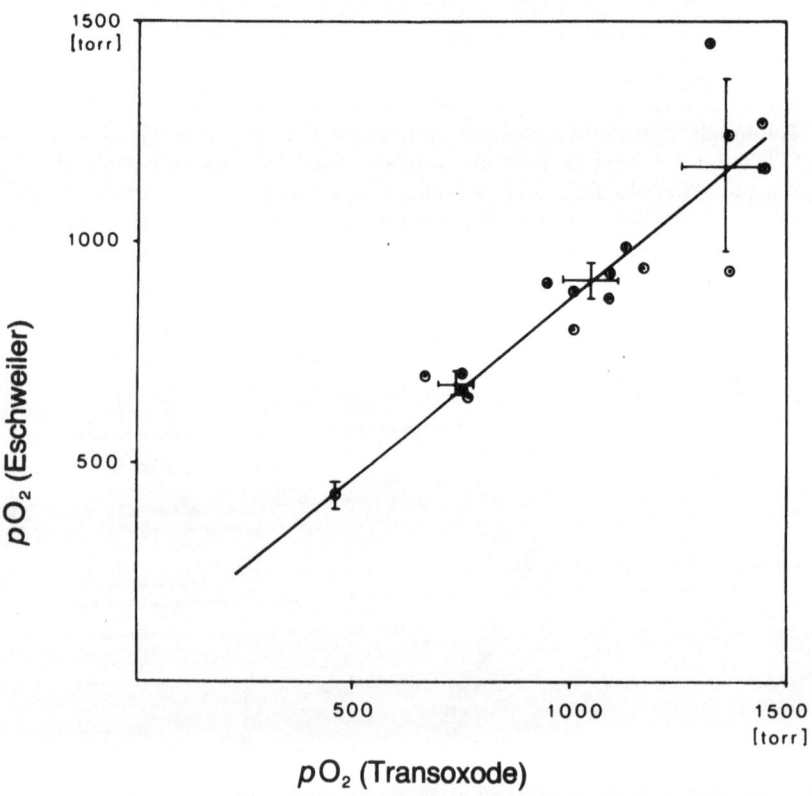

Figure 8 Correlation between transcutaneous data and blood gas data.

188

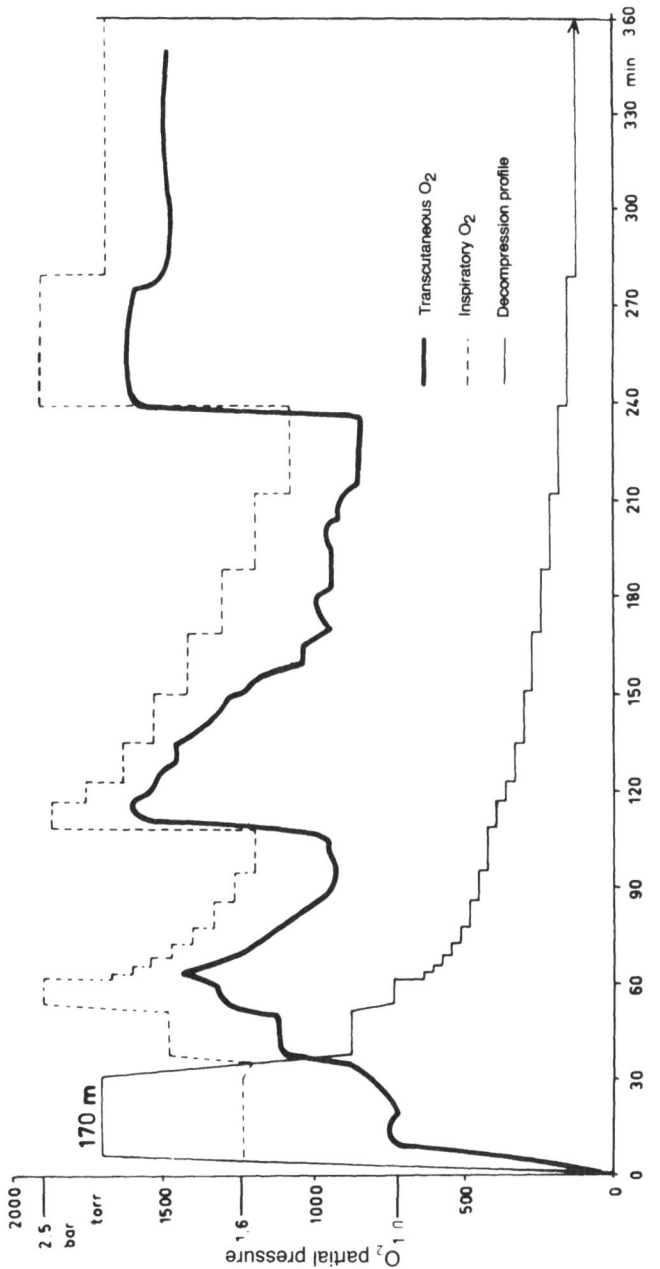

Figure 9   Trend of transcutaneous pO_2 during a 170 m dive (DFLVR).

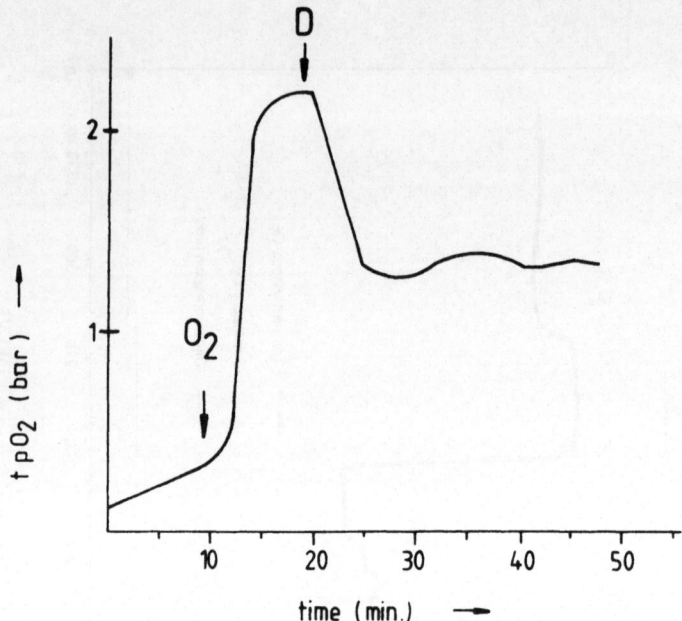

Figure 10   Increase of pO_2 up to oxygen-intoxication (D) and reaction to oxygen decrease.

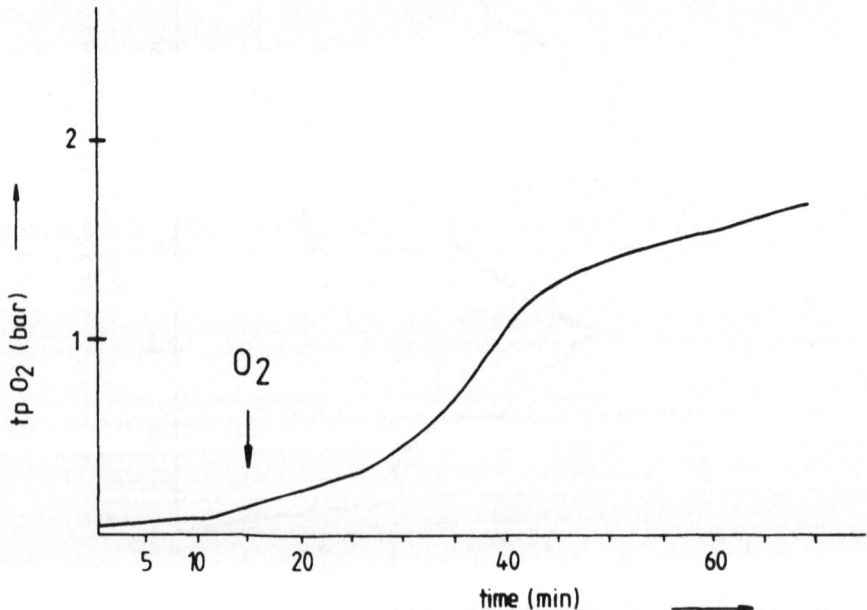

Figure 11   Trend of pO_2 during oxygen treatment in the case of a toxic vasoplegia (gaseous oedema). (Source: Dr Tirpitz).

monitor and recording unit (Fig. 7) is placed outside the pressure chamber. We developed a special accessory set for the connection between sensor and monitor; it consists of a pressure-proof signal feedthrough for the chamberwall and a shielded connection cable between feedthrough and monitor. The monitor is modified by us for extended measuring ranges up to 2400 torr (3.2 bar) oxygen partial pressure.

After the first experiments of Huch *et al.* the "Oxymeter" system was applied intensively for several years at the DFVLR in Bonn–Bad Godesberg in deep dive experiments (Thoma, 1978, 1979; Hampe and Kaiser, 1981), whereas Dr Tirpitz of the hyperbaric center of the St Josephs Hospital, Duisburg, gathered experiences in using it during hyperbaric oxygen treatments (Tirpitz, 1982).

The technical results of all three institutes are similar: the correlation between the oxygen blood pressure determined by other methods and the transcutaneous data shows a fairly good correlation (Fig. 8). The correlation is clearly more individual-dependent in the case of adults than with neonates. The transcutaneous blood pressure data are generally shown to be about 70% of the arterial blood pressure. Taking into account this dependence, the "Oxymeter" can be used as a valuable continuous monitoring trend indicator.

An example for the trend of the transcutaneous pO_2 dive experiments is shown in Figure 9. The clinical utility of the transcutaneous method during hyperbaric oxygen treatment can be seen in Figures 10 and 11. Two special cases of the circulation monitoring should be mentioned. During oxygen treatment (especially in the case of juvenile patients) the onset of oxygen intoxication can be detected by a steep increase of the blood pO_2 (Fig. 10), showing the necessity to lower the dispensed oxygen pressure by decompression or air breathing. During the therapy of shock patients, the very slow increase of the blood pO_2 in extremities indicates from what time the therapy begins to be effective (Fig. 11). The findings from these measurements led to an important improvement of the therapy by a longer, controlled oxygen treatment.

REFERENCES

Hampe, P. and Kaiser, R. 1981. Problems of transcutaneous measurements of oxygen partial pressure under elevated environmental pressure and the influence of sustained acceleration, in *Proc. 2nd Intern. Symp. on Continuous Transcutaneous Blood Gas Monitoring*, Zürich.

Huch, A, *et al*. 1977. Transcutaneous P_{O_2} of volunteers during hyperbaric oxygenation, *Biotelemetry*, **4**, 88.

Thoma, W. 1978. Transkutane Vergleichsmessungen des arteriellen Sauerstoffdrucks bei simulierten Tieftauchgängen, in *Symp. f. Tauchmedizin der Medizinischen Hochschule Hannover*, Hannover.

Thoma, W. 1979. Responses of the transcutaneous oxygen partial pressure in the range of 200 meters during dive experiments and some aspects concerning oxygen toxicity, in *Proc. 5th Congr. European Undersea Biomedical Soc.*, Bergen, Norway.

Tirpitz, D. 1982. Behaviour of P_{O_2} in tissue under hyperbaric condition — first results in transcutaneous measurement, in *Proc. 8th Congr. European Undersea Biomedical Soc.*, Lübeck-Travemünde, Germany, p. 469.

Wiesner, P. 1982. Aspects of modern gas analysis and chamber atmosphere control for diving chambers and diving simulators, in *Proc. 8th Congr. European Undersea Biomedical Soc.*, Lübeck-Travemünde, Germany, p. 57.

18

Thermal Protection Equipment

P. Hayes, AMTE Physiological Laboratory, Alverstoke, UK*

The hazards faced by divers in cold water environments are considered and the basic principles of thermal protection in various environments reviewed. Problems of deep oxyhelium diving are discussed in detail. The need for adequate preparation and training is stressed, both for diver safety and for optimum work results.

SCOPE

It is assumed that the majority of underwater work performed in the Arctic to date has been shallow. The greater part of the additional problem of diving in arctic latitudes as opposed to those more temperate occur as a result of extreme conditions on and above the surface of the water. The extent of the extra difficulty involved with surface support of arctic diving activities is of much greater magnitude than the problems of diver tolerance and diving equipment malfunction when faced with a change in water temperature from 4 to 6 °C (North Sea) to about −2 °C in the Arctic.

Deep helium diving has been performed in the high Arctic (Nuytten, 1973) but in general terms deep activity using helium as well as the use of air saturations has played a minor role in the recovery of resources beneath the Arctic ocean. However, a number of companies have been looking to the needs of the future and are actively engaged in the use and development of arctic drilling rigs capable of working in ice conditions (*The Oilman*, July 1983). Presumably divers will have a part to play in this area of exploration and exploitation.

With increasing demand for and rate of exploitation of undersea resources the development of equipment and procedures for safe diving has lagged behind the required in-water capability, leaving the diver ever more vulnerable to the hazards of the cold water environment.

The basic principles of thermal protection can be applied across the whole depth range from the surface to 500 m and as such are only a reinstatement of the laws of heat transfer. The laws can be stated simply and at the risk of repetition are worthy

---

*Present address: RAF Institute of Aviation Medicine, Farnborough, Hampshire, UK.

of consideration, particularly in a high technology field where it is all too easy to forget the most obvious. Following these statements of the obvious, one can look at how the basic principles can be applied at different depths. The problems of deep oxyhelium diving are discussed in some detail below as well as those of shallow wateg work, and where possible special attention is drawn to the application of protection in the Arctic environment.

BASIC PRINCIPLES

Overall heat balance and the role of thermal protection

The role of thermal protection is represented in Figure 1. In response to a stressful environment the right level of protection is required that allows only as much heat to be lost from the body as is produced. Heat is produced by the body during the processes of cellular metabolism when at rest and increases in response to exercise and/or shivering.

Theoretically, the protection should ensure that the diver is in thermal comfort and can maintain thermal balance indefinitely with a central body (core) temperature of $37 \pm 0.5\ °C$, the variation depending upon the individual and the time of day. Core temperature will rise above $37\ °C$ as a result of heavy exercise. Thermal protection that is good enough to maintain a resting man indefinitely in a comfortable condition when immersed in cold water will undoubtedly lead to over-heating during periods of high exertion.

In reality, in-water protection cannot be constructed both to protect from the cold at rest and to lose enough heat during exercise. The first alone is difficult enough to achieve and still maintain flexibility of movement.

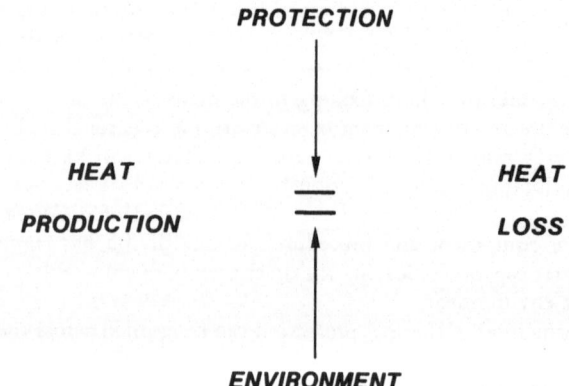

Figure 1   Concept of thermal protection: see text for explanation.

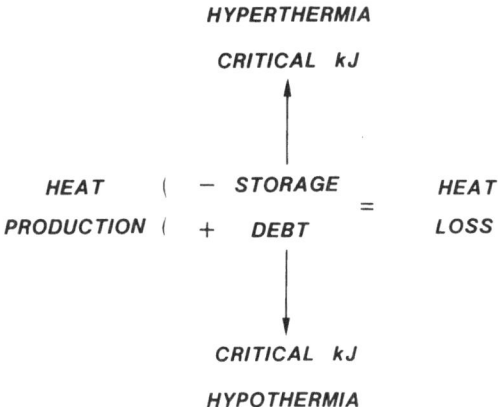

Figure 2   Compromise in protection. The duration of exposure must be limited to prevent too great a change in body heat content; kJ = quantity of heat as kilojoules.

The heat loss/heat gain compromise

The second basic principle is thus the acceptance of a compromise, and can be applied similarly to both heat and cold stress. One is forced to accept a time limitation of the exposure which in effect represents some critical quantity of heat that is either stored or lost from the body (Fig. 2). Extreme critical quantities lead to unconsciousness as a result of either heat exhaustion or hypothermia. Obviously, one chooses a critical quantity of heat that represents a level of core temperature that is neither harmful to the man nor (preferably) is it deleterious to his performance. The time to reach the critical level determines the duration of the exposure (dive). A practical example of this compromise will be a suit that allows the diver to remain comfortable in water at 0 °C for 1 hour if he is working hard, but results in discomfort and shivering if resting for any length of time, thus restricting his time in the water to less than 0.5 hour. The same suit would cause over-heating when working in water at 15 °C.

Heat transfer factors in the balance equation

The heat balance equation summarising the factors involved when the diver is immersed in cold water is shown in Figure 3. The factors on the left represent the heat that is available either as that produced ($\dot{M}$) or removed from stores (the debt, $\dot{S}$), but work must be subtracted from the total heat available as work is performed externally. On the right hand side of the equation are the heat loss terms.

$$\dot{M}_{(\dot{V}O_2)} + \dot{S} - \dot{W} = \dot{E} + \dot{R} + \dot{C} + \dot{K}$$

Figure 3   Heat transfer factors in the balance equation. $\dot{M}(\dot{V}O_2)$ = heat production, + S = heat debt, $\dot{W}$ = work, $\dot{E}$ = evaporative heat loss $\dot{R}$ = radiant heat loss, $\dot{C}$ = convective heat loss and $\dot{K}$ = conductive heat loss. All quantities are a rate (Watts).

Principles of passive thermal protection

In practical terms the heat balance equation for the diver can be simplified because beneath the surface the effects of radiation, conduction and evaporation become insignificant (Fig. 4). Convection from the skin surface ($\dot{C}_s$) and via respiration ($\dot{C}_r$) become the dominant avenues of heat loss. Passive protection can be used to offset both these losses. To prevent heat loss from the skin the diver relies on insulation materials. High respiratory losses can be prevented by some form of thermal regenerator, where heat lost during expiration is reclaimed during inspiration and cold gas is not breathed.

$$\dot{M} + \dot{S} - \dot{W} \quad = \quad \underset{\uparrow}{\dot{C}_s} \quad + \quad \underset{\uparrow}{\dot{C}_r}$$

$$\textit{INSULATION} \quad \textit{REGENERATION}$$

Figure 4   Heat balance equation summarising the factors involved in passive thermal protection in cold water. $\dot{C}_s$ = heat losses through the suit and $\dot{C}_r$ = respiratory heat loss (Watts).

Principles of active supplementary heating

The balance equation describing a diver protected by a hot water suit is modified by the inclusion of a further term on the left hand side of the equation. Excess heat losses through the suit ($\dot{C}_s$) must now be compensated by additional heating at the surface of the man (*HEAT*) (see Fig. 5).

The hot water suit acts as a sacrificial barrier to the surrounding cold water and prevents undue heat loss from the man by maintaining the core to skin temperature gradient at a level compatible with heat balance, as described in Fig. 1.

$$\dot{M} + \dot{S} - \dot{W} + \textit{HEAT} \quad = \quad \underset{\uparrow}{C_s} \quad + \quad \underset{\uparrow}{\dot{C}_r}$$

$$\textit{SUIT} \qquad\qquad \textit{GAS}$$
$$\textit{REPLACEMENT} \qquad \textit{HEATER}$$

Figure 5   Heat balance equation summarising the factors involved in active heating types of thermal protection. Heat is added to the system as skin surface heating as well as heating of the inspired gas.

Prevention of tissue damage

The maintenance of heat balance does not automatically preclude the individual from suffering from cold injury at a specific site on the body surface where protection is totally inadequate. Local sites of poor or non-existent protection may suffer cold injury before any impact at all is made upon the overall body temperature. This highlights the need under particular circumstances to protect every square millimetre of the body surface; the qualifying statement being that this protection needs to be of greater or lesser extent dependent upon level of blood perfusion, activity level of the divers, body shape, size and fatness.

THERMAL PROTECTION AND THE TYPE OF DIVING

The application of the above basic principles is shown in Table 1 and determines the depth and duration of the dive. Body fat acts as an additional protection over and above that of the diving suit and helps to balance the heat loss by reducing the size of the items on the right hand side of the balance equation (as in Fig. 3). Fatness is usually associated with lack of physical fitness but it would be dangerous to sacrifice fitness for fatness. Elevating one's heat production is also a way of restoring the balance of the equation and a less fat person capable of sustained effort may do just as well in terms of cold water survival as a fatter but less athletic colleague. The combination of both fitness and fatness is undoubtedly a good combatant to the cold environment. Long term survival both on land and in the sea is often limited by the inability of the body to continue to produce heat because the thermogenic mechanism is susceptible to fatigue. Other aspects of morphology may also determine the outcome of immersion in the sea, such as size and shape. A low surface area to volume ratio (SA/V) (i.e., as in a short fat man) is helpful as more mass has less area from which to lose heat. Tall skinny divers are at a disadvantage, both because they have a large SA/V ratio and the suit insulation is less efficient when covering surfaces of high curvature. The greater the curvature, as in a finger or on a thin limb, the less effective is a given thickness of insulation.

Other physical factors of the environment itself give rise to the different protection assemblies required (nos. 3, 4 and 5 in Table 1) according to the depth of operation. The clothing assemblies are discussed in detail later, the characteristics of the diving environment immediately below. The type of diving performed and the type of protection used when working underwater in the Arctic may in addition depend upon the surface conditions of weather and the available shelter.

Table 1
Factors affecting the balance equations and the application of the basic principles to the diving environment

1. Fat insulation	— ever present protector
2. Heat production	— susceptible to fatigue
3. Wet suit 0–25 m	— shallow air diving (free swimming)
4. Dry suit 0–50 m	— deeper air diving (surface or bell-supported)
5. Hot water suit 50–500 m	— bell-supported helium diving (bounce or saturation)

CHARACTERISTICS OF THE DIVING ENVIRONMENT AND THERMAL COMFORT

Equivalent comfort ranges for the nude man resting in a number of different surrounding media are portrayed in Figure 6. A man is capable of thermoregulating without undue discomfort in an air environment between temperatures of 22 and 38 °C. The same level of comfort can be maintained only over a smaller and higher temperature range in helium (29–35 °C) and only between 34 and 35 °C in water.

The factors responsible for the shift in mean comfort temperature and a shrinkage of the comfort zone are the increase in the density (ρ) specific heat (C_p) and thermal conductivity (K) when moving from an air to helium to an aquatic environment. A combination of these three factors, ρ, C_p and K, cause a rise in the rate of heat loss and to compensate for this increase in rate the temperature must be raised.

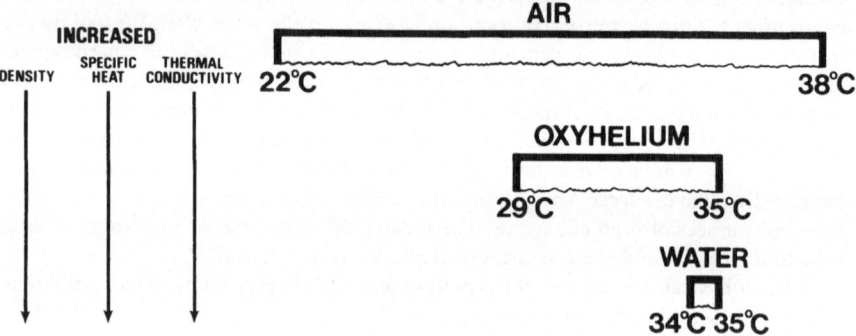

Figure 6   Physical characteristics of the environment determine the comfort zone. Equivalent comfort levels for the nude man.

FACTORS RELATED TO THERMAL PROTECTION — OVERALL VIEW

A summary of the factors related to thermal protection appears in Figure 7.

A given level of protection moves us along a 'notch' from left to right (in Fig. 7) away from the hazards associated with cold and heat stress to an area of more safe diving where there is typically a time limit for the exposure under more tolerable extremes. Still better protection results in a more comfortable and efficient diver working under optimised conditions. Somewhere along this progression we must take into account the testing procedures adopted and risks involved. Such risks (hypothermia, hyperthermia, tissue damage) decrease from left to right. Over-engineering and over-complexity of protection equipment can force the process in the reverse direction because the technology may not be compatible with the human condition. The interaction of man and the machine is most important and,

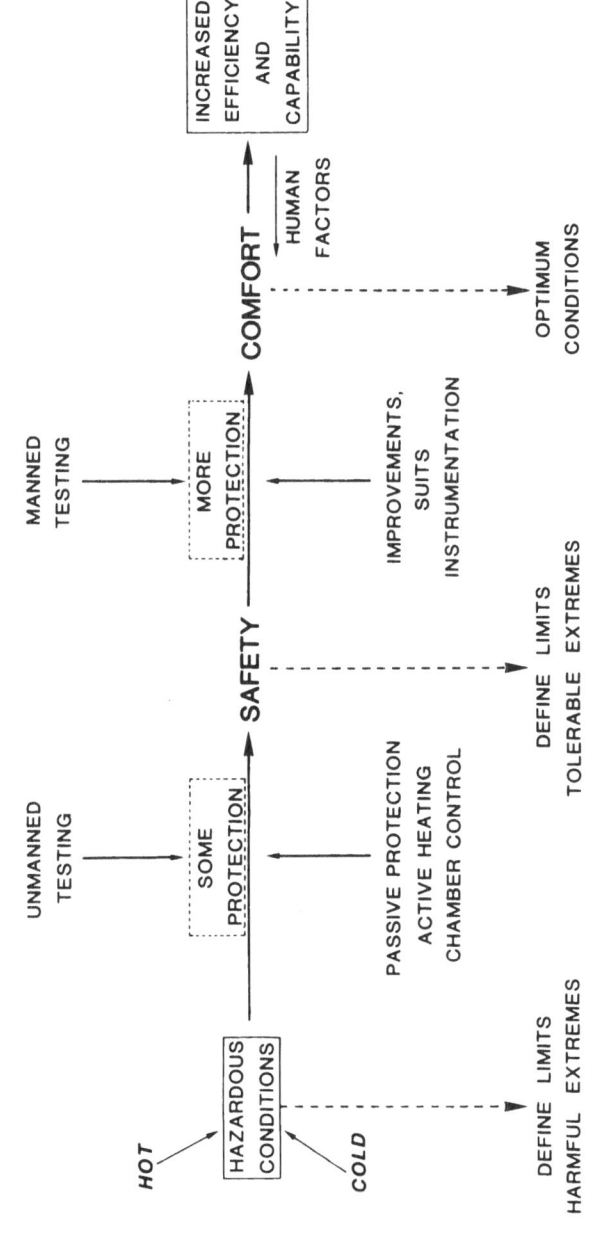

Figure 7 A flow chart of the factors to be considered when increasing the effectiveness of thermal protection (from left to right).

without due allowance for the "human factor", the most sophisticated and highly evolved systems of protection may be worse than their more primitive predecessors.

Provision of adequate thermal insulation

Wet suits

The effect of pressure over the range 0–50 m has a devastating impact on the thermal insulation properties of foam neoprene. The loss of insulation is displayed in Figure 8 and shows that heat flow through the suit has doubled after 30 m and is trebled at 100 m.

For further detailed examination of the effects of pressure on neoprene, see Nuckols (1978) and Wong (1983). The limits of insulation were explored by Beckman et al. (1966) who concluded that true thermal balance could be obtained indefinitely in very cold water (2 °C) with a material of 25 mm (1 inch) thickness. In reality, suits much more than 10 mm in thickness lead to problems of inflexibility and difficulties in controlling buoyancy.

A problem still far from being resolved is what to do about hand protection. The level of insulation required around small radii cylinders (such as fingers) immersed in cold water is prohibitive in terms of thickness (VanDilla et al., 1949). Where delicate and precision work is required by the hands, an alternative solution other than passive insulation is required.

The level of insulation required over the whole body is going to be a function of the activity of the man under the water. If he is working in water (-2 to 0 °C) and producing about 350 W of heat, then he is going to need insulation of about 2 togs

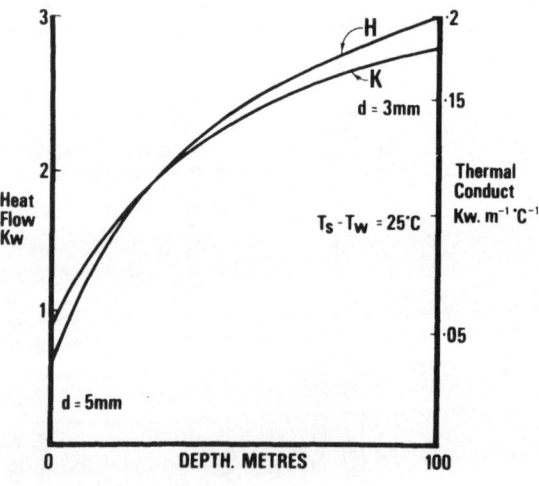

WET SUIT COMPRESSION

Figure 8   The effect of pressure on the properties of foam neoprene (after Bramham et al., 1972).

(1.3 Clo) to maintain comfort. However, once he stops work and relaxes the requirement rapidly exceeds 4–5 togs. Foam neoprene material of 10 mm thickness has a value of about 2.0–2.3 togs when uncompressed. Longjohns and overjacket together can bring the total average insulation of the assembly up to about 2.8–3 togs but as soon as the man gets into the water this will drop to a much lower level (<1.5 togs at 3 m). At 30 m the insulation value is likely to be less than 0.5 togs. By the time the diver has reached 30 m, the thickness of his suit is markedly reduced, the heat flow has increased and he has not become negatively buoyant. The tailoring of the suit has deteriorated and flushing of cold water into the suit will occur more easily.

Dry suits

Any extension of the depth to below 30 m for dives of duration greater than 1 hour require the superior insulation qualities of the dry suit and its underwear. If foam neoprene is used as the outer waterproof layer, then the material will suffer the same fate as that of the normal wet suit upon exposure to pressure. Alternatively, there is little change in the insulation properties of the more solid "dry bag" whose principal purpose is a waterproof outer covering, not insulation. The extent of the insulation offered by the dry suit system is determined principally by the thickness of air trapped within the underwear, usually of an acrylic or woollen type. To offset the squeeze imparted by the hydrostatic pressure of the water suit, inflation, usually with air, is used. This air helps to reinstate an effective insulation layer next to the skin surface and in a few metres of water offers about 50% of the insulation that was previously available at the surface in a typical "Unisuit" arrangement.

Over-inflation causes problems with buoyancy, so further assistance to the diver can be offered as pressure resistant underwear. Pressure differentials (ΔP) of over 2 p.s.i. (14 kPa) exist along the height of a man standing in water, and materials that can withstand a ΔP of 2 p.s.i. without too much crushing offer a real advantage to the diver. Such a material was identified as Thinsulate® insulation. This material, in conjunction with a solid neoprene type dry suit (e.g. Avon, Viking), produces as good an insulation as the Unisuit and conventional underwear together at 1 bar. Just below the surface, both arrangements are of similar insulation value, but at 30 m the conventional arrangement has suffered a further loss of insulation and lost its buoyancy whereas the Thinsulate arrangement has remained unchanged (see Table 2). For a discussion of the development and testing of dry suit systems for the US Navy, see Lippitt and Nuckols (1982) and Piantadosi et al. (1979).

Dry suit assemblies can improve the level of protection beyond that of wet suits for use down to 50 m (75 m in the military) but carry with them the burden of extra bulkiness, repair difficulties, high cost, the occurrence of leaks and the need for spare suits. In arctic conditions the superior protection of dry suits above the surface of the water must also be a point in their favour. Indeed, the air exposure portions of the dive mission often present the critical problems where exposed flesh such as the face and lips may rapidly become numb and swollen, with frostbite being a genuine hazard.

Once under the water a diver equipped with Unisuit, good underwear, three-finger mittens and full face mask can routinely operate in water of -1 °C for

Table 2
Effect of depth on different clothing assemblies. Insulation values for
different passive thermal protection arrangements (modified from
Lippit and Nuckols, 1982)

	6.5 mm wet suit of foam neoprene (togs)	6.5 mm foam dry suit and conventional underwear (togs)	Solid neoprene dry suit and "Thinsulate" (togs)
1 bar air	2.8	3.9	3.9
3 m water	1.2	1.9	1.9
30 m water	0.4	1.2	1.9

1 tog = 0.1 °C m^2 W^{-1} = 0.645 CLO = 0.28 m^2 h °C kcal^{-1}
1 CLO = 1.55 togs = 0.18 m^2 h °C kcal^{-1}
0.1 m^2 h °C kcal^{-1} = 0.86 togs = 0.56 CLO

up to 4 hours less than 10 m deep. However, this cannot be construed as a
particularly comfortable experience.

In spite of the reasonably good record of the dry suit in shallow water, there has
been a move to use active hot water systems, even on shallow work. The diver is
thought to be more comfortable for longer periods, hot water suits are often cheaper
and do not require sich frequent overhaul and maintenance as dry suits.

If hot water heating technology can be used in earnest in the Arctic setting, then,
without a doubt, the most comfortable man would be the diver in the water and
more heed should be taken of the comfort and capability of the support crew left
shivering above. Active heating systems will also offer the advantage of prevention of
freezing of equipment such as the demand valve, where new gas heating systems
could be used to keep the whole assembly warm both in and out of the water (see
shroud gas heaters below).

Active supplementary heating for the diver in excess of 50 m

Skin heating

The complexity and bulk of present surface support equipment necessary for active
heating of the diver may preclude its use in the majority of arctic diving operations.
Alternatives may be the diver-carried heat sources based on closed-circuit tube-suit
systems. These can be either entirely self-contained or need surface support.
Examples are the propane burning systems (Divematics) and the electrical heating
back pack (Andark).

The majority of active heating systems rely on extensive power sources at the
surface to provide large volumes of hot water, which are subsequently pumped down

to the diver via an umbilical and (usually) bell. Surface power sources typically provide 100 to 250 kW and can be used to heat two to three divers. Water is heated at the surface to a high temperature which subsequently cools during its passage in the umbilical down to the diver. The effectiveness of this free-flooding principle relies on a sufficient flow of water at the right temperature reaching the diver. The actual energy requirement of the diver is only about 900 W near the surface in water temperatures of 2–4 °C. At depths from 180 to 300 m the diver needs 1.2–3.4 kW. The free-flooding hot water system is relatively unsophisticated as a concept, although perhaps a wasteful means of keeping the diver warm. However, what it scores on simplicity and acceptability, it loses on the efficiency of temperature control. The temperature of hot water is controlled at the surface and an estimate made of the likely drop in temperature down to the bell and on to the diver. Hopefully, the water reaching the diver will be in the region of 38 ± 2 °C. This method of control is crude but is used in practice on nearly all dives, even down to 300 m. Such a system, however, is reaching the limits of its effectiveness at depths approaching 250 m.

Hot water scalds are a frequent occurrence and are likely to pose a greater problem the greater the depth. For operations within the range 250–500 m the control of the system must be moved to the bell and plausible solutions are automatic mixing of hot and cold water to make sure the diver gets the correct temperature or, alternatively, the use of close-circuit hot water systems based on having the energy source mounted on the bell. Prototype systems also exist for mixing of hot and cold water at the inlet to the suit. Also available now is a system for the control of topside water temperature dependent on the sensing of temperature at the diver's suit inlet. A disturbing feature of the present hot water systems is the number of instances of scalds that occur without the diver being aware of any accompanying tissue damage. This appears, at least in part, to be due to an inadequacy in the surface supply, where in response to a general feeling of cooling the temperature is turned up; in reality, a greater volume flow rate would and should compensate.

Respiratory gas heating

Breathing cold gases of high specific heat and density can cause a marked loss of heat from the respiratory tract and central areas of the body. Respiration of cold HeO_2, particularly at high levels of ventilation, can jeopardise the safety of the man in the water. Gas breathed in is both heated and humidified by the respiratory tract. During expiration, the heat transfer mechanisms within the tract are not quite so good at reclaiming the heat and water; as a consequence, expired temperatures and humidity are considerably higher than those inspired. The rate of heat loss can be calculated from the following equation, neglecting the relatively minor evaporative component:

$$H_{resp} = C_p \rho \dot{V} (T_e - T_i)/60$$

H_{resp} = respiratory heat loss W $\qquad$ $\dot{V}$ = minute ventilation $\quad$ l min^{-1}
C_p = specific heat $\qquad$ J g^{-1} °C^{-1} $\qquad$ T_e = expired temperature $\quad$ °C
ρ = density $\qquad$ |g l^{-1} $\qquad$ T_i = inspired temperature $\quad$ °C

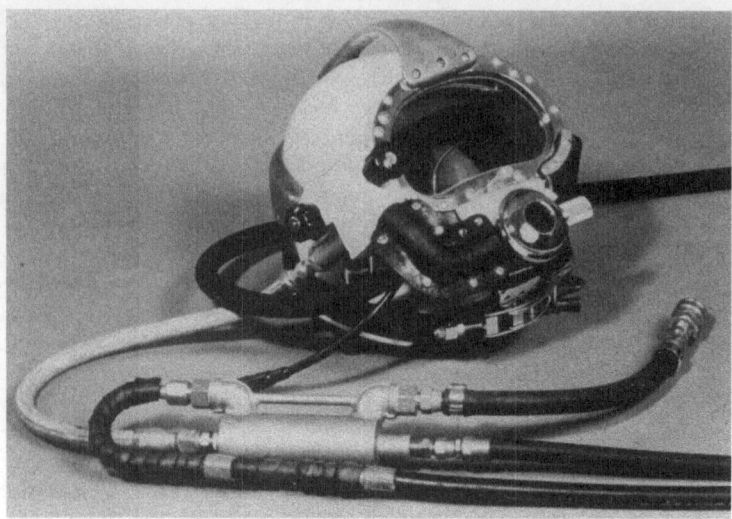

Figure 9   In-line gas heating systems within the umbilical and a shroud system mounted on the helmet.

If the rate of respiratory loss exceeds approximately 200 W, then the diver may begin to experience difficulties in breathing and a feeling of tightness in the chest may occur (Hayes *et al*, 1982). Within 5 to 15 min of breathing the very cold gas, the respiratory tract begins to produce copious quantities of mucus, in some cases sufficient to fill an oronasal mask and make breathing impossible. A solution is to heat the gas before it reaches the diver, and this is done with moderate success using the same hot water supply as for the suit heating. The majority of the inline respiratory heaters readily available are adequate for use down to 180 m but alternative "shroud" systems mounted on the mask or helmet are being used for deeper activity. In-line gas heaters effectively heat the gas on its passage through the exchanger but the majority of the heat is then subsequently lost in the delivery pipes and the demand valve. A flow of hot water over the delivery tube, keeping side-block and demand valve warm appears to heat the incoming gas to the desired level. New and old type gas heating equipment is shown in Figure 9.

Environmental control in the high pressure environment

Living chambers

As mentioned previously, increases in the density, specific heat and thermal conductivity of the surrounding fluid environment result in the need for a progressively higher mean comfort temperature and a narrowing of the comfort zone. The deeper the saturation exposure, the higher the temperature needed to retain comfort and the more precise and accurate the environmental control system needs to be. Figure 10 demonstrates the mean comfort temperature during a dive to 540 m.

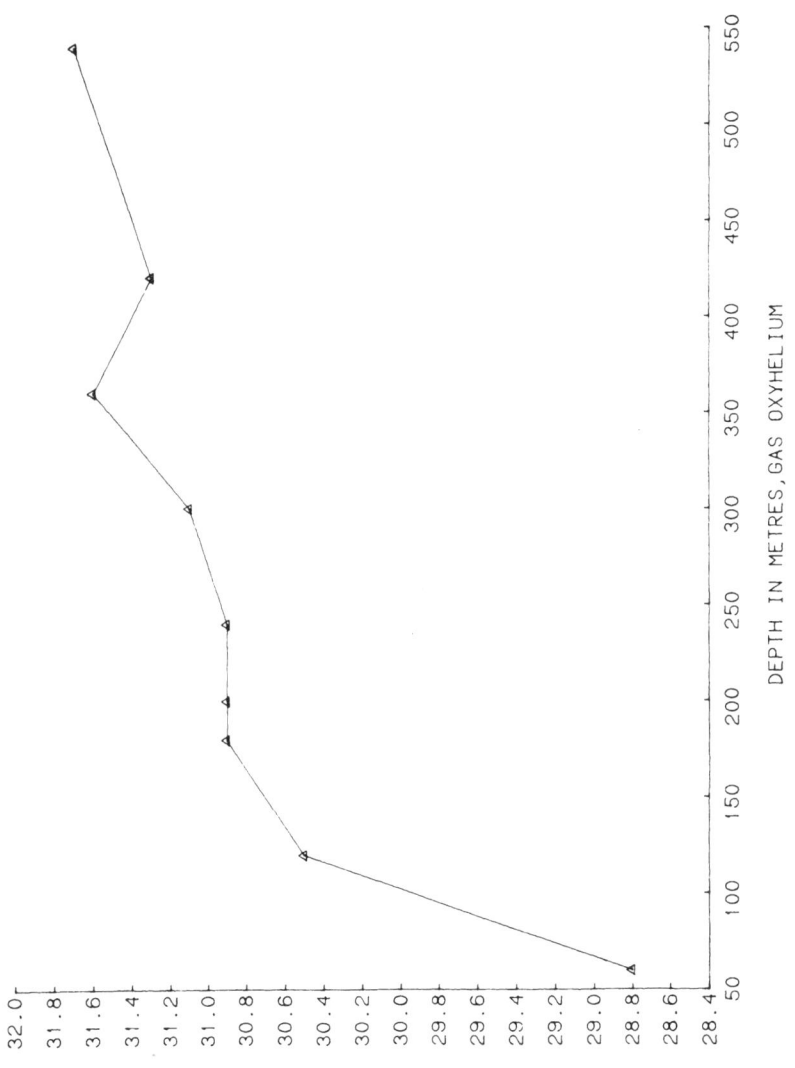

Figure 10 Selected comfort temperature of gas (HeO₂) during a dive to 540 m.

Limits of comfort at the greater depths can be as little as ±0.5 °C with ±1.0 °C, giving rise to profuse sweating or violent shivering.

Environmental gas temperature is not the only factor affecting comfort; the close proximity of the chamber walls means that the radiant heat transfer (and thus mean radiant temperature) must also be taken into consideration. Traditional heat stress indices such *WBGT* (wet bulb globe temperature, Yaglou and Minard, 1957) take into account radiant losses but cannot be readily transferred to hyperbaric surroundings, though attempts have been made to produce a psychrometric chart for use in chamber systems (Nishi and Gagge, 1977). In practice, we have found that a combination of background electrical panel heating of the chamber walls used in conjunction with gas temperature conditioning affords the best means of producing comfortable surroundings for the diver. A two-system approach prevents many of the inter-individual wrangles over thermal comfort (induced by different radiant transfer rates), due to the proximity and orientation of different divers to the chamber walls. Such heating systems aid more restful sleep and generally make the conditions more comfortable.

Another of the consequences of the increase in the magnitude of the physical characteristics of the environment is the ability to transfer high levels of heat "to" as well as "away from" the diver. It is thus prudent to consider the risks of hyperthermia as well as those of hypothermia, for once either of these states is initiated they progress at a much faster rate in the high pressure environment compared with air at 1 bar.

Another factor of the high pressure environment not always appreciated is that a given level of relative humidity, for instance 60% RH, does not mean the same level of evaporative heat loss or comfort at different depths. The pertinent factor in terms of comfort and capacity for work is the rate at which water can evaporate from the skin surface. The effect of density is to reduce the rate of water vapour transfer, so the same level of evaporative heat loss is accomplished at 35% in HeO_2 at 250 m as by 60% RH in air.

Hyperbaric welding habitats

With the advent of pre-heating pipes to high temperatures to ensure the quality of the weld, the diver becomes exposed to both high radiant temperatures and high gas temperatures. The combination of light work in the hot environment, high sweat loss, with stooping and bending is likely to cause instances of fainting. In the absence of any recognised thermal stress index catering for the highly specialised habitat, it would be sensible to consider a number of practical procedures to prevent accidents.

The welder will be working in a position somewhere along steep radiant and thermal gradients between the pipe and the habitat walls, and will experience high local skin temperatures as well as an elevated core temperature. Provision should be made for active cooling of the man should he need it and a store of water inside the habitat should be available for cooling as well as an ample supply of drinks. In fact, the welders should be encouraged to drink more than they feel is their actual requirement. Welding operations lasting longer than 2 hours should adopt a regimen of 50% work and 50% rest if temperatures rise above 38 °C. Such temperatures should be measured inside a black globe within 1 m of the heated pipe. Supervisors would do well to consider the advantages of previous acclimatisation to heat by the men involved in extensive welding activity.

The risks involved

An awareness of the risks involved requires education about the effects of cold upon the body and the diving equipment. Reducing the risks requires better and improved equipment as well as proper preparation for the dive. Minimising the risks during operations, should anything go wrong, is achieved by training and experience.

Diving in the Arctic carries a greater risk during entry and exit into the water than in more moderate climates. Even when ice is not present, the dangers from freezing and non-freezing cold injury are to be taken seriously both above and below the surface, for the diver and for the support team. Hypothermia is a constant threat to the diver who does not have active heating and the insidious nature of its occurrence means one cannot always rely on the subjective awareness of the man to tell you if he has become too cold. Skin freezes at a temperature of -0.5 °C but, provided the freezing is both superficial and lasts only a few minutes, the changes are reversible. The reversibility of this so-called "frostnip" distinguishes it from the more serious damage to tissues known as "frostbite". Frostnip can be treated on the spot; the exposed parts should be rewarmed by reducing the cold insult and moving to a warmer environment, or covering the area with warm hands, or applying warm water in a flannel or cloth. With immediate treatment frostnip will not progress to frostbite. Divers who work in water at temperatures less than zero and experience a tear or major leak in their suit should therefore attempt to warm the affected area without delay, and preferably within 10 min.

Changes induced by frostbite are not immediately reversible and can lead to effects lasting many months. Cases of frostbite can be treated at a well-equipped base camp but will probably require evacuation to hospital. A description of and a code of practice for the treatment of frostbite has been produced by Ward (1974).

Injury to nerve and muscle can occur at higher water temperatures if the exposure continues for any length of time. A skin temperature of 6–8 °C for as little as 30 min will give rise to an injury in some cases. This non-freezing type of cold injury has been typified by "immersion foot" or "trench foot" and should be treated in hospital. It is likely to have lasting, often permanent, effects. Non-freezing cold injury may occur in more moderate water temperatures in divers with a previous history of tissue damage and may perhaps manifest itself as decompression sickness (Leitch and Pearson, 1978).

As well as the dangers of peripheral cold injury, divers should also be aware of the risks involved when the body becomes generally cold and this should cover a working knowledge of the signs and symptoms of developing hypothermia. It is important to be aware of the insidious nature of hypothermia and the important role of exhaustion in the loss of body temperature. Descriptions and recommendations for working in cold water are to be found (Arthur, 1980; also the UEG publication entitled "Thermal Stress in Divers", 1983).

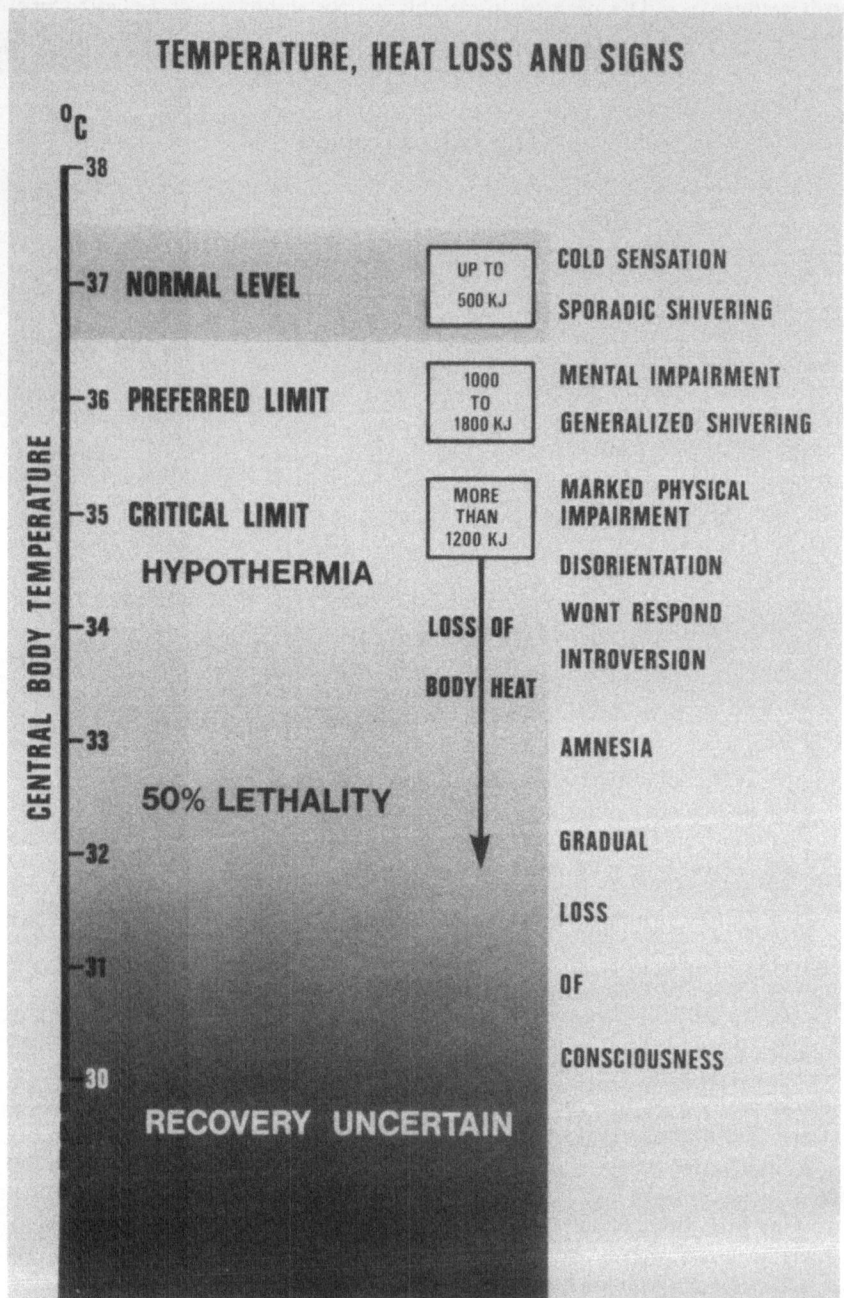

Figure 11   Signs and symptoms associated with loss of body temperature. Figures for heat loss as kilojoules (kJ) correspond to net heat lost from body stores.

At a central body temperature of 35 °C, only 2 °C below normal, a man is considered to be suffering from hypothermia and will show impaired muscular and cerebral function. Signs and symptoms associated with the changes in central body temperature are shown in Figure 11. In the event of hypothermia, the diver must be rewarmed. For serious cases, involving body temperatures probably below 34 °C and where the victim is not fully conscious, hospital procedures are required. This may not always be feasible in the diving environment or in arctic conditions, and a number of alternative suggestions are made by this author in *Thermal Stress in Divers* (UEG, 1983) and by Golden (1983).

Rewarming of less serious cases, as might be necessary during repeat dives, may be just as essential in the Arctic but from a logistical point of view may be difficult to achieve or monitor. Neither the diver nor the observer is likely to know when rewarming is complete The need for rewarming during a repeat dive series is essential, as the heat debt incurred by one dive can be added to the next and, although he might feel fit to dive at the onset, the man may very quickly become incapacitated.

Being accustomed to cold conditions may not necessarily be an advantage when it comes to either recognition of the level of cold or the extent of the rewarming. The process of habituation as one becomes accustomed to the cold tends to reduce the subjective suffering and severity of the physiological response. Cooling can occur more readily but without so much discomfort. Habituation may offer some local advantage as an increased (or less reduced) capability of the diver when working in the cold. Hands become less painful and dexterity and sensitivity of the fingers can be maintained in the cold, but perhaps at the expense of more heat loss from the body.

Rewarming of the diver between dives can be performed by immersing him in a hot water bath (about 40 °C) or drenching with warm water inside his suit while the diver is resting. Alternatively, he can be wrapped up in a warm area, given warm drinks and allowed to rewarm from his own heat production. Rewarming can be assumed to be complete when sweating is initiated but this may not occur until the man is too hot. Splashing of cool water on to hands and feet will often illicit a much stronger response than normal with obvious shivering and discomfort if the person is not adequately rewarmed. During either shallow water work or deeper bell lockouts, body temperatures can have a marked effect on the success of the decompression. If the diver cannot be kept warm the whole time, every attempt should be made to rewarm him before the start of the decompression period and keep him warm throughout its course.

Training is intimately involved with the use of equipment, dive procedures and successful team work in hostile conditions. In addition, it also involves a contingency for personnel injury and equipment failure. Within the considerable range of depth now exploited, the training procedures must cover a plethora of possible equipment failures and accidents, ranging from a leaking suit to free flow under the ice to survival procedures in the lost bell (Hayes *et al.*, 1981) or during hyperbaric evacuation. Preparation and training go hand in hand and many of the practical aspects of both are covered by Jegou (1972), Jenkins (1976), Langerman (1979) and MacInnis (1976), whose articles must constitute the major sources of knowledge prior to ICEDIVE '84. Other sections of the community, such as government departments (e.g. fisheries), also undertake diving in extreme cold conditions and provide a source of practical know-how. As an example of the care and attention to procedure and preparation undertaken by the "biological diver", two extracts from the University of Newfoundland Guide for Diving Safety are quoted below. The most basic principle, that of self preservation, is evident.

Special Equipment and Procedures
   (a) Listed below is the special equipment required for ice diving.
      (1) Ice saw or axe.
      (2) Double-hose regulators (preferred).
      (3) Safety line for each diver (both the same length).
      (4) Safety line for standby diver shall be (a) of highly visible colour, (b) twice the length of the lines used by other divers, and (c) made of a material which floats.
   (b) Procedures;
      (1) Before the divers enter the water a safety line shall be tied to each diver's waist.
      (2) Only one pair of divers shall dive through the same hole at any one time.
      (3) Ice diving should not be attempted in an ice field composed of separate pans of ice.
      (4) The hole through which the divers enter and exit the water shall be at least four feet square and shall be well marked around the perimeter.
      (5) On completion of the dive, if possible, the ice which was removed from the hole should be replaced and the site visibly marked as a warning of dangerous ice to all persons.

Emergency Procedures (Lost Diver)
   If a diver should become separated from his line the following procedure should be followed:
      (1) The lost diver immediately goes to the surface and stays in that position under the ice.
      (2) The second diver returns to the hole and leaves the water.
      (3) When the topside attendant realizes that no one is attached to the line he sends in the standby diver who has a line attached to him which is twice as long as that of the lost diver.
      (4) The standby diver does a 360° swim immediately under the ice at the end of his line.
      (5) When the lost diver sees the line he swims back to the hole on this line.
      (6) The standby diver is recalled.
      (7) If this procedure fails after two attempts a similar bottom research is carried out by the standby diver.

Test procedures

The value of any unmanned test procedure is only as good as the accuracy and closeness of the test to the manned operational circumstances. Unmanned testing is prevalent in the underwater field particularly on UBA and has crept into the testing of thermal protection suits and gas heaters. From an ethical point of view, this is to be encouraged, particularly in the early stages of prototype testing where the limits of tolerance and safety may be exceeded during a manned test. What is often lacking now, and has been missing in the past, is any form of specification on which to base the unmanned test. Instead we have been forced to adopt the "let's see what the diver has to think about it" approach. The diver may well think about it but his impression of thermal stress can so easily be wrong. What may also be lacking if one does try to adopt an unmanned approach is any consideration for the many ergonomic aspects and physiological transients that occur when the apparatus is strapped to the diver. Aspects of the ergonomics and a specification for the design of thermal protection systems are discussed below.

Caution must be exercised when making an extrapolation from an unmanned test to the reality (e.g. thermal manikin to man), as it is difficult to simulate many of the factors associated with human movement: effects of pumping and flushing water in and out of suits, increased heat production as a consequence of the exercise, extra heat loss due to forced convection, and the effect of fatigue and fit between different personnel.

Human factors

The need for thermal protection cannot be viewed in isolation, distinct from the work the man has to do or the other equipment with which he must interact. This somewhat obvious comment is easy to make but difficult to implement or describe in general terms. Although increasing the thickness of insulation would provide better protection, it is necessary to sacrifice the need for total protection to the need for flexibility, control of buoyancy, reduction in fatigue and better use of tools. The need for thick protection on the hands means that locks, latches, handles, tools and cameras require modification, and the overall more cumbersome clothing used in the Arctic will limit the performance of the man both above and below the water.

Performance will also be affected by fatigue, and life in a high pressure environment, even without the addition of frequent cold exposures, appears to cause a general feeling of lethargy. Add to this the problems of operating in cold harsh conditions, and the time expectancy for any given task is likely to be considerably extended. Swimming in bulky dry suits is fatiguing and, at greater depths, the drag of an umbilical and increased work of breathing exacerbate the problem of fatigue. The weight, balance and noise levels are important aspects of helmet design and become only slightly less significant than obtaining an adequate gas supply over a long dive.

Another particularly human factor is the diuresis inevitably associated with immersion in water and enhanced by becoming cold. Urine in dry suit underwear dramatically reduces the insulation and urine collection systems are now an integral part of some military dry suit arrangements.

There has until relatively recently been the tacit assumption that on shallow dives the man uses passive insulation for protection and active heating at depths in excess of 50 m. Either concept could be used in the alternative depths range where a combination of both principles may provide the best overall approach to protection. Such combinations are worthy of consideration and have been explored by Thornton (1980).

Even with the established procedures and heavy reliance on technology to provide diver safety and comfort, mild hypothermia is not uncommon in the apparently adequately protected diver (Vaughan and Anderson, 1973; Keatinge *et al.*, 1980). Small decrements in central body temperatures have given rise to poor vigilance, less effective verbal memory and decreased arithmetic skills. Alternatively, cold has been shown to have no effect on a number of performance tasks such as reaction times, navigation and verbal reasoning. The outcome of a number of studies is summarised by Ellis (1982), who states that cold does have a deleterious effect on the performance of certain cognitive skills. Peripheral cold (hands and feet) will also have a distracting effect on the performance of some tasks, as well as causing the loss of sensitivity, dexterity and strength.

In summary, mild body core cooling will give rise to poor performance of continuous and/or complex tasks and may produce errors of judgement. Errors of judgement compounded by unfamiliar and cumbersome equipment and clothing can lead to mistakes and panic. Sometimes these mistakes become dangerous, even fatal.

The need for education, preparation and training in order to assess the risks, reduce their size and provide a contingency is essential for safe working in the undersea Arctic.

Physiological design criteria for thermal protection equipment

A summary of recommendations appears in Table 3. These statements can be supplemented by additional comments concerning hot water suits.

Table 3
Physiological design criteria for thermal protection suits

Body heat loss and temperature	1 Do not exceed net heat loss (dept or change in enthalpy) of 15 kJ per kg body weight
	2 Central body temperature no lower than 36 °C or a decrease of 1 °C (whichever lower; preferred limit); critical at 35 °C
Skin temperature	1 Mean value no lower than 25 °C
	2 No local skin site less than 20 °C
	3 Except hands and feet down to 15 °C

Table 4

Minimum and maximum inspired gas temperature in HeO_2 as a function of depth

Depth (msw)	Gas temperatures		
	Minimum permitted (°C)	Minimum preferred (°C)	Minimum hazardous (°C) (more than few minutes)
100	0	0	−2
150	10	11	8
200	15	19	10
250	18	25	12
300	20	28	14
350	22	30	15
	Maximum permitted (°C)	Maximum preferred (°C)	Maximum hazardous (°C)
50–300	42 (wet gas > 70% RH)	40 (wet)	46
	39 (dry gas < 30% RH)	36 (dry)	41

(a) Water should be maintained between 35 and 40 °C at the point of suit entry to avoid either cooling or scalding of the diver.

(b) Hot water suits should be designed to accommodate a range of flows (7 l min^{-1} or 1.5 gall min^{-1}) up to 27 l min^{-1} (6 gall min^{-1}) to allow for differences in demand arising from the fit of the suits, the size of divers as well as their level of activity and depth of working.

Active heating systems also need provision for gas heating, as shown in Table 4. On certain push–pull configurations over-heating of the gas may be a problem and maximum levels are suggested based on the limited practical experience available. Such maxima appear highly dependent upon the water content of the gas.

ACKNOWLEDGEMENTS

My thanks to Dr Robert Elner (Canadian Fisheries and Oceans) for his comments; to Dr Martin White (BAS) for his help and to Dr Val Hempleman for his encouragement during the time much of this work was done.

REFERENCES

Arthur, D. C. 1980. *Hypothermia: An Educational Manual for Instruction of the Fleet Duty Corpsman*, Naval Submarine Medical Research Lab Report No. 943, Groton, Connecticut 06340.

Bramham, E., Dineley, J. L. and Wharmby, B. 1972. Heat loss compensation in deep diving, *Hydrospace*, August, p. 20.

Ellis, H. D. 1982. The effects of cold on the performances of serial choice reaction time and various discrete tasks, *Human Factors*, 24(5), 589–598.

Golden, F. 1983. Rewarming, in *The Nature and Treatment of Hypothermia*, R. S. Pozos and B. Wittmers (eds), University of Minnesota Press, Minneapolis.

Hayes, P. A., Padbury, E. H. and Atherton, P. J. 1981. *Astronaut Trial Section One. Passive Systems. AMTE (E)*, Report R81 403, Alverstoke, Hants.

Hayes, P. A., Padbury, E. H., Florio, J. T. and Fyfield, T. P. 1982. Respiratory heat transfer in cold water and during rewarming, *J. Biomech. Eng.* 104, 45–49.

Jegou, A. 1972. Deep Diving and Cold Water, some practical results, in *The Working Diver Symposium, 1972*, Marine Tech Society, USA, pp. 127–143.

Jenkins, W. T. 1976. *Guide to Polar Diving*, Naval Coastal Systems Lab, Panama City.

Keatinge, W. R., Hayward, M. G. and McIver, N. K. I. 1980. Hypothermia during saturation diving in the North Sea, *Br. Med. J.*, 280, 291.

Langerman, N. 1979. Mountain Divers in Winter, in *Proc./XIth Int. Conf. On Underwater Education*, Nat. Assoc. of Underwater Instructors, Colton, California, pp. 184–187.

Leitch, D. R. and Pearson, R. R. 1978. Decompression sickness or cold injury, *Undersea Biomed. Res.*, 5(4), 363–367.

Lippitt, M. W. and Nuckols, M. L. 1982. *The Development of an Improved Suit System for Cold Water Diving*, Tech. Memo NCSC TM 336–82, Panama City, Florida 32407.

MacInnis, J. B. 1976. The Underwater Arctic: Earth's most hostile frontier, in *The Working Diver Symposium, 1976*, Columbus, Ohio.

Nishi, Y. and Gagge, A. P. 1977. Effective temperature scale useful for hypo- and hyperbaric environments, *Aviat. Space Environ. Med.*, 48(2), 97–107.

Nuckols, M. L. 1978. Thermal considerations in the design of divers suits, in *Hyperbaric Diving Systems and Thermal Protection*, C. E. Johnson (ed.), ASME OED, Vol. 6, pp. 83–100.

Nuytten, P. 1973. Deep Helium Diving in the High Arctic, Fifth Annual Offshore Technology Conference, Houston, 29 April to 2 May 1973, PAPER No. OTC 1779.

Piantadosi, C. A., Ball, D. J., Nuckols, M. L. and Thalman, E. D. 1979. *Manned Evaluation of the NCSC Thermal Protection (DTP) Passive System Prototype*, NEDU Report No. 13–79, Panama City, Florida 32407.

The Oilman. 1983. Statoil seeks arctic rig design, July, p. 9.

Thornton, A. G. 1980. *The Potential of Passive Methods for Thermal Protection of Divers*, AMTE(E) R80522, AMTE Experimental Diving Unit, HMS Vernon, Portsmouth, England.

UEG Technical Note 28. 1983. *Thermal Stress on Divers in Oxy-helium Environments*, CIRIA, London.

Vaughan, W. S. R. and Anderson B. G. 1973. *Effects of Long-duration Cold Exposure on Performance of Talks in Naval Inshore Warfare Operations*, Tech. Report ONR Contract N00014–72–C–0309, Washington DC, USA.

Ward, M. 1974. Frostbite, *Brit. Med. Journal*, 1, 67–70.

Wong, P. 1983. Measurement and Analysis of Thermal Protection in Diving, PhD Thesis, University of Salford, Dept of Chemical Engineering.

Yaglou, C. and Minard, D. 1957. Control of heat casualties at military training centres, *AMA Arch. Ind. Health,* 16, 302–316.

19

Mechanical Design and Operation of Thermal Protection Equipment

L. E. Virr, Head of Research, Admiralty Marine Technology Establishment, Experimental Diving Unit, HMS *Vernon*, Portsmouth, UK

Criteria for avoiding physical injury due to hypothermia caused by deep diving are considered. Techniques for thermally protecting divers are covered, including passive protection, active protection, electrical heating. The difficulties of monitoring the physiological states of divers are discussed. Implications for diving to depths in excess of 300 m are considered.

INTRODUCTION AND BRIEF HISTORY OF HYPOTHERMIA IN DIVING

One of the earliest recorded diving fatalities in which hypothermia was identified as a major factor occurred in 1969 during the Sealab III experiment. At a depth of 189 m and water temperature of 7 °C, one of two divers experienced a convulsion, lost his mouthpiece and was found to be dead on reaching the surface via the PTC. He was not wearing a heated suit. In 1973, the Johnson Sea-Link submersible became fouled at 100 m. Two observers were compressed to 24 m eq. depth on air and subsequently 107 m on He/O_2. Both lost their lives, and this was attributed to the high pressure He/O_2 environment at approximately 6 °C. In another early incident, on the Borgny Dolphin installation in the North Sea, a diver at 140 m lost his communications and life-line, was brought to the SCC by another diver but was found to be dead on reaching the surface. In this case, death was believed to have been caused by hypoxia compounded by cold and over-exertion.

In the Royal Navy, no deaths attributable to hypothermia have occurred. However, RN incidents of physical injury due to cold and dangerous states of hypothermia are on record. In one such incident, actual physical injury to a diver's hands resulted from diving in cold water wearing $\frac{3}{8}$ inch neoprene mittens. In a later incident, a near-fatality occurred when a diver, wearing the same $\frac{3}{8}$ inch neoprene mittens, was unable to operate his equalising valve due to cold-impairment. In another incident, a dangerous state of hypothermia was reached by a diver who continued diving in 6 °C water despite a leaking dry-suit.

Table 1
Physiological criteria for thermal protection in diving

(a) Deep body core temperature should not fall below 35.5 °C. This gives a margin over the temperature at which cooling is considered debilitating (35°).

(b) The maximum corresponding deficit from heat stores (i.e. net heat loss from the body) should not exceed 12 kJ kg^{-1} body weight (3.3 Wh kg^{-1}). This assumes a starting core temperature of 37 °C or above. Special care should be taken to ensure complete rewarming between dives. The figure of 12 kJ kg^{-1} corresponds to 825 kJ (0.23 kWh) for 70 kg body weight.

(c) Mean skin and local head temperature should not fall below 25 °C, with no local measurement below 20 °C except for hands and feet which should be maintained above 15 °C (for useful work in the fingers) and above 10 °C to prevent pain and possible cold injury over long dives.

(d) Maximum local skin temperature should not exceed 42 °C.

(e) Core temperature should not exceed 39 °C.

(f) Maximum inspired temperature of dry gas ($<$ 8 mg H_2O l^{-1}) should not exceed 35 °C for long periods of breathing, and not exceed 45 °C when humidified ($>$50 mg H_2O l^{-1})

(g) Minimum inspired gas temperatures should not result in a respiratory loss in excess of 175 W. In practice, it is worth using a heat exchanger whenever a hot water suit is worn and a level of humidity in the breathing gas over 40% RH would be desirable. For deeper diving around 300 m, then the current exchangers and arrangements are not sufficient. Recommended minimum inspired temperatures:

Depth (m)	Temperature (°C)
150	6
200	16
250	22
300	26
350	28

(From Hayes, 1980, and UEG Technical Note, No 28, 1983, based on work carried out at AMTE/PL and INM)

The above incidents have been selected, from a large number of reported incidents, to underline the serious consequences of cold and the fact that cold may affect the diver directly or indirectly. In many instances, although cold cannot be identified as a prime cause of accident or injury, it is nevertheless believed to have been an important factor. The operational implications of cold-impairment are self evident.

Experience of hypothermia in commercial diving prompted the Chief Inspector of Diving to issue "Diving Safety Memorandum No. 4" dated Feb 1977. This document made suit heating below 50 m and gas heating below 150 m mandatory in commercial diving. Following this, and a research programme carried out by the

Admiralty Marine Technology Establishment (AMTE/PL) and the Institute of Naval Medicine (INM), the recommendations listed in Table 1 were published (Hayes, 1980, and subsequently, with minor modifications, in UEG Technical Note 28, 1983).

The above criteria raise two important issues: achievement of the required thermal conditions in a diving environment, and monitoring of the thermal state of a diver to ensure that the required thermal state is maintained. The first of these issues is the main consideration of this paper, following a brief discussion of the mechanisms by which the body loses heat through clothing and by respiration. In particular, some of the recent work carried out by AMTE(EDU) on electrical heating of diving suits is described.

HEAT LOSS MECHANISMS

Heat is lost from the surface area of the body by the physical processes of conduction, convection, evaporation and radiation, and via the respiratory system. These mechanisms are listed in Table 2 with comments regarding pressure and temperature dependence. For a detailed mathematical treatment of heat-loss mechanisms, the reader is referred to Wissler (1978).

Table 2
Nature of pressure and temperature dependence of body heat transfer mechanisms

Mechanism	Formula	Pressure dependence	Temperature dependence
conduction	$\dot{Q} = -KA \dfrac{dT}{dx}$	independent	$\alpha(T_1 - T_2)$
convection	$\dot{Q} = h_c A (T_1 - T_2)$	$\alpha p^{0.5}$ to $p^{0.6}$	$\alpha (T_1 - T_2)$
radiation	$\dot{Q} = AF\sigma (E_1 T_1{}^4 - A_1 T_2{}^4)$	independent	$\alpha (T_1 - T_2^4)$
evaporation	$\dot{Q} = NA \Delta H$	$\alpha p^{-0.4}$ to $p^{-0.5}$	$\alpha (T_1^2 - T_2^2)$
respiratory exchange	$\dot{Q} = \dot{M} C_p (T_1 - T_2)$	αp	$\alpha (T_1 - T_2)$

Symbols

A	Surface area of body losing heat (m,^{2})	T_1	Temperature of body losing heat K
C_p	Specific heat of breathing gas at constant pressure (J kg^{-1} K^{-1})	T_2	Temperature of surrounding environment K
E_1	Emissivity of radiating body	ΔH	Enthalpy change (J kg^{-1})
A_1	Absorbtivity of radiating body (assumed $A_1 = E_1$)	σ	Stefan's constant (W m^{-2} K^{-4})
		$\dot{M}$	Rate of inspiration ((mass) kg^{-1})
F	Numerical factor to allow for human body form	N	mean evaporation rate (kg m^{-2} s^{-1})
		$\dot{Q}$	Heat transfer rate (W)
h_c	Film coefficient, derived from Nusselt number (Wm^{-2} K^{-1})	$\dfrac{dT}{dx}$	Temperature gradient (K m^{-1})
p	Pressure (Nm^{-2})		

The conductive loss through a medium, for example, the material of a diving suit, is proportional to the temperature gradient across the medium, and essentially independent of pressure for solids, liquids and gases. Convective losses occur within liquid or gaseous media and are due to movement of the medium with respect to the body. This movement may be caused by density gradients due to temperature differences (natural convection) or by some external force (forced convection). Convective losses are proportional to the temperature difference between the body and the convecting medium. In the case of conduction, the constant of proportionality is the conductivity of the medium, whereas for convection the constant of proportionality is the film coefficient. The thermal conductivity of the gas in which thermal insulation is worn thus determines, to a great extent, the insulation value, a higher thermal conductivity giving rise directly to an increase in conductive loss, and to higher convective loss by increasing the film coefficient. Unlike heat capacity, discussed below in the context of respiratory loss, thermal conductivities vary considerably with different gas mixtures but are virtually independent of pressure. The thermal conductivities of helium, air, argon and freon II, for example, are 1415.10^{-4}, 241.10^{-4}, 162.10^{-4} and 60.10^{-4} W m^{-1} K^{-1} respectively. Convective losses increase with pressure, but are not proportional to pressure, in a gaseous environment. Radiative losses are generally considered unimportant in diving, owing to the low temperatures encountered. Evaporation loss reduces with pressure, and this can lead to "thermal runaway" in a diver subjected to a hot hyperbaric environment. The reason for this is that sweat, which evaporates under normabaric conditions, thereby extracting latent heat of vaporization from the body, may evaporate more slowly than sweat is being produced under hyperbaric conditions and thus remain on the skin.

Respiratory losses arise because gas is inhaled at a low temperature and exhaled at near body temperature. Although this causes only a small heat loss at atmospheric pressure, the increased mass of each lungful of breathing mixture gives rise to a heat loss proportional to pressure. Thus, at a breathing rate corresponding to a respiratory loss of 20–30 W at the surface, the loss at 30 bar could become 600–900 W.

It should be noted that respiratory loss is due to the mass of gas which is passed through, and heated by, the lungs, and not to the high thermal conductivity of helium. This is because of the nature of the lungs and the fact that gas remains in the lungs for long enough to permit thermal equilibrium to be attained. It is thus the heat capacity, rather than thermal conductivity, of the breathing gas which gives rise to respiratory loss; the heat capacities (per unit volume) of breathing gas mixtures do not, in fact, vary greatly: values per unit volume (J l^{-1} K^{-1}) for air and 5/95 O_2/He being 1.29 and 0.94 respectively at 1 bar, and 41.3 and 28.5 respectively at 31 bar. At any depth, therefore, breathing air (if this were possible) would lead to greater respiratory loss.

Strictly, heat removed from the body by water vapour during respiration should be added to the heat loss due to the heat capacity of the gas. This heat loss due to water vapour can be ignored at depths where respiratory loss becomes important, but the humidity level of inspired gas is important from the point of view of diver comfort, as low humidity leads to an exaggerated feeling of dryness in the throat (Hayes, 1980). Again, respiratory loss can lead to "thermal runaway", since shivering (or working) requires an increase in breathing rate and the consequent increase in respiratory loss can exceed the heat produced by the activity.

In summary, conduction and convection are the most important heat-loss mechanisms through clothing; respiratory heat-loss is unimportant at shallow depths but increases with depth, and becomes the major heat-loss avenue at 300 m. In order to limit heat loss by conduction, insulation must be kept dry. The major contributions of both conduction and convection to total heat loss mean that suit design and fit and material insulation value are important.

The need to maintain thermal undergarments in a dry state calls for the use of a vapour barrier between the insulating material and diver's body. This has the added advantage of preventing heat loss by evaporation during a long dive.

THERMAL PROTECTION TECHNIQUES

Methods for providing thermal protection can be categorized under two headings, namely "passive" and "active". Passive methods do not involve any external energy supply, and involve clothing and heat exchangers or other heat storage devices, whereas active methods imply an external supply of energy, for example, hot-water or electricity.

Improvements in passive protection techniques can be anticipated, but it is unlikely that passive protection alone will be sufficient for all diving operations, and thus, in general, a combination of passive and active techniques will always be required. However, better clothing reduces the energy demand on an associated active system and by so doing increases the scope for design of the latter.

Passive protection

Thornton (1981) reviewed the "present" position regarding passive thermal protection, in relation to particular dive missions, and also estimated the effect of improved passive protection making "reasonable" assumptions regarding the feasibility of clothing and heat-exchanger development. Table 3(a) gives estimates of heat losses by respiration and from the body surface, wearing either a conventional wet-suit or "woolly bear"/dry suit combination, at various depths of 300 m. The effect of pressure in reducing the insulation value of wet-suit materials (e.g. neoprene) is evident. Table 3(b) shows the expected heat-loss figures using the dry suit/woolly bear passive protection for four different dive missions. It should be noted that the physiologically permissible maximum heat loss of 0.23 kW (based on the figures given in Table 1 and assuming average body weight) is exceeded in all cases. Table 3(c) gives the corresponding heat-loss figures, making the "reasonable" assumptions of 50% efficient breathing gas heat exchangers and an increase in insulation value of clothing by a factor of between 2 and 3 (depending on depth). It should be noted that the 300 m dive for 1 h is still borderline from the point of view of total heat loss. A metabolic heat production of 150 W was assumed in Tables 3(b) and 3(c).

Table 3
Passive protection (present* and projected positions)

(a) *Typical heat losses, conventional thermal protection*

Depth (m)	Respiratory loss (W)	Wet suit loss (W)†	Dry suit/'woolly bear' loss (W)††
0	10	240	170
10	20	290	200
30	40	550	300
100	110	1,200	450
300	250	1,600	500

(b) *Estimated heat losses, dry suit/woolly bear*

Depth (m)	Time (h)	Respiratory loss (W) (22.5 l min⁻¹)	Suit loss (W)	Total loss (W)	Net loss (W)	Total loss (kWh)§
10	6	20	200	220	70	0.42
50	1	60	350	410	260	0.26
100	1	110	450	560	410	0.41
300	1	250	500	750	600	0.60

(c) *Estimated heat losses, improved protection (see text)*

Depth (m)	Time (h)	Respiratory loss (W)	Suit loss (W)	Total loss (W)	Net loss (W)	Total loss (kWh)§
10	6	10	100	110	0	0
50	1	30	170	200	50	0.05
100	1	55	200	255	105	0.105
300	1	125	225	350	200	0.20

*1981
†Nuckols, 1978
††Wattenbarger and Breckenridge, 1978
§Maximum permissible loss (Table 1, 0.23 kWh |
Assumptions: body surface area, 1.8 m²; Sea temperature, 4 °C; internal suit temperature, 25 °C; RMV, 22.5 l min⁻¹; metabolic heat production (Tables 3b and 3c), 150 W.

Table 4
Insulation resistance of thermal protection systems at different depths

	Insulation resistance togs (clo*)		
Environment	6.5 mm foam neoprene wet suit	6.5 mm foam neoprene dry suit + conventional u/wear	Solid neoprene dry suit + "Thinsulate" undergarment
Surface (in air)	2.8 (1.8)	3.9 (2.5)	3.9 (2.5)
3 m (in water)	1.2 (0.8)	1.9 (1.2)	1.9 (1.2)
30 m (in water)	0.4 (0.26)	1.2 (0.8)	1.9 (1.2)

*1 clo = 0.155 m² K W⁻¹. (This is the level of insulation required to maintain comfort in a resting, sitting, individual at 21 °C, 50% RH with normal ventilation. A naked person is comfortable at approx 30 °C; 1 clo is the level of clothing insulation which would produce similar comfort at 21 °C, i.e. 1 clo would compensate for a drop of 9 °C. The level of insulation provided by the clothing normally worn by men is 1 clo. (Wattenbarger and Breckenridge, 1978))

togs × 0.645 = clo.

The point has been made recently (UEG Technical Note, No. 28, 1983) that recent developments in insulating materials, in particular "Thinsulate", have been such that further improvement is not likely for some considerable time. Table 4, extracted from the UEG paper and based on Lippitt and Nuckols (1982), summarizes the present position regarding insulating clothing. The UEG document also states that in order to limit heat loss to 150 W (the figure assumed for heat production by the body), an insulation value (thermal resistance) in excess of 4 togs (see Table 4) is necessary. Table 4 shows that, although this is virtually attained with modern insulating materials in air at atmospheric pressure, this is far from the case in water even at shallow depths.

In designing thermal protection systems the hands present a particular problem, owing to their large surface area to volume ratio and the need to maintain manual dexterity while preventing cold-impairment. Insulation materials must also be able to withstand up to 2 p.s.i. bulk crushing, which can arise because of "suit squeeze".

A recent paper by Lippitt and Nuckols (1983) describes the development of, and the philosophy behind, a diver passive protection system which probably represents the best compromise between thermal and operational performance which can be achieved using materials presently available. The system consists of specially designed thermal underwear worn under an outer suit which incorporates inflation and deflation facilities. Dry gloves, with insulating liners, are included. Following a study of available materials, a crushed neoprene material sandwiched between layers of knitted nylon fabric was selected for the majority of the outer suit, a micro-fibrous polyolefin batt (e.g. "Thinsulate") with a vapour barrier for the insulating undergarment and a similar material for the glove liners. It proved necessary to use different materials for the hood, neck and wrist seals though these were integral with the outer suit. Design considerations included the outer suit, buoyancy control, weight distribution, thermal insulation of undergarments, gloves and diver urine collection system.

Lippitt and Nuckols claim that the variable volume dry suit system with thermal underwear described above is not affected by working depth provided the suit inflation gas is not changed. This was shown to be the case in an initial series of test dives to 10 fsw and 70 fsw, and consequently later test dives were carried out at 15 ft or less. Following the series of tests, it was concluded that the passive protection system could maintain a resting diver within acceptable thermal limits for 6 h in water at approximately 4 °C (the physiological criteria adopted were very similar to those given in Table 1). Thus, for shallow diving at least, a situation between Tables 3(b) and 3(c) appears to have been reached. This is still far from the case for deep diving where respiratory loss becomes important, and the presence of helium in the diver's clothing can reduce the insulation value to as little as 20% of the value with air.

Further improvement of passive protection in the form of clothing seems unlikely in the near future. Where helium is used for breathing and suit inflation, passive protection for the working diver must be regarded as totally inadequate. The greatest potential for improvement of passive protection for helium diving appears to be in respiratory heat exchangers. In a series of experiments using helium at IATA, McKay (1979) has investigated the performance of simple gauze-type heat exchange devices in which either copper or nylon gauze discs with different mesh geometries were used as the heat exchange medium. Efficiency, that is, the fraction of exhaled heat returned on inhalation, and pressure drop across the heat exchange devices were determined for different heat exchanger geometries and the two different heat exchange materials. One conclusion was that the maximum efficiencies observed, approximately 90%, showed little dependence upon heat exchange mesh geometry or material. Pressure drop, however, was strongly dependent upon heat exchanger and heat exchange material geometry. McLean and Thornton (1982) have investigated the performance of simple heat exchangers under hyperbaric conditions, using nylon mesh as the exchange medium. The heat exchanger investigated took the form of a PVC tube, internal diameter 8.0 cm and length 5.0 cm, containing up to 20 nylon discs, each of thickness 0.7 mm and estimated surface area 10 360 mm^2. Efficiency and pressure drop were measured for different ambient pressures to 10 bar and flow rates of 22.5, 40 and 62.5 l min^{-1} RMV with different numbers of nylon discs up to a maximum of 20. A pressure drop of 8 mbar across the heat exchanger was considered reasonable and this pressure drop was observed, at 10 bar ambient pressure, using 20 nylon discs, at 40 l RMV. Under these conditions, 60% efficiency was observed. It was concluded that greater thermal efficiency for a given pressure drop would require re-design of the exchanger housing and optimization of mesh geometry.

Active Protection

For deep-diving and long duration diving in cold shallow water, it is evident that an external supply of heat energy to provide the diver with a warm layer to limit heat loss from the body is essential. For deep diving below 150 m, a means of heating breathing gas is also necessary.

Deep diving for which a SCC is a statutory requirement (>50 m in commercial diving) is normally carried out from an extensive surface facility where sufficient space and electrical power are available to provide hot water for suit and breathing gas heating for the SCC and excursion diver(s). In order to illustrate the principles and disadvantages of conventional hot-water suit/gas heating, it is convenient to consider a simplified system supplying one diver only at depth. Seawater temperature, T_s, is assumed uniform; water is heated from T_s to T_o (typically 50–90 °C) at the surface, pumped down to the diver at a flow rate $G \, \mathrm{l s}^{-1}$, entering the diver's suit/breathing gas heater at temperature T_d (typically 35–40 °C), and discharged into the sea at a temperature T_2. A heat loss rate, P_H, occurs along the umbilical hoses, a heat loss rate, P_D, at the diver and heat is "dumped" into the sea at a rate, P_2, by virtue of the "dump" temperature, T_2, which must be high enough (e.g. >25 °C from Table 1) to ensure diver thermal comfort and safety. The hose power loss, P_H, at any point is, to a first approximation, proportional to the temperature difference between the hose (at that point), and the sea, and is therefore likely to be greatest near the surface (in practice, sea temperature reduces with depth). The diver power loss, P_D, is dependent on the outer suit material but can be considered "useful" power since the presence of the hot-water around the diver prevents heat loss from his body (it is important to note that the hot-water in the diver's suit does not heat the diver, but, ideally, establishes a near-zero temperature gradient next to the body so preventing heat loss).

Regarding P_D as useful power, the efficiency of the system (assuming 100% efficiency for the boiler) can be expressed (simply) as

$$\text{Efficiency } \% = \frac{100 \, P_D}{P_D + P_H + G \rho_w S_w (T_2 - T_s)}$$

where S_w = specific heat of seawater, ρ_w = density of seawater, ($\rho_w \, S_w = 4187 \, \mathrm{J \, l^{-1} \, K^{-1}}$).

P_D and P_H are dependent upon diver and hose inlet and outlet temperatures, as well as the insulation afforded by the diver's outer garment and the hose material. Increasing flow rate, G, will enable T_o and T_D to be reduced, thereby reducing P_H and P_D and providing a more even temperature within the diving suit. However, an increase in flow rate gives rise to a proportionate increase in the "dumped" power loss, P_2, and a reduction in efficiency (although an optimum flow rate is mathematically possible). At a flow rate of 15 l min^{-1} per diver (10 l min^{-1} for the suit and 5 l min^{-1} for the gas heater), an outlet temperature of 25 °C, and sea temperature 6 °C, P_2 becomes approximately 20 kW (per diver).

Typically, between 100 kW and 200 kW at the surface is required to heat the SCC and excursion divers' suits and breathing gas. Hayes (1980) has considered the power requirements at the diver (P_D) in some detail, and concludes that (using a free-flooding hot-water suit) between 500 W and 1000 W are required in 0 °C water up to 50 m, and 3000 W in 0 °C at 300 m for the excursion diver. For the SCC diver, Hayes gives power requirements of 800 W and 1500 W for gas temperatures of 10 °C and 0 °C respectively at 100 m, increasing to 1200 W and 2000 W for 10 °C and 0 °C gas temperatures respectively at 300 m. An overall efficiency of a few per cent is to be expected.

A detailed study of the protection provided by hot water suits has been carried out by Kuehn and Zumrick (1980). Laboratory studies were carried out to a depth of 427 msw, and showed that hot-water heating, with proper control of inlet water flow and temperature and inspired gas heating and present suit technology, affords thermal comfort and safety at this depth for up to 1 h. Field trials proved less satisfactory, and this was attributed to greater difficulty in maintaining the required hot-water flow parameters.

In addition to low efficiency, the surface supplied hot-water system suffers from the disadvantages of temperature control problems, inadequate heat distribution over the diver, high cost and the physiological problem of skin disorders caused by prolonged exposure of the body to hot water (Hayes, 1980). A major advantage of the conventional hot-water system, however, is that leaks or small tears in a diver's suit are not likely to cause abortion of the dive mission. This is not the case with the more efficient closed circuit hot water or electrical systems discussed below.

A conventional hot-water heating system is feasible only where sufficient space and electrical power are available. This is not the case, in general, for shallow diving operations in cold water or for lock-out submersible diving.

Clearly, a more efficient method of providing a supply of hot-water for diver "heating" would be to heat water at the SCC, or on the diver himself, by chemical or electrical means, and circulate this hot water via tubes in the diving suit and via a return hose to the heater. Such closed-circuit systems are under development and promise to overcome some of the disadvantages of the conventional open-circuit systems.

Electrical heating

Electricity is the most efficient method of transmission and distribution of energy, and offers a means for providing diver heating from a lock-out submersible or limited surface facility.

An electrical suit-heating system for divers is under development at AMTE/EDU. The original objective was to provide a viable alternative to hot-water heating for deep saturation diving but, although this remains a long-term intention, the emphasis has now shifted to shallow diving operations in cold water for which no alternative method of heating exists. A prototype system, operating at 24–30 V DC has been in existence for some time and is presently undergoing sea trials with the RN. This system is shown schematically in Figure 1.

The electrically heated suit itself consists of a woollen under-garment and "woolly bear". Sandwiched between these two layers of thermal insulation is the heating element, approximately 50 m of multi-stranded PET-covered resistance wire, insulated overall and enclosed in an outer braided screen of tinned copper (monel

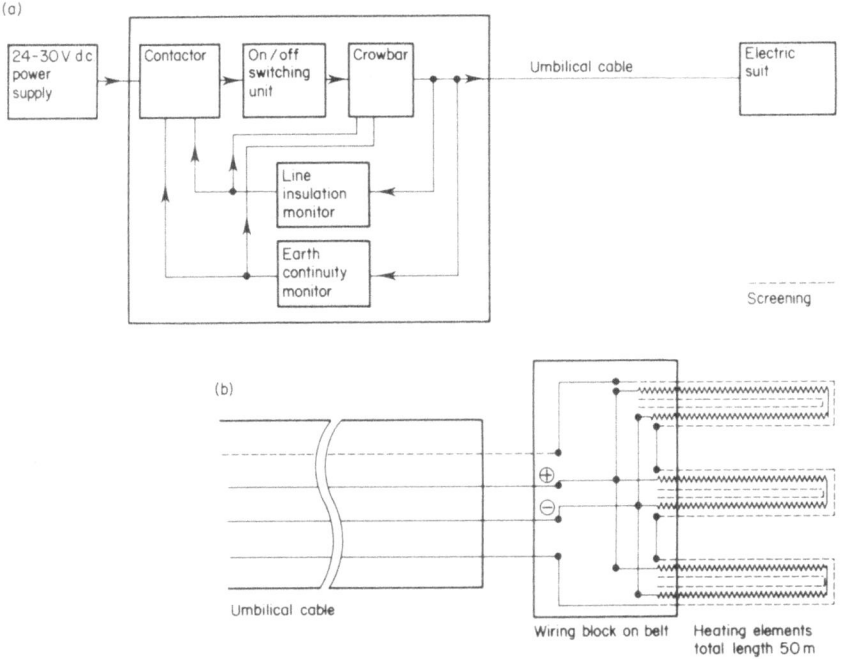

Figure 1   Schematic diagram of (a) electric suit heating arrangement and (b) umbilical/suit wiring.

metal is now under consideration for this purpose, being relatively immune to attack by seawater). The function of the outer screen will be described later. The suit is wired in three sections of approximately 3 Ω resistance, connected in parallel to give a complete suit resistance of 1 Ω; all connections are made at the waist belt from where a short length (2 m approx.) of electric cable (terminated with one half of an underwater connector) is led through a "spout" fitted in an outer dry suit for connection to the supply umbilical. The suit entry point for the supply cable is made watertight by means of a "Jubilee" type clip which tightens around the entry spout onto a rubber bung fitted to the cable. An electric umbilical, typically 30 m in length, connects the suit to a control unit on the surface (or SCC); a 24–30 V DC supply, conveniently provided by lead-acid batteries, is connected to the control unit. The control unit combines the functions of power control and electrical protection. Power control is effected by on-off switching (electronic), the mark–space ratio being determined by a control-knob setting. The conducting screen around the suit heating wire is continued, via the suit connector and supply umbilical, to the surface, and the "remote" end of the suit screen is connected to the surface by means of a separate conductor, thus providing an "earth -loop". One protective function of the control unit is to monitor the resistance of this earth-loop to check that it is continuous. The main protective function of the control unit is

Table 5

Results of electric and passive thermal protection trials (see temperature 8 °C, breathing gas 20/80 O₂/He, max depth 10 m)

Dive/Diver	Thermal protection*	Temperature fall over 1 h period (°C); upper figure (in brackets) initial temp ±1 °C; (−ve sign indicate rise in temperature)										
		R Foot	R Thigh	L Foot	L Thigh	R Bicep	L Bicep	Nape of Neck	Abdomen	R Wrist	L Wrist	Sternum
1 A	Thermal u/wear† El suit§ P ≏ 120 W	(33) 9¶	(33) 0	(31) 2	(29.5) 1.5	(35) 1	(33) 1	(34) 1	(32.5) −1.5	(33.5) −0.5	(30) 1.5	(36) 1
2 B	Thermal u/wear† El suit§ 1st 30 min 123 W 2nd 30 min 178 W	(27) −1	(30) 1	(26) −1	(30) 1	(24.5) −10.5‖	(35) −2	(34.5) 3.5	(33) −1	(33.5) 2	(32) 0	(36) 0
3 C	Thermal u/wear† and "Thinsulate" undergarment	(27) 5	(27) 4	(26) 2	(28.5) 7.5	(33) 7	(34) 4.5	(33) 3	(33.5) 5.5	(28.5) 1.5	(34.5) 8	(35) 2
4 A	Thermal u/wear* and "Thinsulate" undergarment	(28) 2	(31) 7	(29) 3.5	(30) 5	(32) 3	(32.5) 4.5	(30) 5	(28) 0	(32) 5	(32) 5	(32) 3
5 B	Thermal u/wear† and "Thinsulate" undergarment	(26) 2	(29) 3	(25) 3	(28) 3.5	(32.5) 1.5	(32.5) 1.5	(34.5) 2.5	(35) 2	(32) 5.5	(33) 2.5	(36) 3
6 C	"Usual" u/wear‡ "Woolly bear"	(29) 4.5	(30) 4	(30) 4.5	(29.5) 3.5	(29.5) 0.5	(33) 4	(32) 1.5	(30) 3	(34) 8	(34.5) 6.5	(34) 0.5

*Dry rear entry oversuit worn in each case.
†Standard RN issue, worn next to skin.
‡As above, plus 'T' shirt.

§Incorporating thin woolly bear as supplied with electric suit.
¶Slight leak in this area towards end of dive.
‖Initial temperature very low.

that of a line insulation circuit breaker (LICB). In the event of a single insulation fault (usually taken as <20 kΩ) between the heating element circuit and the earth-screen, or discontinuity in the earth-loop circuit, the suit is de-energised within 1–2 ms. This switching speed is achieved by electronic "crowbar" action which precedes operation of the main contactor. At 24–30 V DC operating voltage, such protection is considered essential (Department of Energy, 1982).

Following an initial period of subjective trials of the electrically heated diving suit described above, objective trials in open water were carried out to compare electric and passive techniques for diver (suit) thermal protection. A total of six dives were carried out over a period of two days using three divers, the electric suit (with standard RN issue thermal underwear) being worn for the first two dives and "passive" protection clothing only on the remainder (see Table 5). Temperatures were measured at 11 skin sites (using thermistors taped to the skin) and rectum (however, no significant changes in rectal temperatures were observed during the dives).

Actual maximum depth was 10 m, but a 60 m dive with 10 min bottom time, and total in-water time of 1 h, was simulated, with appropriate "stops" in each case. Breathing gas was 20/80 O_2/He changing to pure O_2 at "15 m" after approximately 30 min. Sea temperature was 8 °C (46°F). Dry-suit inflation gas was air.

A 50 m neutrally buoyant umbilical was used, with 4×4.1 mm^2 aluminium power conductors, total resistance of each 50 m conductor 0.33 Ω, two conductors being used in parallel in each direction (i.e. total umbilical resistance 0.33 Ω). The umbilical was screened overall and also contained an extra conductor for earth-loop monitoring.

The actual amount of power supplied to the diver could be calculated from the dial setting on the front of the control unit. Thus, although the power was varied in response to dive subjective comment, the system of heat control was essentially "open-loop" in the sense that no hard-wired feedback loop was incorporated for temperature control. The results obtained are summarised in Table 5.

For the two dives using the electric suit, it should be noted that standard RN thermal underwear only was worn in addition to the electrically heated undergarment, and this underwear was worn beneath the electric suit. The suit itself incorporates a "woolly bear" over the heating element. For dive 1, an average power of approximately 120 W was supplied and for dive 2 average power supplied was 123 W for the first 30 min and 178 W for the remainder of the dive. With the particular power supply, control unit, 50 m umbilical and electric suit used, the maximum power which could have been supplied was approximately 300 W. However, the suit was designed to dissipate up to 1 kW when connected to the power supply by means of a 30 m length of (copper) umbilical cable.

With regard to the distribution of electric power in the suit, Table 6 shows the power (% of total) applied to the different parts of the body in the present design.

It may be seen from Table 5 that the electric suit performed well in comparison with the conventional "woolly bear" and "Thinsulate" suits. Subjective comment from the divers was to the effect that they felt warm in the electric suit but rather cold after 1 h in the other (passive) clothing.

A slight leak in the region of the right foot in dive 1 caused a sudden rapid fall in temperature in this area towards the end of the dive but apart from this a

Table 6
Electric suit power distribution

Body area	% total power
head	0
torso	17
abdomen	20
thigh	21
calf	22
Foot	7
upper arm	8
lower arm	5

near-equilibrium state appears to have been reached soon after entering the water during dives 1 and 2 (using the electric suit). In dives 3–6 there is evidence that temperatures were continuing to fall even after 1 h. The implication is that in colder water or for a longer dive duration, divers wearing passive protection only could become extremely cold. It should be noted that the "Thinsulate" undergarments themselves failed to provide adequate protection in 8 °C water over a period of 1 h. The observation reinforces the comments made earlier concerning the importance of a systems approach, as adopted by Lippitt and Nuckols, and accurate tailoring of insulating clothing to provide a close fit to the diver.

One important conclusion of the trials is that, in the dry condition and in conjunction with passive insulation ("woolly" bear), a relatively small amount of electric power, between 100 and 200 W in shallow water at a temperature of 8 °C, is effective in maintaining the diver in a state of thermal comfort (with air as the inflation gas). Such a low level of power, for a limited period of possibly to 1 h, could conceivably be provided from a diver-borne (or diver-associated) power supply. By way of example, a reasonable figure for the energy supply capability of conventional lead–acid batteries is 14 W h lb^{-1} and, on this basis, to provide 200 W for 1 h would only require a supply of 14 lb.wt approx in air. In-water weight would be considerably less. A disadvantage of a diver-borne supply is that adequate electrical protection (e.g. LICB) would be difficult, if possible, to incorporate.

In addition to the shallow water trials reported above, the electric suit in its present form has been taken to 300 m under simulated diving conditions. In the helium/water environment, the suit was found to be capable of maintaining skin temperatures in general at an acceptable level with an electrical input of up to 1 kW. However, divers reported some cold areas, notably feet and hands (electrical hand heating was incorporated at this stage), and considerable difficulty was experienced on occasions in separating the two halves of the connector joining the supply umbilical to the suit. The connector problem was attributed to temperature changes in the high pressure environment.

Disadvantages of electrical suit heating are that it is difficult to provide heat over certain areas of the body, for example, knees and particularly hands, and an electric suit, though electrically safe provided adequate electric protection is incorporated, becomes very much less effective (as would be expected bearing in mind the low power levels involved) if a leak develops in the outer suit (in common with purely passive protection). A hot water suit, on the other hand, remains effective in the event of leakages in the outer suit; however, this benefit is extremely costly in terms of energy and heating facilities. There is little doubt that a hot water suit is an extremely effective and reliable method of heating a diver at 300 m where the provision of heating (and pumping) facilities does not present a problem. At present, the electric suit is seen not so much as an alternative to the hot water suit, but as a means of providing thermal protection where hot water cannot be used, in particular when diving from a lock-out submersible or limited support facility. However, as saturation diving depths increase, electrical heating for both suits and breathing gas may become the preferred solution for thermal protection, because of the high efficiency and also because of the need for more accurate temperature control at the diver as hyperbaric pressure is increased.

With regard to power levels required for electric suit and breathing gas heating (where required), Hayes (1980) quotes figures of 400 W at the surface (4 °C water), 780 W at 80 m, 870 W at 180 m and 900 W at 250 m. These figures include an assumed 25% power loss in the supply umbilical.

As stated earlier, hand heating presents a particular problem. Attempts to produce an effective design for electrically heated gloves have so far been unsuccessful. However, for depths beyond 150 m, for which breathing gas heating is required, a successful hybrid sytem has been developed at EDU. This system comprises an electrically heated suit and separate (electrical) breathing gas heater, the latter also providing a supply of hot water for hand heating. The gas heater consists of a water-filled cylinder (approximately 14 $''$ × 6 $''$ diameter), vented to the sea to prevent pressure differentials, containing gas supply tubes and an electric heating element rated at 1 kW (minimum). Hot water thus provides a heat-exchange medium between the electrical heating element and gas supply tubes. By means of one-way valves fitted to the gas heater, which allow outward flow of hot water only, a supply of hot water for hand heating is made available through rubber tubes leading to gloves. Hand movement provides the necessary pumping action. Initial trials in very cold water (approximately 2 °C), albeit at very shallow depths ($\simeq$ 4 m), have been very successful. The difficulty in designing an electrical hand heater arises because heating elements must be confined to the palm and back of the hand, and cannot be extended to include the fingers because of the need to maintain manual dexterity. The gloves must therefore be water-filled in order to provide a heat-transfer medium to the fingers. This leads to temperature differentials and, in addition, the warm water is expelled from the gloves and replaced by cold water by the action of the hands.

To date, insufficient operational experience has been gained with the electrical suit and gas heating system for an assessment of long-term reliability to be made. Weaknesses in the present DC system include the electrical connectors, which can prove difficult to separate under some circumstances, and the need for complex protective circuitry. The use of high frequency AC, at 10–30 kHz, promises to overcome both these difficulties, and high frequency suit and gas heating systems are now under investigation. The advantages of high frequency for diver suit/gas heating are:

(a) HF is safer from the point of view of electric shock.
(b) Different voltages can be used for transmission along umbilicals and for heating at the diver.
(c) Magnetic couplers, rather than electrical, can be used.

A disadvantage of HF is that the inductance of the load reduces the voltage available for heating. However, the supply can be converted to DC by rectification at (or near) the diver and the DC voltage so obtained can be low enough (because of the ability to "step-down" using HF) to permit operation without LICB or ELCB protection.

Diver monitoring

The physiological criteria listed in Table 1 raise the issue of thermal monitoring. Although monitoring of thermal and other physiological data is necessarily standard practice in experimental work, such monitoring of the working diver is difficult, if practicable. On the other hand, the diver is not always the best judge of his own thermal state. The diver, and indeed a person in a cold normabaric situation, can unknowlingly become dangerously hypothermic; at the other extreme, there have been many reported instances of divers receiving hot-water scalds of which they have been unaware at the time. The reasons are unknown, but it appears that subjective sensation is affected in such a way that the diver is unaware of his predicament (Hayes, 1980).

Pearson (1981) has considered the philosophy of diver monitoring, and makes the point that it would be unreasonable to expect that a dive supervising team, and the diving team itself, should include the necessary expertise for fitting and interpreting the results of physiological monitoring devices. Pearson makes the point that, if a case is to be made for monitoring of selected operational dives, this must be regarded as an extension of the research situation.

Skin temperature is commonly measured using thermistors taped in close contact with the body. Kuehn (1977) has reported the development of heat-flow pads for the purpose of skin temperature measurement. However, both thermistors and heat-flow pads and their associated wiring are extremely delicate and a large number of such sensors are required for sensible analysis (Hody and Kacirck, 1972).

The usual device for measuring core temperature is a rectal probe but, again, this type of monitor and its associated wiring is delicate and unlikely to be acceptable in operational diving. Alternatives to the rectal probe and thermistor or heat-flow pad for core and skin temperature measurements are the radio pill and radio "tab" respectively (Kuehn and Ackles, 1978). These devices avoid the problem of wiring transducers on the diver; their transmission range being a few feet, signals can be received by a small aerial at some convenient point on the diver and subsequently multi-plexexd to the surface. However, most of the other objections to monitoring in the operational situation remain. A major criticism of the radio pill is that although it gives a useful indication of core temperature and thus of approaching hypothermia, it is a poor indicator of thermal "stress" as hypothermia is approached and consequently of the ability of a diver to work effectively (Kuehn and Ackles, 1978).

Although physiological monitoring of the operational diver may be contentious, monitoring of his immediate environment, for example, suit, hot water and inspired gas temperature, is certainly desirable, the more so as diving depth is increased. Such monitoring can provide the surface support team with information which can be analysed readily and used as a basis for either human or automatic action.

As a final word on the subject of diver monitoring, there can be no doubting the importance of voice communication between the diver and the surface; adequate communications can provide the experienced listener with a very effective means of assessing many aspects of a diver's physiological state.

Thermal implications of deeper diving

The present depth record for simulated diving is well in excess of 600 m (660 m, AMTE/PL, Alverstoke, 1981, and shortly afterwards 690 m, Duke University, USA). Operation at such depths in the open sea will call for a higher degree of temperature control, particularly with regard to inspired gas temperature, as reference to Table 1 will show. Surface-supplied hot-water systems would be considerably less efficient even than the present 300 m systems which have efficiencies of only a few per cent. The heat requirement to the diving suit would not be expected to increase significantly, but breathing gas heating requirements would (approximately) double, the supply hose loss would be greater and a higher flow rate would almost certainly be necessary in order to lower the required water temperature at the surface. Thus P_D, P_H and P_2 (defined earlier) would increase considerably and efficiencies of the order of 1% and surface heating power approaching 500 kW could be expected. Diver monitoring, or at least monitoring of his immediate environment, would become more important than at 300 m.

If diving beyond 300 m is to be contemplated, alternative forms of heating should be developed. One possibility is an electrical system similar to that described earlier. Apart from the high efficiency of an electrical system, electricity offers the facility for relatively rapid response automatic temperature control from the SCC. Electricity may well prove to be the only viable means of providing thermal protection at the maximum depths attainable.

CONCLUSIONS

With regard to passive protection, the present state of the art is such that for shallow diving, using air for breathing and suit inflation, the situation envisaged in Table 3(c) is approached, although zero net heat loss after 6 h in 4 °C water, suggested as a feasible target by Thornton (1981), making "reasonable" assumptions regarding the development of passive protection, has not yet been achieved. The passive system recently developed and described by Lippitt and Nuckols (1983) probably represents the best compromise between thermal and operational performance which can be achieved at present and in the immediate future. For maximum effect, attention must be paid to choice of materials, system design and tailoring to provide closeness of fit. Both conduction and convection play major roles in the process of heat loss through clothing. Insulation must be kept dry; this requires careful attention to oversuit design, and calls for the incorporation of vapour barriers and urine collection (for long-duration diving).

Where the suit inflation gas is helium, it is unlikely that passive protection techniques can be developed to provide adequate protection for much longer than 1 h. For depths at which respiratory loss is significant, the greatest potential for improvement of passive techniques lies with breathing gas heat exchangers.

Conventional surface-supplied hot-water heating systems, despite their extremely low efficiency, high cost and associated temperature control problems, offer the only viable means at present for providing thermal protection for divers operating at depths down to 300 m.

Electricity is a means of providing efficient, low cost thermal protection at all depths, augmenting passive protection for long-duration shallow diving and offering an alternative to hot water for deep saturation diving. Electricity also offers a means of providing thermal protection for deep lock-out submersible diving for which passive protection alone is totally inadequate (unless the presence of helium next to the body can be avoided) and hot water (certainly on a free-flow basis) cannot be made available. In designing electrical systems, careful consideration must be given to operating voltage and supply frequency, and electrical protection.

With regard to physiological monitoring, the general consensus of opinion is that, at present, this is appropriate only for research purposes. The development of radio pills and in particular radio "tabs" (for skin temperature measurements) may result in acceptable monitoring for operational diving. Monitoring of the diver's thermal environment, that is, suit, hot water and breathing gas temperatures, is considered extremely important. The diver cannot always be considered the best judge of his own thermal state.

Diving to depths greatly in excess of 300 m would present major problems of temperature control and high power requirements if surface-supplied hot-water were to be used for diver heating. Electricity may prove to be the only solution.

ACKNOWLEDGEMENTS

The author acknowledges the help of numerous colleagues and friends: in particular, the Officer in Charge of The Admiralty Marine Technology Establishments (Experimental Diving Unit) and Surgeon Captain R. R. Pearson, RN, with whom he conducted the experiments summarized in Table 5. Copyright, Controller HMSO, London, 1984.

REFERENCES

Department of Energy. 1982. *Draft Code of Practice for the Safe Use of Electricity Underwater*, Document No. ISBN 086017/191/4.

Hayes, P. 1980. Hazards of diving — cold and heat, *Oceanology International*, 33–40.

Hody, G. and Kacirck, K. 1972. Combined skin temperature and direct heat flow measurements in a thermally stressful environment, *Proc. Aerospace Medical Association Annual Meeting*.

Kuehn, L. A. 1977. *Assessment of Convective Heat Loss from Humans in Cold water*, ASME, No. 77-WA/B10-5.

Kuehn, L. A. and Ackles, K. N. 1978. *Thermal Exposure Limits for Divers*, ASME OED-6, 39–51.

Kuehn, L. A. and Zumrick, J. 1980. *Assessment of Thermal Protection Afforded by Hot Water Diving Suits*, ASME 80-WA/OCE-4.

Lippitt, M. W. and Nuckols, M. L. 1982. *The Development of an Improved Suit System for Cold Water Diving*, NCSC Report No. TM 336–82.

Lippitt, M. W. and Nuckols, M. L. 1983. *Development of Passive Diver Thermal Protection System*, NCSC Report No. TM 378–83.

McKay, W. 1979. Unpublished MOD report.

McLean, A. and Thornton, A. G. 1982. The hyperbaric performance of a simple passive respiratory gas heat exchanger, *J. Soc. Underwater Tech*, Autumn, 13–16.

Nuckols, M. L. 1978. *Thermal Considerations in the Design of Divers Suits*, ASME OED-6, 83–99.

Pearson, R. R. 1981. Why do we need diver monitoring?, *Divetch 1981: Aids to Underwater Operations*.

Thornton, A. (1981) Potential of passive methods for protection of divers, *Divetch 1981: Life Support Systems*.

UEG Technical Note, No. 28, 1983.

Wattenbarger, J. F. and Breckenridge, J. R. 1978. *Dry Suit Insulation Characteristics Under Hyperbaric Conditions*, ASME OED-6, 101–116.

Wissler, E. H. 1978. *An Analysis of Heat Stress in Hyperbaric Environments*, ASME OED-6, 53–74.

20

Hand Protection

J. A. Adolfson, L. Sperling and *M. Gustavsson*, FOA 58 Naval Medicine Division,
Naval Diving Center, National Defence Research Institute, Harsfjärden,
Sweden

The human hand is designed as a mechanical tool but it is also an important sensory
organ. The normal hand of a diver in arctic water must be protected against the cold
and also against injuries of different kinds. The aim of this experiment was to study as
a matter of principle thermal and functional properties of a thin and flexible glove
system for divers. Two prototypes were evaluated, one with an inner layer of polyester
and one with a double inner layer of natural silk. Subjective ratings of perceived hand
temperature, hand moisture, difficulty in performing the manual dexterity test and
degree of exertion in the test of grip strength were obtained. Both were warmer than a
previously tested five-fingered prototype but colder than previously tested conventional
deep-diver gloves.

INTRODUCTION

Most of us have sometimes tried to do up a button with cold fingers and found that
this simple act costs a lot of trouble. If the cold hands are covered by gloves, the
trouble will be still more pronounced, and if the gloves are heavy and stiff it will be
impossible to use the hands. This is — in a figurative sense — exactly what happens
to a diver working in cold water.

The human hand is designed as a mechanical tool meant to be used with force and
precision. It is able to form itself round objects and hold and handle objects of
different shape, surface structure, consistence and resistance. It is also able to twist
and turn and move objects in different directions. The human hand is not only a
mechanical tool, however. It is also a very important sensory organ. The shape of an
object, the surface structure, the consistence and the temperature are parts of the
information to the brain mediated through the sensory organs of the hand. It is a
prerequisite of doing a job under water that the diver, by means of hands and
fingers, is able to examine and handle objects which he most often cannot see. Lack
of tactile sensation or sensation of touch of the hand makes it a very rough tool,
useful only in situations where the visual preception is unaffected, and that is a very
rare situation for the working diver. The sensation of touch and most of the other

functional properties of a normal hand of a diver in arctic water is affected, *inter alia*, by cold and the rough, thick and stiff protecting gloves.

Gloves

Gloves are necessary not only for protection against the cold in the oceans but also against injuries of different kind. It is important, then, that the hand wear is designed in such a way that it gives best possible protection against actual risks in combination with best possible comfort and with smallest possible disturbances of the functions of the hand.

The skin is an ideal "glove" which adjusts to the movements of the hand because of its flexibility and general structure. This quality can be sufficiently kept when the demands of protection can be fulfilled by means of thin, soft and extensible material. However, if the work demands protection against physical injuries and high or low temperatures, the material of the gloves must be relatively thick and, as a consequence, less elastic. Therefore, the suitability and design of the hand wear is in that case of decisive importance as well for the grip function as for the comfort.

The gloves should be designed basically from the functional resting position of the hand. Gloves formed after the extended hand have a negative influence on the strength. Thick and stiff seams should be placed in such a manner that they will not disturb the grip. The stiffness of the material is also important. Some materials become stiffer in cold, and compressible materials become certainly thinner but also stiffer at increased ambient pressure.

The friction of the material is of great importance, especially on motion of rotation, for instance, while turning round a screw driver. Smooth materials which become slippery in the water will cause the diver to use much of his force just to hold the object to be handled.

The optimal function of the hands wearing gloves is determined to a great extent by how well the glove fits the hand. Too big gloves produces excess material, which cause folds in the palm during grasping. The tactual sense deteriorates, and it will be hard to get a firm grasp of the object.

Heat regulation and functional properties of the hand

A basic condition of the optimal function of the hands is to preserve the temperature. Cold, but also heat, acts on the thermal balance of the hand and might result in changes of the tissue temperature. A normal sense of touch, a normal motor performance and a normal grip strength are of essential importance to the working capacity of a diver. However, protective gloves necessarily bring about a deterioration of important functions of the hand. Therefore, a hand protecting system must be a compromise between the climatic protection and retained functional capacity of the hand.

The hand is warmed by circulating blood from the body core and is dependent on the heat balance of the body and the conditions of the surroundings. The heat production of the hand even at hard work is of secondary importance. The bloodflow to the hand is the important factor and varies greatly due to the vasomotor influence regulated by the nervous system (Montgomery and Williams, 1977).

The skin temperature of the resting hand is about 32–34 °C at thermal comfort (Olesen and Fanger, 1973) but there are individual differences due to biological, medical or other circumstances. The heat regulation in the hands (and fingers) is determined by a complicated interplay of external physical and internal physiological factors, not only locally in the hand but also in the body as a whole. It is difficult to maintain a more considerable heat flow to the hands if the body is cold.

The heat loss from the hands to the surroundings takes place through conduction, radiation, convection and evaporation. At a constant temperature the heat flow to the hands is equal to the heat loss from the hands. The big surface of the hand compared with the mass is favourable from an exothermic point of view, but this ratio makes it much more difficult to protect the hand against cold. The isolation certainly increases with hand wear but the additional contribution becomes small compared with the required thickness of the gloves (van Dilla et al., 1949). The differences between a five-fingered glove of cotton and a mitten of fibre pile is relatively small. The isolation is primarily brought about by air — a bad thermal conductor — which can be tied in layers on the outside of the skin, and this can be done much better if the hand protection is built up of several layers with an outer wind- and water-proof surface. Additionally, condensation problems and discomfort caused by the natural perspiration of the hand can also be mastered by means of a hand wear consisting of an inner glove and a tight outer glove.

Hand protection in diving

The protection of the hands in diving constitutes a problem that is hard to solve. Most divers seem to be unsatisfied with their gloves. About 90% of Swedish divers, interviewed in 1977, stated that their grip strength was highly reduced, fine motor work became difficult — sometimes impossible — and they experienced very little tactile sensibility when wearing gloves under water (Adolfson, 1978; Günther and Ivergård, 1978). The diving gloves which are commercially available are anatomically and functionally badly designed, according to Lewin (1977). The material is thick and rigid, and it is a strain just to clench the hand. It is hard — often impossible — to handle an adjustable wrench, for instance. These circumstances lead to risks for a diver in situations where he is bound to drop his weight belt, to open his reserve air valve, to cut himself loose or where he in some other way has to be able to use the entire strength and swiftness of his hands.

Preliminary studies

The interviews of the Swedish professional divers led to a pilot study of six types of commercially available diving gloves. Ten naval divers were tested with respect to grip strength, manual dexterity and finger sensibility while wearing the gloves. The dives were performed at 30 msw (metres of sea water) in a pressure chamber wet pot. The water temperature varied between 1.7 °C and 3.7 °C, and each dive lasted about 45 min (Adolfson et al., 1980).

It was found that the skin temperature on the hands dropped to 19.5 °C and on the forefinger to 10.0 °C as an average. The grip strength was reduced by 30–35% and the grip strength endurance by about 50%. The manual dexterity and the tactile sensibility of the fingers were both reduced by about 40% compared with the results bare-handed. Only small differences were noted between the gloves, of which five were made of neoprene foam and one was a five-fingered cast glove of crude latex rubber. This rubber glove was not designed as a diving glove but was used and preferred by the divers because they perceived it as more flexible and easier to work with than commercially available diving gloves. However, that glove was also perceived as cold and the divers usually put a glove of wool underneath which, they said, made it more comfortable but also more difficult to work with.

The results of this pilot study indicated that an improvement in diving gloves was necessary with respect to functional as well as thermal qualities. Therefore, a comparative study of a thin and flexible glove system and a common deep-diver glove was performed in 0.1 °C water at a depth of a 5 msw in a pool. Twenty dives were performed by 10 divers wearing the glove systems, one at a time (Adolfson *et al.*, 1981).

The glove system prototype to be evaluated consisted of a five-fingered cast outer glove made of 1.1 mm thick crude latex rubber and formed after the natural human hand in resting position. It fitted well to the hand and was very flexible. Underneath was a glove made of PVC (polyvinylchloride) yarn which did not absorb moisture and which also fitted well to the hand. It was knitted with a special technique to prevent air movements in the microclimate (Fig. 1). The deep-diver glove was an ordinary three-fingered 7 mm thick glove made of neoprene foam and underneath a five-finger woollen glove (Fig. 2). Both glove systems were sealed to the diving suit

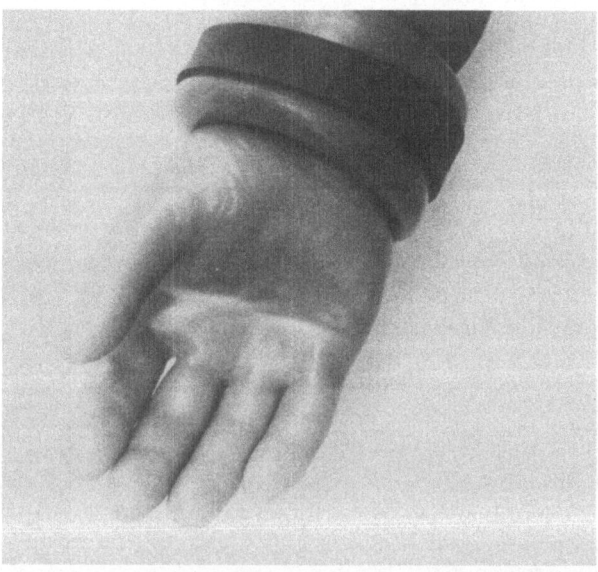

Figure 1   Five-fingered glove of 1981 sealed to the diving suit over cuff-ring.

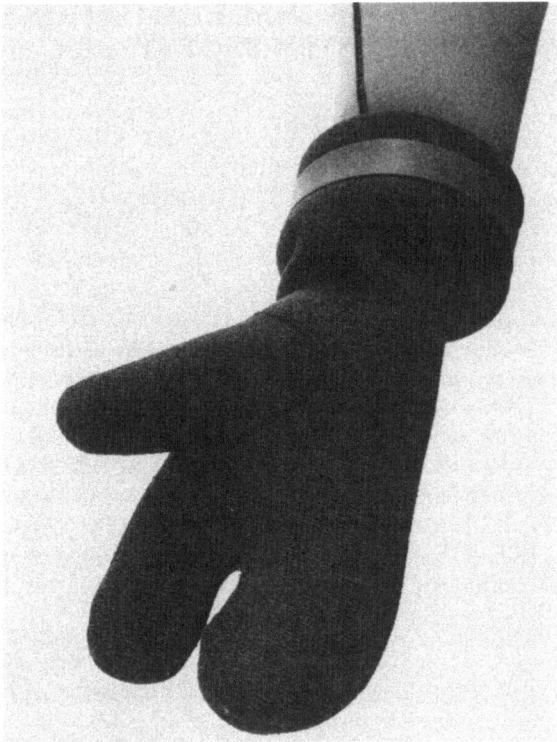

Figure 2   Deep-diver glove of 1981 sealed to the diving suit over cuff-ring

over cuff-rings inside the sleeves of the suit in order to avoid occlusion of the blood flow to the hands.

Measurements used for calculation of mean skin temperature, mean hand temperature, mean body temperature and thermal insulation of the two glove systems were made. Performance testing of maximum grip strength, grip strength endurance, manual dexterity and tactile discrimination ability was made bare-handed and while wearing either of the two glove systems.

Results indicated that the insulation properties of the conventional deep-diver glove system were better than those of the prototype, but also that the performance with the prototype was better in all the tested performance parameters. Subjectively, the divers perceived the prototype as excellent and far superior to the deep-diver gloves from a functional point of view, but they also regarded the thermal properties as not good enough for underwater work in cold water. It was considered necessary, then, to continue the work of developing a more comfortable diving glove system, and the suggestion was that it should be possible to find a compromise between the insulating and the functional properties based on the principles of the tested prototype design.

THERMAL AND FUNCTIONAL EVALUATION OF TWO GLOVE SYSTEM PROTOTYPES

The aim of the latest Swedish investigation of possible glove systems designed for divers (and workers in cold and wet environments) was to design and evaluate two glove systems of which at least one should fulfil the demands of functional capability as well as sufficient thermal insulation.

These glove systems consisted of an outer and an inner layer with the outer layer common to both and designed as a cast three-fingered glove made of 1 mm thick latex rubber formed after the resting position of the human hand (Fig. 3). It was made in two sizes, small and medium, and the subjects were told to choose the most suitable size. The inner layer of glove system 1 consisted of a five-fingered glove of polyester (the polyester gloves). Polyester was chosen as a material which does not absorb moisture. The inner gloves were knitted with a special technique to minimize air movements in the microclimate. The inner layer of glove system 2 consisted of double five-fingered gloves of natural silk (the silk gloves). Silk is traditionally said to have unique climatic properties, and this was the only reason why it was chosen. The inner gloves were knitted with the same technique as the polyester gloves but, as the yarn was thinner, they were worn in double layers to be comparable to the polyester gloves. Both types had a long cuff and were made in one flexible size.

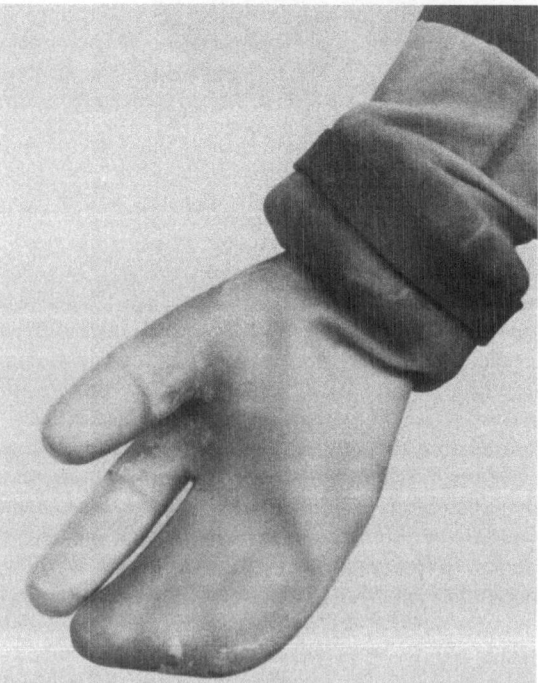

Figure 3   Three-fingered glove of 1983 sealed to the diving suit over cuff-ring.

METHODS

Nineteen 40 min air dives at 6 msw in a pool and eighteen 40 min air dives at 30 msw in a pressure chamber wet pot were made by 10 professional Navy divers. The dives were performed in freshwater with a mean water temperature of 1.7 °C at 6 msw and 1.5 °C at 30 msw. The diving suits were Unisuit pressure-compensated constant volume dry suits and undergarments of synthetic fibre pile.

Temperature measurements

Subjects were fitted with a temperature measurements system consisting of eight skin thermistors and a rectal probe. Thermistor location is shown in Figure 4. All the temperatures were monitored and printed at 1 min intervals with a microcomputer. Except for dorsal hand and forefinger temperatures, rectal, mean skin and mean body temperatures were determined. In addition, the thermal insulation of the two glove systems was measured on an electrically heated hand model (Elnäs and Holmér, 1980).

Performance testing

The functional properties were tested with respect to grip strength, manual dexterity and tactile discrimination ability. The grip strength was tested by means of a strain-gauge dynamometer (Sperling et al., 1980) which was effectively insulated against the ambient water. Contractions were performed at approximately 90° elbow

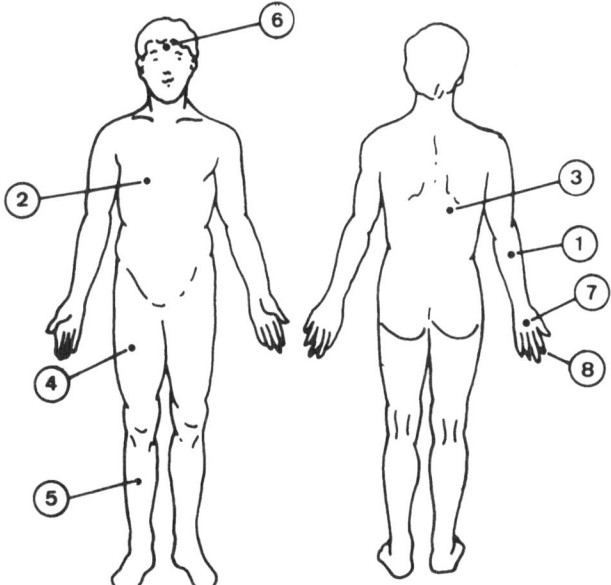

Figure 4 Thermistor location.

angle, and each contraction lasted 5 s and was repeated six times at 5 s intervals. The maximum value in each series was registered as MVC. Grip strength endurance was calculated by comparing the last value with the maximum value in each series and expressed as a percentage.

Manual dexterity was tested by means of a screw-bolt test (Adolfson, 1965). Nut-case-nut-bolt had to be shifted from one side to the other of a vertical board in 4 min without using tools. Each finished detail counted as one point, and the subject was not allowed to start a new screw-bolt until he had finished the previous one. Changes in performance were calculated as a percentage of the performance bare-handed.

Tactile discrimination ability was tested by means of a tactile identification test (Sperling *et al.*, 1980) in which the blind-folded subject had to identify pellets, barrels, cubes, hexagon bars, flush screws and cheese-head bolts of different size and mounted in random order.

Subjective ratings of perceived hand temperature, hand moisture, difficulty in performing the screw-bolt test and degree of exertion in the test of grip strength were made by the subjects by means of a modified category scale with ratio properties (Borg, 1982).

Statistics

Standard statistical methods were used to calculate means (M) and standard deviations (SD). The results are expressed as M ± 1 SD. Student's t-test and paired t-test were used to compare means. Correlation of measurements were determined by the correlation coefficient (r).

TEST PROCEDURE

During a one-day training period the subjects had to perform the tests in order to minimize learning effects. The subjects were tested on land, first bare-handed and then wearing gloves in order to investigate the influence on performance of the two glove systems *in situ*. Once the subject was fitted with the measuring instruments, he donned the underwear and the Unisuit over the thermistors from which cables were fed through a waterproof penetrator in the front of the suit to the monitoring device outside the pool or the pressure chamber. The gloves were sealed to the diving suit over cuff-rings inside the sleeves of the suit in order to avoid occlusion of the bloodflow to the hands. He was then tested in cold water at 6 msw — which means a slightly increased ambient pressure — and at 30 msw — a common depth for a working diver. The temperatures were recorded continuously during the dives. Performance testing was made three times during each dive except for the grip strength, which was recorded twice — when the diver arrived at the bottom and just before the dive ended. In that way it should be possible to compare performance at different hand temperature levels. The subjects were monitored by a closed circuit TV and communication was maintained by underwater telephone.

RESULTS AND DISCUSSION

Thermal properties

When the dives started the rectal temperature averaged 37.4 °C and when the dives ended 37.3 °C. The mean skin temperatures were all higher than 25 °C, with no individual point below 20 °C except the hands. The forefinger and dorsal hand temperatures at 6 msw during the course of the dives are shown in Figure 5. There were no significant differences between the two glove systems, but there was a substantial temperature drop during the 40 min in the water.

The corresponding temperatures at 30 msw are shown in Figure 6. The picture is about the same as at 6 msw (Fig. 5). The lowest forefinger temperatures were at both depths about 9 °C with the polyester gloves and about 10 °C with the silk gloves, and the lowest dorsal hand temperatures were about 18 °C with the polyester gloves and about 19 °C with the silk gloves. Laboratory studies of the thermal properties showed that both glove systems had a heat transfer coefficient of 21.5 W/m^2 K, which means that the thermal insulation was the same for both.

A desirable condition for operational diving is that "core temperature (rectal) does not go below 36 °C, or 1 °C below starting temperature, whichever is lower. Thus, a diver starting at 36.8 °C could go down to 35.8 °C" (Hamilton, 1983). The condition is also that skin temperatures should not reach a mean value lower than 25 °C, with no individual point below 20 °C except hands and feet. In this study the drop in the rectal temperatures averaged only 0.1 °C during the course of the dives, and all the divers were well within the limit. Also, the other condition was

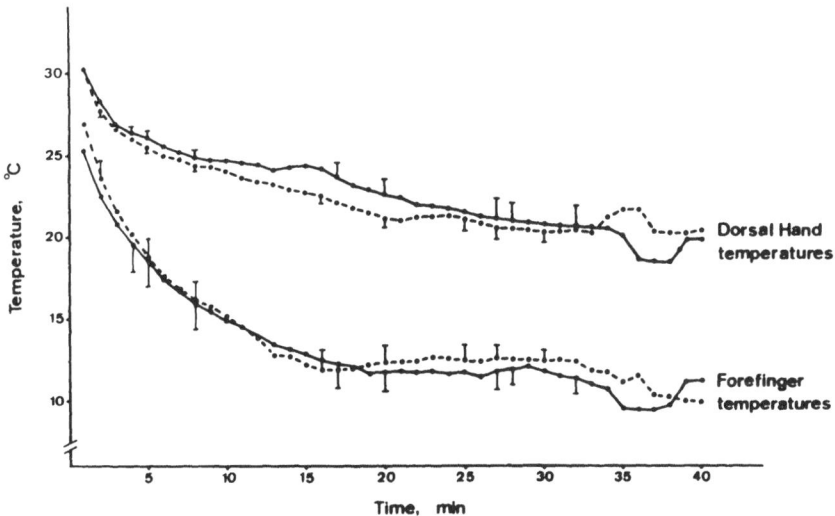

Figure 4 Dorsal hand and forefinger temperatures at 6 msw. Mean water temperature 1.7 °C. ——, the polyester gloves; . . ., the silk gloves. Means ± 1 SD.

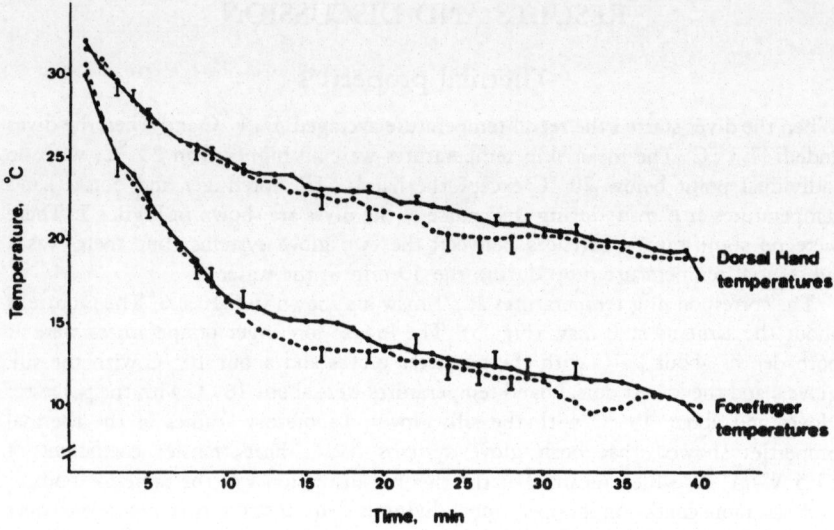

Figure 6 Dorsal hand and forefinger temperatures at 30 msw. Mean water temperature
1.5 °C. ——, the polyester gloves; . . ., the silk gloves. Means ±1 SD.

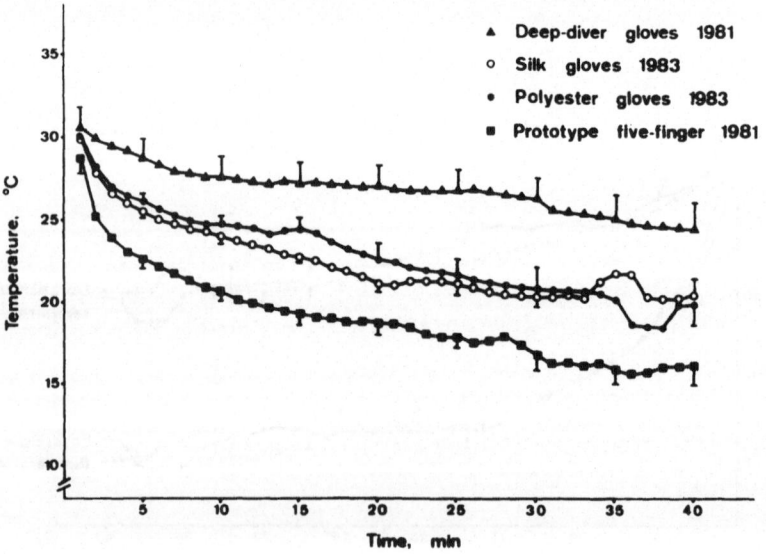

Figure 7 Dorsal hand temperatures during 40 min in cold water at 5–6 msw while wearing
deep-diver gloves, five-fingered prototype gloves, polyester gloves and silk gloves. Means ±
1 SD.

fulfilled. It may be concluded, then, that the diving suits used gave sufficient protection against the cold during the experiment and that no cooling of the body core could have influenced the cooling of the hands.

The use of the rating scale showed that the majority of the subjects perceived the silk gloves as warmer and dryer than the polyester gloves. According to the recorded temperatures, no significant differences were found between the two studied glove systems, however.

As can be seen in Figure 7, the dorsal hand temperatures of both glove systems lie somewhere in between those of the deep-diver gloves and the five-fingered prototype studied in 1981 (Adolfson et al., 1981). The forefinger temperatures (not shown in this Figure) did not differ from those of the five-fingered prototype of 1981. Thus, the three-fingered design did not significantly improve the forefinger temperatures compared with the five-fingered design.

The hands of the subjects were kept dry in both the tested glove systems, and the sealing to the diving suit over cuff-rings allowed them to fill the gloves with air now and then. Besides, the heat production in the fingers is almost entirely dependent on the bloodflow to the fingers (Burton, 1939; Newburgh, 1949), and it seems important to preserve the bloodflow by using cuff-rings as a routine.

Performance

Maximum grip strength (MVC)

Bare-handed, the subjects produced 617 ± 31.6 N as an average on the dynamometer. They did not differ from the sports divers (611 ± 34.3 N) investigated previously (Adolfson et al., 1981). Figure 8 shows percentage performance of the maximum force bare-handed and while wearing the glove systems. Dry on land, the MVC was reduced by 23% with the polyester gloves and by 24% with the silk gloves compared with the force bare-handed, and this reduction was statistically significant ($p < 0.001$). The results of the series of contractions under water, at 6 msw as well as at 30 msw, did not differ from the results obtained while wearing gloves dry on land. No significant differences were found between the two glove systems, but MVC was slightly better ($p < 0.05$) at 30 msw than it was at 6 msw with both the tested glove systems.

Thus, the maximum grip strength was significantly reduced while wearing either of the glove systems on land under dry conditions. The reduction averaged 23–24%, apparently due to slippage between the inner and outer layer of the glove systems. This is in accordance with previous findings in pilots and flying cadets wearing standard heavy flying gloves consisting of wool liner inside and a leather shell (Hertzberg, 1955). Also, stiffness of the material (Adolfson et al., 1981) and insecure grasp because of difficulty in closing the fist firmly caused by disturbed sensory feedback mechanisms in the fingers (Tichauer, 1976) may have contributed to the reduced MVC. A 20% reduction was also found while wearing the five-fingered prototype in the study of 1981.

Coppin et al. (1978) found that grip strength decreased as a consequence of a 30 min immersion of the naked forearm in 10 °C water. In the present study the forearm temperature decreased significantly during 30 min at 6 msw and at

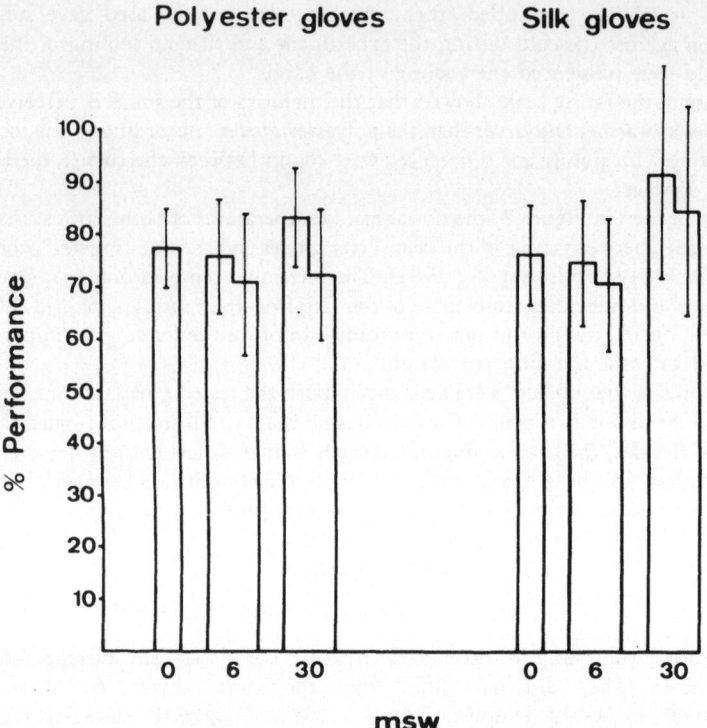

Figure 8 Maximum grip strength (MVC) as percentage of the maximum force bare-handed and while wearing the polyester gloves and the silk gloves dry on land, at 6 msw and at 30 msw. Means ± 1 SD.

30 msw, but the cooling of the hands under water was apparently not enough to cause a further significant deterioration of the MVC. However, the forearms were well protected, and the loss of forearm skin temperatures was at the most less than 5 °C, which might explain the divergence. The increased ambient pressures had no influence on the grip strength, which is in accordance with previous findings (Lambertsen *et al.*, 1978).

Grip strength endurance

As can be seen in Figure 9, the grip strength endurance averaged on land 76% bare-handed, 74% with the polyester gloves and 78% with the silk gloves. Thus, the reduction of force after six contractions was 24%, 26% and 22% respectively. No significant differences were found between the endurance bare-handed and with either of the glove systems. Under water the endurance did not change, neither at 6 msw nor at 30 msw.

Functionally, the grip strength endurance is of great importance. A reduction of MVC can be compensated by a relatively better endurance (Sperling, 1979). This

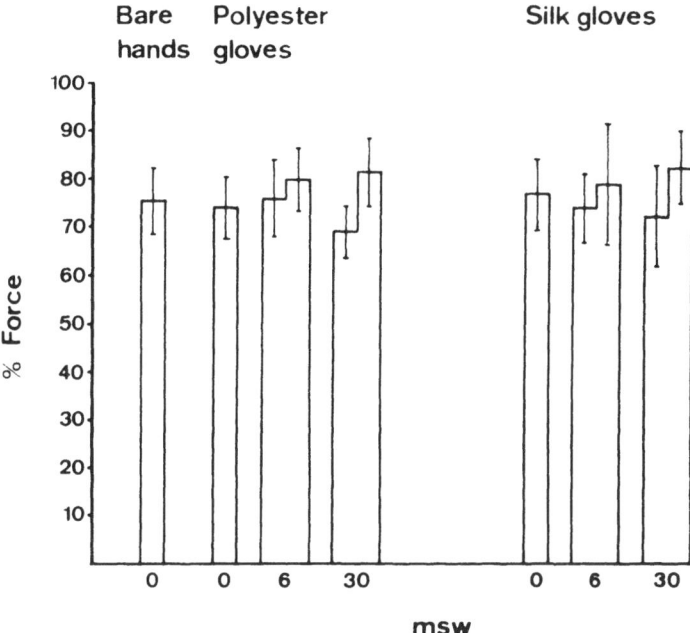

Figure 9 Grip strength endurance. Percentage of force after six maximum contractions bare-handed and wearing the polyester gloves and the silk gloves dry on land, at 6 msw and at 30 msw. Means ± 1 SD.

was the case in the present study, showing a decreased MVC in combination with a high degree of endurance in the second series of contractions at 6 msw and 30 msw with both glove systems. Previous studies of conventional divers' gloves showed a high degree of deterioration of MVC as well as endurance, which means a great handicap to the diver.

It could be expected that a series of maximum contractions is matched by a rated maximal exertion. This was the case when the grip strength was tested bare-handed and with both glove systems on land. However, in the first series of contractions during each dive at both depths the perceived exertion was lower compared with the corresponding results on land. This can be explained by the compression of the air layer in the gloves, making the slippage between the inner and outer layer less prominent and the grasp more secure and comfortable. In the second series of contractions during each dive, the degree of exertion was again higher, probably due to fatigue.

Manual dexterity

Bare-handed, the subjects achieved 76 ± 3.1 points on average on the screw-bolt test. Figure 10 shows percentage performance of the results obtained bare-handed and while wearing the two glove systems. On land under dry conditions, the

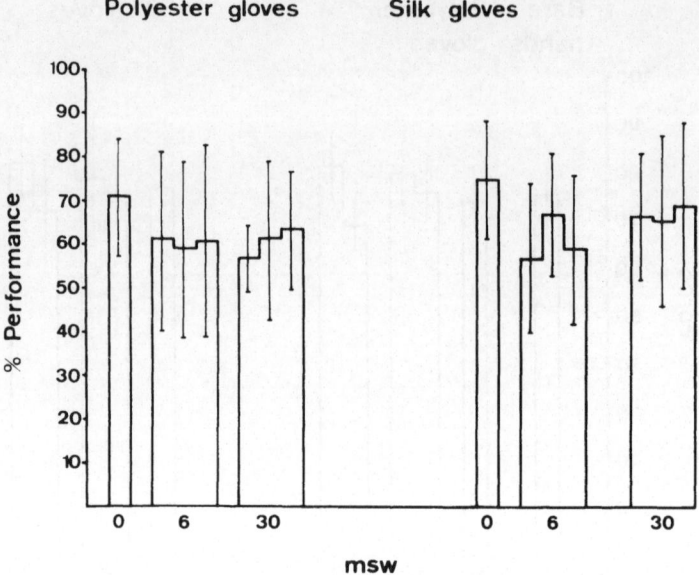

Figure 10 Screw-bolt test. Manual dexterity as percentage of the performance bare-handed and while wearing the polyester gloves and the silk gloves dry on land, at 6 msw and at 30 msw. Means ± 1 SD.

performance was reduced by 29% with the polyester gloves and by 25% with the silk gloves compared with the performance bare-handed, and these reductions were statistically significant ($p < 0.001$).

The results obtained in the test series under water at 6 msw and at 30 msw did not differ from the results obtained while wearing gloves dry on land. No statistically significant differences were found between the two glove systems, neither on land, nor under water, and the performance at 30 msw did not differ from the performance at 6 msw.

The perceived difficulty in the first and second test series in each dive at 6 msw was stronger with the polyester gloves compared with the ratings bare-handed ($p < 0.01$). With the silk gloves, no significant differences were seen, neither compared with the ratings bare-handed, nor with the polyester gloves. During the third test series, in each dive at 6 msw the perceived difficulty was larger than with bare hands for both the glove systems ($p < 0.02$). At 30 msw, the perceived difficulty was larger only during the third test series and only with the polyester gloves ($p < 0.02$). Thus wearing either of the glove systems on land under dry conditions caused the performance to decrease significantly. That was also the case in the study of the five-fingered prototype of 1981. However, manual dexterity with the five-fingered gloves of 1981 was significantly better than either of the glove systems in this study ($p < 0.001$ in all comparable parameters). It cannot be excluded that the three-fingered design may have caused this functional impairment.

It is well known that manual performance is impaired when hand skin temperature drops below a certain level although the rest of the body is kept warm (Gaydos, 1958; Gaydos and Dusek, 1958; Clark, 1961). Severe impairments of manual dexterity in divers have also been demonstrated earlier (Bowen and Pepler, 1967; Bowen, 1968). In the present study the small differences between the performance as well as perceived difficulty while wearing the glove systems on land and under water are statistically insignificant. It might be suggested that this deterioration is caused by the gloves and *not* by cooling of the hands or by the wet environment at depths down to 30 msw. However, a possible training effect during the test series in the present study may explain the divergence between these results and the results obtained by Bowen and Pepler.

Tactile discrimination ability

Bare-handed, the subjects had no difficulties in identifying all the items of the tactile discrimination test. As can be seen in Figure 11 the tactile discrimination ability was not affected by the glove systems or by the stay in cold water. On land under dry conditions, it was reduced by 4% with the polyester gloves and unaffected with the silk gloves. The ability while under water was of about the same magnitude. No statistically significant differences were found between the two glove systems, neither dry on land, nor under water. The results obtained with either of them did not differ from the results obtained bare-handed, and the ability at 30 msw did not differ from the ability at 6 msw.

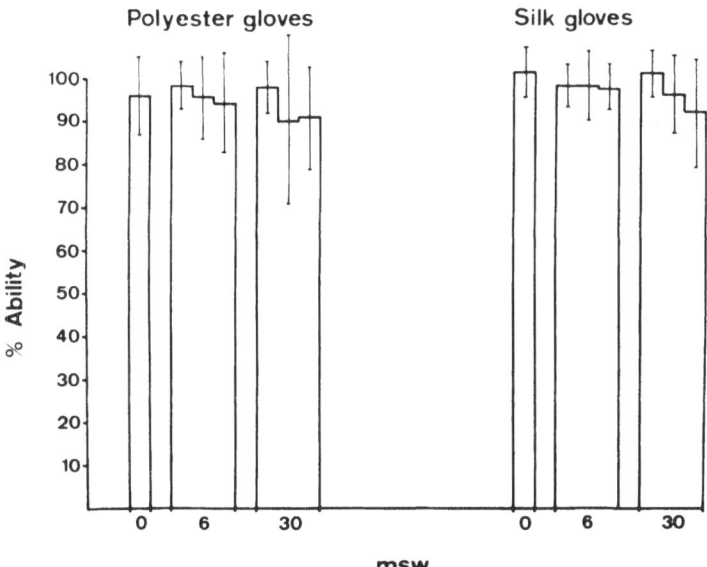

Figure 11 Tactile discrimination test. Tactile discrimination ability as percentage of the ability bare-handed and while wearing the polyester gloves and the silk gloves dry on land, at 6 msw and at 30 msw. Means ± 1 SD.

The cooling of the hands during the course of the dives did not cause any statistically significant deterioration of this performance, nor did the ambient pressures at depths or the wet environment. The "cold effect" described by Bowen (1968) was not found in this study. Bowen's divers reported that contact or pressure produced more pain in their hands than tactile sensation. No such reports were delivered in this study. The cooling of the hands was apparently not great enough to bring about pain.

CONCLUSIONS

The performance properties of both the polyester gloves and the silk gloves were far superior to those of the deep-diver gloves of 1981, demonstrated in Figure 12. The opinion of the subjects was that both these glove systems were very much easier to work with than any other deep-diver gloves they were used to, and the majority of them perceived the silk gloves as a bit better than the polyester gloves. Figure 7 illustrates that the thermal properties of both glove systems were better than those of the five-fingered prototype but poorer than those of the conventional deep-diver gloves both tested in 1981. However, all the subjects considered both the polyester gloves and the silk gloves to provide sufficient protection against the cold water

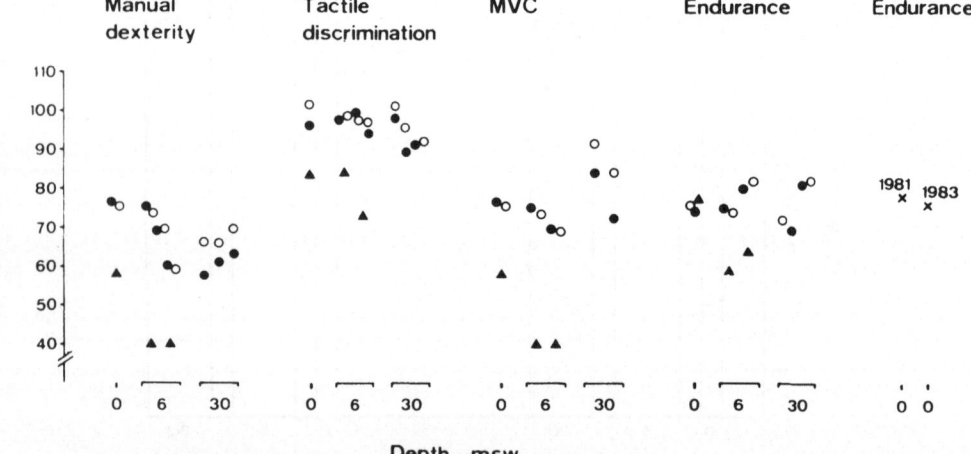

Figure 12 Manual dexterity, tactile discrimination ability and MVC as percentage of the performance bare-handed. Grip strength endurance as percentage of the maximum value in each series of contractions. Means ± 1 SD.

under the circumstances of these experimental dives (depths, bottom time, water temperature). The majority of them perceived the silk gloves as warmer and dryer than the polyester gloves. There were no statistically significant differences between the two glove systems, regarding neither the thermal nor the performance properties.

The aim of this experiment was to study, as a matter of principle, thermal and functional properties of a thin and flexible glove system for divers consisting of an inner layer to keep the hands sufficiently warm and comfortable and an outer layer to keep the hands dry. The conventional diving gloves commercially available are considered anatomically and functionally badly designed, and the material is thick and rigid. These are circumstances which lead to risks for a diver in situations where he is bound to use the entire strength and swiftness of his hands. The results of this investigation indicate that the tested principle can be used as a base for future design of diving gloves in different fields of application.

The material of which the outer gloves are made should be chosen with regard to the kind of work in which they will be used. Heavy work with the hands demands heavy gloves, and reconnaissance, inspection, underwater photographing and that kind of underwater work demands more flexible gloves. It seems crucial, however, that the gloves fit well to the hands in each work situation and that they are correctly designed anatomically and functionally.

ACKNOWLEDGEMENT

This work was supported by the Swedish National Defence Research Institute, Project No. H2 36, and by the Swedish Work Environment Fund, Contract No. ASF Dnr 82–0330.

REFERENCES

Adolfson, J. A. 1965. Deterioration of mental and motor functions in hyperbaric air, *Scand. J. Psychol.*, 6, 26–31.

Adolfson, J. A. 1978. *Den militäre dykarens arbetsmiljö, utrustning och arbetsförhållanden under vattnet. En intervjuundersökning,* Nat. Swedish Defence Research Institute, Report No. C 58004–H9. Stockholm. (In Swedish).

Adolfson, J. A., Sperling, L. and Pettersson, S. 1980. *Funktionstudie av dykarhandskar,* Nat. Swedish Defence Research Institute, Report No. C 58007–H9. Stockholm. (In Swedish)

Adolfson, J. A., Elnäs, S. and Sperling, L. 1981. *Utveckling av funktionsdugliga dykarhandskar,* Nat. Swedish Defence Research Institute, Report No. C58011–H2. Stockholm. (In Swedish)

Borg, G. 1982. A category scale with ratio properties for intermodal and interindividual comparisons, in *Psychophysical Judgement and the Process of Perception,* H.-G. Geissler and P. Petzold (eds), VEB Deutscher Verlag der Wissenschaften, Berlin, pp. 25–34.

Bowen, H. M. 1968. Diver performance and the effects of cold, *Human Factors,* 10(5), 445.

Bowen, H. M. and Pepler, R. D. 1967. *Studies of the Performance Capabilities of Divers: the Effects of Cold,* ONR–N0014–67–C–0263, Washington, DC.

Burton, A. C. 1939. Range and variability of bloodflow in human fingers, *Am. J. Physiol.*, 127, 437.

Clark, E. R. 1961. The limiting hand skin temperature for unaffected manual performance in the cold, *J. Appl. Physiol.*, 45, 193.

Coppin, E. G., Livingstone, S. E. and Kuehn, L. A. 1978. Effects on hand grip strength due to arm immersion in a 10 °C water bath, *Aviat. Space Environ. Med.*, 49(11), 1322–1326.

van Dilla, M. A., Day, R. and Siple, P. A. 1949. Special problem of hands, in *Physiology of Heat Regulation and the Science of Clothing*, L. H. Newburgh (ed), W. B. Saunders Co., Philadelphia.

Elnäs, S. and Holmér, I. 1980. *Termisk utvärdering av handbeklädnad med en elektriskt uppvärmd handmodell*, Nat. Swedish Board of Occupational Safety and Health, Report No. 1980:38. Solna. (Abstract in English).

Gaydos, H. F. 1958. The effect on complete manual performance of cooling the body while maintaining the hands at normal temperatures, *J. Appl. Physiol.*, 42, 373.

Gaydos, H. F. and Dusek, E. R. 1958. Effects of localized hand cooling versus body cooling on manual performance, *J. Appl. Physiol.*, 42, 377.

Günther, C. and Ivergård, T. 1978. *Rapport om yrkesdykares arbetsmiljö*. Ergolab AB, Report No. R 78:08. Stockholm. (In Swedish).

Hamilton, R. W. 1983. Personal communication.

Hertzberg, H. T. E. 1955. Some contribution of applied physical anthropology to human engineering, *Annals of the New York Academy of Sciences*, 63, 616–629.

Lambertsen, C. J., Gelfand, R. and Clark, J. M. 1978. Work capability and physiological effects in the He–O_2 excursions to pressures of 400–800–1200 and 1600 feet of sea water, *Int. Environ. Med.*, Univ. Pennsylvania Med. Center, Philadelphia.

Lewin, T. 1977. Personal communication.

Montgomery, L. D. and Williams, B. A. 1977. Variation of forearm, hand, and finger blood flow indices ambient temperature, *Aviat. Space Environ. Med.*, 48, 231–235.

Newburgh, L. H. 1949. In *Physiology and Heat Regulation and the Science of Clothing*, L. H. Newburgh (ed), W. B. Saunders, Co., Philadelphia.

Olesen, B. W. and Fanger, P. O. 1973. The skin temperature distribution for resting man in comfort, *Arch. Sci. Physiol.*, 27, 385–393.

Sperling, L. 1979. The perception of force in hand grips in subjects with normal and impaired hand function, in *Greppfunktion i dagligt liv* (L. Sperling, Thesis), Department of Handicap Research, Univ. Göteborg.

Sperling, L., Jonsson, B., Holmér, I. and Lewin, T. 1980. *Testprogram för arbetshandskar*, Nat. Swedish Board of Occupational Safety and Health, Report No. 1980:18. Solna. (In Swedish)

Tichauer, E. R. 1976. Biomechanics sustains occupational safety and health, *Industrial Engineering*, Feb., 46–56.

21

What A Diving Team Needs To Know About Hypothermia

H. J. Manson, Assistant Director, Center for Offshore & Remote Medicine, Memorial University of Newfoundland, St. John's, Newfoundland, Canada

The training given to divers on the subject of hypothermia is often confusing and perplexing. Detailed knowledge of the prevention and treatment of hypothermia appears to be dangerously lacking among some divers. The author provides details of a training scheme aimed at overcoming this lack.

INTRODUCTION

The coast of Newfoundland and Labrador is affected by the Labrador Current which flows down from the Arctic to meet the Gulf Stream south of the Grand Banks of Newfoundland. This results in surface water temperatures ranging from approximately 10 °C in summer to -2 °C in winter and early spring. Principal industries of Newfoundland and Labrador are fishing, forestry, mining, and to these has recently been added exploration for offshore oil. Since the raw materials produced by mining and forestry are exported by sea, this results in a great deal of marine activity in this region. In addition, several substantial hydroelectric projects have been undertaken to provide electricity for the region.

A consequence of all this marine activity is that a surprising amount of diving is undertaken in this region for commercial reasons, in addition to which, in spite of the cold water, there is a large and active sport diving community. One might think this would result in a diving community with a large practical experience of the management of hypothermia, but this is not necessarily so. In fact, this author has seen a number of serious errors in safe diving practice with respect to hypothermia and it is too early yet to say whether educational efforts in this respect have been successful. The number of cold related problems seen is very large but perhaps three examples will serve to set the scene.

A 30-year-old police diving instructor, teaching a course in ice diving, was totally unaware of the dangers of non-freezing cold injury. Repeatedly, over a 12 h period, he exposed his lower extremities to -2 °C salt water, albeit whilst wearing a dry suit. The consequence of this was that he suffered a classic non-freezing cold injury complicated by influenza-like symptoms, gross haemoglobinuria and, somewhat amusingly, a positive Wassermann reaction. More seriously, a company engaged in diving in support of offshore oil exploration at depths between 600 and 700 feet complained of problems with two of their divers who developed signs and symptoms consistent with bronchitis. The symptoms manifested themselves following a dive and closer investigation revealed that the divers were using hot-water-heated suits, but, although the company possessed heat exchangers for breathing gas, the supervisors and safety officers had not considered it necessary to use these.

An incident also came to light when a young diver, not long graduated from his training school, refused to leave the diving bell when the hot water supply to his hot-water suit failed. For this excellent piece of judgment, he was dismissed.

Such experiences led me to question closely the divers I see on their knowledge of hypothermia. Although it appears that most local divers realize the importance of cold with regard to decompression schedules, the majority are confused and perplexed by the training they have received with respect to hypothermia. Many have been the recipients of contradictory instruction from different instructors at different times and detailed knowledge of the prevention of and treatment of hypothermia appears to be dangerously lacking in this diving community.

PREVENTION

It is obviously better to teach diving crews to prevent hypothermia than to treat it, (Cooper, 1976). Up to a certain level, this is quite easily done and there is general agreement on the procedures to be taught. For example, it is clear that commercial diving operations in eastern Canada cannot be realistically carried out using wet suits. Therefore the use of dry suits for shallow operations and heated suits of the conventional type for deeper operations is appropriate. Once it is pointed out, the need to heat breathing gas for deeper diving operations is also obvious to all, as is the need to control closely the temperature in hyperbaric chambers.

A recurring problem in diver instruction is the importance of hypothermia in repetitive shallow diving. Much of the shallow ship and harbour diving and smaller construction projects are undertaken by small diving companies in eastern Canada. This is an area not covered by any Canadian regulations, and the Canadian Standards Association standard for commercial diving is frequently ignored by these operators. The question arises as to how the diver or diving team should monitor the onset of hypothermia in repeated dives in very cold waters.

It has been widely taught that, provided a diver exercised, drank hot liquids and rewarmed himself so that he was sweating before his next dive, he would not suffer a cumulative problem with hypothermia. The available evidence suggests it is impossible for subjects to sweat while still experiencing significant negative heat balance (Snellen, 1966). Suggestions that a diver's core temperature be measured under these circumstances are impractical since it is most unlikely that they would be complied with. There is some anecdotal evidence to suggest that the problem of insidious onset of hypothermia in repetitive shallow diving may have been involved in several near-miss incidents attributable to lack of judgment in this region.

Also of interest in the prevention of hypothermia is the question of providing thermal protection for divers in a diving bell which has lost its heat supply from the surface. It is now becoming clear that commercially available thermal protection systems for use in such lost bell incidents can provide acceptable limitations in the rate of core temperature fall, compatible with realistic rescue times (Tonjum *et al.*, 1980a, 1980b). There have been some attempts at computer modelling of heat loss in these circumstances (Silcock and Flook, 1982) but uncertainties remain, and for the present perhaps it may be wiser to rely on human testing until more validation of computer simulations has been carried out. We have recently carried out some experiments in air at one atmosphere which indicated that the estimation of airway heat loss is not simple. Our preliminary results indicate that sensible heat loss in the airway may be overestimated by 5–15% if maximum exhaled air temperature is multiplied by tidal volume (King *et al.*, 1984) instead of taking the integral of exhaled air temperature with volume. The pattern of respiration also appears to influence airway heat loss with the greatest heat loss being found at high tidal volumes of respiration, but there is another factor perhaps related to frequency of respiration which we have not yet been able to quantify.

Another frequently asked question in the area of prevention is whether a suitable diet can help prevent hypothermia when diving in cold water. Since many of the diving population have to guard against obesity, it would perhaps be unwise to advocate a very high calorie or high fat diet indiscriminately. In this context, a modest amount of subcutaneous insulation is perhaps an advantage in our waters.

Now that investigation is under way into the use of theophylline compounds to increase heat production (Lin *et al.*, 1980; Wang and Anholt, 1982), perhaps we shall in the future be required to make up our minds as to whether the use of these compounds is reasonable in the diving environment. We emphasize the cognitive effects of hypothermia and the possible loss of normal warning sensations when the periphery is warmed while the core is cooled via the lungs.

TREATMENT

The treatment of hypothermia continues to have controversial areas and is a field in which the last few years have seen a number of changes. There are some areas which may be particularly relevant to the handling of the hypothermic diver and it is these which we shall consider.

The first area of instruction of importance to a diving team is perhaps the awareness of hypothermia as a problem to divers. It is important that someone who may be confronted with the necessity of recoving and treating a hypothermic diver realize the possibility that the patient may be hypothermic, otherwise the diagnosis may be missed and the patient handled in an inappropriate way. Thus, it is important to draw the students' attention to circumstances likely to lead to diver hypothermia.

The hazards of rescue of the hypothermic victim and of post-rescue death have recently received a good deal of attention and it seems that this is an area which we must emphasize in training divers. Quite apart from the problem of the hypothermic diver, it is by no means unknown for a diving team to be pressed into service in a rescue role.

It may be useful to list the putative causes of post-rescue death and to consider briefly the importance of each one in the context of diving. The most commonly suggested causes of post-rescue death are, not necessarily in the order of importance: further fall in core temperature caused by further exposure particularly to wind chill, further fall in core temperature consequent upon the afterdrop, mechanical stimulation of the hypothermic heart leading to ventricular fibrillation, and shock caused by relative hypovolemia in the hypothermic victim.

The problem of continuing exposure and wind chill will affect the hypothermic diver in the same way as anyone else except that there is perhaps one special case, and that is the case of a hypothermic diver recovered into a dense heliox atmosphere where it might be possible for continuing heat loss to occur, say, in a lower bunk although the temperature in other parts of the chamber, say, in an upper bunk was comfortable for the diver's companions.

There has been a large volume of literature developed over the years on the problem of the afterdrop. This phenomenon is extremely familiar to a practising anaesthetist, both from former times when surface cooling was used for neurosurgical operations, and from the modern practice of core cooling for cardiac surgery. The phenomenon of the afterdrop used to be due in large part to equilibriation between the different tissues of the body and therefore between the "core" and the "shell". This means that the core temperature continues to drift downwards when surface cooling has ceased, the warmer core dropping towards this mean temperature by losing heat to the shell. Indeed, in core rewarming after cardiac surgery, one can see the converse phenomenon where the shell lags behind the core and core temperature falls after active heating ceases. Because rapid cooling produces greater temperature differentials within the body, afterdrop is seen most markedly in those who have been rapidly cooled. It is to be anticipated that a diver who has become hypothermic from such causes as failure of his hot water supply will have been rapidly cooled, and this is also likely to be true for a wet suit diver or one whose hypothermia is due to major leakage in his dry suit.

A diver becoming hypothermic while unprotected in a heliox atmosphere will also likely be rapidly cooled. On the other hand, a diver in a well-functioning dry suit or perhaps a diver who had become hypothermic over a prolonged period of time while well-protected in a heliox atmosphere could be expected to have cooled slowly.

The problems of mechanical stimulation leading to dysrhythmia in the hypothermic heart are present for everyone, and the only likely special cases in the diving environment would come about because of the difficulty of handling a diver in bulky equipment or of rescuing an unconscious diver into a diving bell.

The onset of shock due to relative hypovolemia in hypothermia may have a number of causes. These include loss of circulating volume due to diuresis secondary to immersion (Arborelius et al., 1972; Boening et al., 1972), diuresis secondary to the effects of cold, and loss of oedema fluid into the tissues of the shell, as well as impaired control of the peripheral circulation, bradycardia and limitation of cardiac output.

When we consider the hypothermic diver, he may or may not suffer diuresis as a result of immersion. This will depend upon the attitude in which he has spent the majority of his time working underwater. To take the extreme example, if he has been working in a head-up position, one would expect that he would be subject to the same central displacement of blood volume as one sees in a subject immersed up to the neck at the surface but, if he has been working in a horizontal position, as is often the case, such effects might tend to be minimized.

There seems no reason why diuresis due to the effects of cold should not occur to the same extent as in other hypothermic subjects. This should also be true of loss of fluid into the shell except in the case of the diver core cooled by breathing heliox while he remains warmed from the surface by a hot water suit.

As the heart becomes hypothermic, bradycardia is observed with limitation of cardiac output and this is a factor in the development of hypovolemic shock in the hypothermic patient. We do not know to what extent the bradycardia seen at great depths may affect the situation.

We have based our teaching on diver rescue on these concepts but we find continuing difficulty with the practical problem of rescuing a diver into the diving bell. In this situation a vertical lift is almost inevitable and it is virtually impossible to guarantee the absence of mechanical stimulation to the chest wall. Where it is possible, partial flooding of the bell during the rescue might help to minimize the problems of a vertical lift.

The next area in which we encounter confusion in our diving population is in regard to the question whether or not cardiopulmonary resuscitation should be practised on the hypothermic victim. Some instructors have gone so far as to say that cardiac massage should never be carried out on hypothermic victims because of the fear of inducing ventricular fibrillation in circumstances in which the diagnosis of cardiac arrest is rendered difficult by hypothermia. They argue that on the one hand there is some protection for the hypothermic brain with its diminished oxygen uptake and on the other adequate cardiac output for these circumstances may be present even though normal clinical signs are difficult to detect. This subject was discussed at length at a recent conference held in Alaska, and the consensus reached was that the only reliable sign of cardiac arrest in hypothermia is careful palpation of the carotid artery with the rescuer's warm hand for upwards of a minute and that if no pulse can be detected after that time cardiac massage should be instituted (Doolittle *et al.*, 1982). This is what we teach, and we point out that many of the signs of cardiac arrest have become difficult to detect by 28–30 °C core temperature whilst at this temperature only about eight or ten minutes of cerebral protection can be expected. We also point out that, once mouth to mouth breathing has been initiated in the hypotherhmic victim, it is unlikely that spontaneous respiration will easily return before the patient is rewarmed, and we point out that it is unlikely that spontaneous recovery of cardiac rhythm will occur before substantial rewarming has taken place.

This brings us to our greatest area of difficulty, and that is the controversy over the relative merits of rapid and slow rewarming (Kaufman, 1983). North American physicians are in some difficulty here since the guidelines published following the recent Alaskan conference (Doolittle *et al.*, 1982) could well be pointed to as standards of best practice, which would have considerable medico-legal implications. In brief, the consensus of this meeting was that rewarming in the field was undesirable and should not be practised if avoidable, and that rapid rewarming in field conditions was unacceptably dangerous. We have some trouble with these conclusions in that in the context of diving it is clearly not possible to rapidly transfer all divers to a medical facility, both because they may be subject to considerable decompression requirements and because of the great distances, difficult weather, problems with communication and transportation in our region (Bartlett, 1983). Therefore we are forced to teach diving teams how to carry out rewarming in the field. Our approach has been to teach that, if rapid rewarming is to be carried out, it must be so accomplished as to increase core temperature rapidly and that attempts at rapid rewarming which result in perhipheral rewarming without core rewarming carry the danger of precipitating hypovolemic shock and of returning volumes of cold acidotic blood from the periphery to the core. We also teach that if rapid rewarming is to be practised sufficient personnel must be present to ensure gentle handling of the patient and minimize the risk of accidental mechanical stimulation of the hypothermic heart.

There has been considerable debate as to the efficacy of core rewarming by heated humidified air or oxygen and it is our view that, although core rewarming is certainly accomplished by this means, it is an extremely slow method since only quite small amounts of heat can be transferred in this way. It is surprising that research has not been directed towards the possibility of using this method of rewarming in the hyperbaric heliox environment where much larger amounts of heat could be transferred. Perhaps the considerable body of knowledge regarding the protection of divers from heat loss by this route could be applied to the question of diver rewarming.

We teach that rapid rewarming by hot water immersion can be practised safely in mild to moderate hypothermia and that a practical means of determining this is by assessing the victim's level of consciousness. We teach that this method of rewarming may be dangerous in those particularly susceptible to hypovolemic shock, and here one thinks of those who have been made hypothermic slowly or who have been hypothermic for extended periods of time, allowing full physiological changes to take place. One also thinks of those with other factors predisposing to hypovolemia such as trauma and intercurrent illness, a good example of which might be decompression sickness. We teach that patients not in these categories should be rewarmed by surface warmth augmented, if desired, by airway rewarming with the aim of achieving a rise in core temperature of 1 °C, and not more than 2 °C, per hour. We emphasize the importance of measuring the rectal temperature with a low-reading thermometer in order to monitor this. Methods which are suggested for surface rewarming include the use of hot packs, hot blankets, the rescuer's body heat. The danger of burns from heating apparatus is emphasized.

Lastly, we emphasize the dangers of diagnostic confusion in the hypothermic patient and, in particular, the dangers of confusion between signs and symptoms of hypothermia and of decompression sickness or indeed arterial gas embolus. This again highlights the high priority that must be given to measuring the core temperature of the stricken driver.

CONCLUSION

This paper cannot answer all the questions raised by its title. While recognizing that the best answers may not be available at this time, it is necessary to attempt to provide a rational and coherent body of knowledge for the diving team in order that they may respond reasonably and without panic in situations which are often remote and difficult. It is the hope of the author that this account of his training practices may provoke the exchange of information in the area of the various rewarming methods.

REFERENCES

Arborelius, M., Balldin, U. I., Lilja, B. and Lundgren, C. E. G. 1972. Haemodynamic changes in man during immersion with the head above water, *Aerospace Med.*, 43, 592–598.

Bartlett, R. 1983. Problems in providing medical support for remote diving operations, in *Proceedings of the Canadian Ocean Technology Congress II*, Toronto.

Boening, D., Ulmer, H. V., Meier, U., Skipka, W. and Segeman, J. S. 1972. Effects of multi-hour immersion on trained and untrained subjects, I. Renal function and plasma volume, *Aerospace Med.*, 43, 300–305.

Cooper, K. E. 1976. Hypothermia, in R. H. Strauss (ed), *Diving Medicine*, Grune and Stratton, New York, pp. 211–226.

Doolittle, W., Hayward, J., Mills, W., Nemiroft, M., and Samuelson, T. 1982. *State of Alaska Hypothermia & Cold Water Near Drowning Guidelines*, Emergency Medical Services Section, Division of Public Health, Alaska Department of Health and Social Services. Juneau.

Kaufman, W. C. 1983. A reconsideration of rewarming techniques, in *Proceedings of the Canadian Ocean Technology Congress II*, Toronto.

King, F. G., Manson, H. J., Snellen, J. W. and Chang, K. S. 1984. Demonstration of a problem in estimating sensible heat loss from the respiratory tract by thermometry, *Can. Anaesth. Soc. J.*, in press.

Lin, M. T., Chandra, A. and Liu, G. G. 1980. The effects of theophylline and caffeine on thermoregulatory functions of rats at different ambient temperatures, *J. Pharm. Pharmacol.*, 33, 204–208.

Silcock, S. and Flook, V. 1982. A computer model designed to make rapid predictions of diver temperature changes, Paper presented to the Advanced Course in Diving Medicine for Medical Practitioners, sponsored by the Institute of Environmental and Offshore Medicine, Aberdeen; Comex S. A., Norwegian Petroleum Directorate, Marseilles.

Snellen, J. W. 1966. Mean body temperature and the control of thermal sweating, *Acta. Physiol. Pharmacol. Neerl.*, 14, 99–174.

Tonjum, S., Hamilton, R. W., Brubak, A. O., Peterson, R. E. and Youngblood, D. A. 1980a. *Project Polar Bear: Testing of Diver Thermal Protection in a Simulated "Lost Bell"*, NUI Report, 2–80, Bergen.

Tonjum, S., Pasche, A., Furevik, D., Holand, B., Brubak, A. and Olsen, C. 1980b. *Subproject: Polar Bear II: Test of Survival Systems for Stranded Bell Divers at 31 bars*, NUI Report, 40–80, Bergen.

Wang, L. C. and Anholt, E. D. 1982. Elicitation of supramaximal thermogenesis by aminophylline in the rat, *J. Appl. Physiol: Respirat. Environ. Exercise Physiol.*, 53, 16–20.

22

The Case of the Lost Bell

S. Tønjum, A. Påsche and *J. Onarheim*, Norwegian Underwater Technology
Centre, (NUTEC) Ytre Laksevåg, Norway,
and
P. Hayes and *H. Padbury*, RAF Institute of Aviation Medicine, Farnborough,
Hampshire, UK

Two passive survival systems were tested in a simulated lost bell. The subjects were
monitored for core temperature, skin termperature and inspiratory gas temperatures.
The results indicate that one survival had a poorer passive thermal protection than the
other. However, individual morphological characteristics are of importance.

The conditions in a stranded bell in the North Sea are mainly defined by pressure and
temperature. Pressure may range between 7 and 20 bars, and temperature can drop
as low as 4–6 °C. The representatives of the diving industry indicate that rescue
might take as long as 24 h under typical North Sea conditions. Therefore, divers
need a protection system enabling survival for at least that long, while they are
waiting to be rescued.

Life support in a stranded bell requires oxygen to breathe, removal of carbon
dioxide (CO_2) and protection against excessive heat loss. Sufficient oxygen for an
extended period without surface supply is built into the life support system of the
bell and is usually not a problem. The build-up of CO_2 can be prevented if the divers
breathe through a CO_2 absorbent material. The heat produced by this biochemical
CO_2 reaction can be used to heat the breathing gas, and accordingly prevent
respiratory heat loss. Thermal protection of the stranded divers' bodies can be
achieved using heavy insulative clothing. Based on the results of two former tests at
16 and 31 bars in a cold heliox environment, both of which lasted for approximately
10 h, the Norwegian Underwater Technology Centre concluded that, by using only
passive thermal protection, survival for 24 h in a stranded diving bell would be
possible (Brubakk *et al.*, 1982; Tønjum *et al.*, 1980). However, in 1981 AMTE-PL
in conjunction with the Royal Navy (Hayes *et al.*, 1981) validated three different
passive survival systems at 26 bars under realistic conditions in a diving bell, but
had to abort the test after 6 h because of a critical drop in body core temperature in

two of the test divers. Due to this, it was decided that another test was necessary. The same gas cooling profile was used for the AMTE-PL test for the first six hours, and a cooling profile extrapolated from that curve was used for the following hours of the test.

MATERIALS AND METHODS

The experiment was performed at 16 bars in NUTEC's hyperbaric chamber complex. The experiment was performed in one of the living chambers which had been modified, so that it was easy to cool and heat. A canvas sheet was stretched inside the chamber, approximately 80 cm above the deck plates, and the test subjects lay on this in a supine position for the whole test period.

The two divers who took part in this experiment had, before the dive, been through a test which gave information of their response to gradual cooling. They were immersed in water at 32 °C, and the water was slowly cooled to 28 °C over the next 130 min. During the immersion period, the skin and rectal temperatures were measured. Mean skin temperatures were almost the same for both subjects; the rectal temperature remained approximately constant, at a slightly decreased level, for one subject, whereas that for the other subject fell slowly to a subnormal level.

Both test subjects were SCUBA divers with limited diving experience, one being a medical doctor and the other a physiologist. The physical characteristics of the two subjects are given in Table 1.

A limited food intake during the test period was established and consisted of chocolate biscuits and high-energy bars (not candy). Approximately 1 liter of cold water was available to drink. It was planned that the subjects would stay in the survival systems as long as possible, a maximum of 24 h, or until their core temperatures dropped to 35.5 °C.

The divers' core temperatures were assessed directly by thermistor rectal probes. Skin temperatures were measured by six thermistors and in addition three heat flow discs were used. These measured skin temperatures as well as heat loss in W/m^2 of body surface. Inspiratory gas temperatures were measured inside the oro-nasal masks, and temperatures were also measured in the CO_2 absorbent canisters in addition to several temperature registrations from the experimental chamber.

Two survival systems were tested. Both are passive systems and designed specifically for the purpose of providing protection for a diver in a stranded bell

Table 1
Physical characteristics of test subjects

Age (years)	Height (cm)	Weight (kg)	Surface area (m^2)
36	178	79	1.97
40	185	73	1.96

situation. Both systems consist of high quality sleeping bags, and combined thermal regenerators and CO_2 absorbent canisters. In addition, both systems had individual thermal protection clothing, i.e. hooded survival vest without sleeves, underwear, boots and mattress.

RESULTS

The test was performed for 24 h as planned. The temperature at the bottom of the experimental chamber was, at the termination, 6.2 °C. Thirty minutes before the test was terminated, the chamber heating for the experimental chamber was put on. Both subjects were able to leave their survival systems without assistance, but complained about some temporary dizziness. They disconnected all the monitoring equipment themselves before they left the experimental chambers. They entered the warm living chamber, but had no need to go into the sleeping bags, which were prepared for rewarming procedures. The rectal temperatures of the two test subjects never dropped below 36 °C (Fig. 1), had a tendency to fall during the night, but increased again in the morning. At the end of the test there was a small drop in rectal temperature in both divers: 36.2 °C and 36.6 °C at the time of termination.

The curves in the middle of Figure 1 show the subjects' rectal temperatures during the test. The curve drawn with the open circles is from the subject with the highest heat loss per m^2 body surface. The curves at the bottom of Figure 1 show the inspired gas temperatures measured in the subjects' oro-nasal masks. Again, the curve with the open circles is from the subject with the highest heat loss per m^2 body surface.

The mean skin temperatures were based upon readings from chest, thigh, front calf and upper arm. One subject had a much higher mean skin temperature during the 5–6 h of the test than the other subject. This is obviously due to the fact that this diver put on the complete survival system immediately when the test started and felt uncomfortably warm during the first few hours, whereas the other subject was not in his "complete system" until 2–3 h later. However, the subject who had the highest mean skin temperature for the first half of the test had the lowest mean skin temperature for the second half of the test. The survival system he used had no pants or boots, whereas the other had. Toe temperature was measured as low as 13 °C for the test subject in that system; toe temperature measurements never went below 27 °C for the subject testing the other system.

The inspired gas temperatures were measured in the oro-nasal masks. The inspiratory gas temperatures (Fig. 1) were on average higher in one subject's oro-nasal mask than in the other. The subject with the highest temperature complained several times that the gas temperature felt too hot, and that he therefore either had to go off the mask for some minutes, or breathe without keeping the oro-nasal mask tight to the face. The temperature in this mask was relatively stable during the whole test and varied between 31.5 °C (lowest) and 34.5 °C (highest). The temperature in the other mask was as high as 36 °C at the beginning of the test, but dropped to 27 °C during the following 10 h. For the rest of the test it varied between 24 and 28 °C. However, the subject testing this system never complained about cold breathing gas. The heat flow measurements were made from thigh, chest

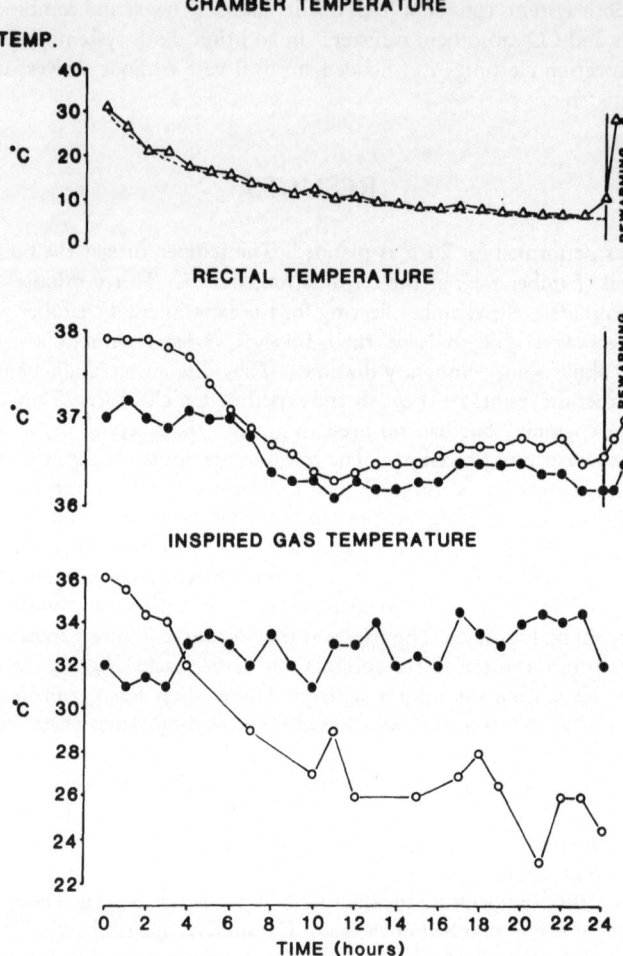

Figure 1 *Top*: Chamber temperature curves: the temperature cooling profile extrapolated from the AMTE-PL test, and the temperature cooling profile used in the 24 h test. *Centre*: Subjects' rectal temperatures during the test. Open circles represent data from the subject with the highest heat loss per m² body surface. *Bottom*: Inspired gas temperatures measured in the subjects' oro-nasal masks. Open circles represent data from the subject with the highest heat loss per m² body surface.

and back, and 48 recordings were done from each of the mentioned places in both divers. The measurements represent heat loss in W/m² of body surface. Both subjects had the greatest heat loss from their backs with mean values of 81 W/m² and 69 W/m² respectively. Both had the lowest heat loss from their chests, with mean values of 32 W/m² and 31 W/m² respectively. The mean value of the heat loss from the thigh was the same in both subjects: 45 W/m².

DISCUSSION

Data concerning the gas cooling profile of a manned stranded diving bell in cold water is limited. The AMTE-PL test (Hayes *et al.*, 1981) of survival systems gave valid data for the first six hours of the gas cooling profile in a manned bell without heat supply in 7–8 °C water, and a pressure equivalent to 250 msw. To complete a 24 h cooling profile, extrapolation was necessary. It is, of course, possible that the cooling of a stranded bell could be faster than what was used in this test. The humidity in the test chamber was lower (60–75%) than it would have been in a real "lost bell" situation, where 100% is to be expected.

During the test one subject had a higher mean skin temperature than the other, and had as well, for more than half of the second part of the test (the coldest part), higher inspiratory breathing gas temperatures than the other. The same subject had a smaller average heat loss (48 W/m^2) than the other (53 W/m^2). In spite of this, the subject with the higher heat loss had a lower average inspiratory breathing gas temperature, a lower mean skin temperature, and a higher rectal temperature during the whole test.

This should indicate that the subject with the greater heat loss was testing a survival system with poorer passive thermal protection than the other. It indicates however, that individual morphological characteristics are present and of great importance in such tests. This indication is supported by the fact that their physical characteristics (Table 1) show a significant difference. The shorter subject is the one who had the greater heat loss during the test but, in spite of that, had a higher rectal temperature than the other. In the pre-dive test, when both divers were immersed in 32–28 °C water for 130 min, the same subject had a higher rectal temperature than the other, whereas skin temperatures were similar. The night–day fluctuations of both subjects' rectal temperatures in normal hyperbaric temperatures were, however, similar and within the range which is normal at the surface (36.4–37.4 °C). These observations support the view that heat production (metabolism) is individual for each subject and the fact that core temperature depends on the balance between heat production and heat loss. No observations were made that should indicate that hyperbaric heliox environment is different from surface conditions, especially taking into account that neither of the subjects were shivering during the test period.

REFERENCES

Brubakk, A. O., Tønjum, S., Holand, B., Peterson, R. E., Hamilton, R. W., Morild, E. and Onarheim, J. 1982. Heat loss and tolerance time during cold exposure in heliox atmosphere at 16 ATA, *Undersea Biomed. Res.* 9, 81–90.

Hayes, P. A., Padbury, E. H. and Atherton, P. J. 1981. Validation of Emergency Procedures — The Lost Bell Situation. International Conference DIVETECH '81, Workshop B. Aids to Underwater Operations, Session 4.

Tønjum, S., Påsche, A., Furevik, D., Holand, B., Brubakk, A. and Olsen, C. 1980. *Polar Bear II: Test of Survival Systems for Stranded Bell Divers at 31 bars*, NUI Report No. 40.

Part III

Underwater Operations

Underwater Operations

23

Arctic Operations

R. Goodfellow, President of the Engineering Committee on Ocean Resources,
Vice-President of the Society for Underwater Technology, Goodfellow
Associates, London, UK

Climatic factors of crucial importance to offshore operations in the Arctic are described.
Environmental conditions in the Beaufort Sea. Arctic Islands and the east coast of
Canada are described together with the various techniques and equipment which have
been developed to exploit the gas and oil reserves.

INTRODUCTION

The search for hydrocarbons (oil and gas) has taken the operators into many strange
and hostile areas, but perhaps none more so than the arctic waters. Locations in the
Canadian Beaufort Seas and high Arctic Islands provide a challenge for the aggressive
and innovative nature of the offshore engineers to harness these resources (see
Fig. 1).

For shallow water exploration, artificial islands have already been widely used as
drilling platforms. Now, with the moves into deeper water, there is the need to
build bigger islands to support drilling and production facilities (see Fig. 2).

There has also been assessment of arctic engineering needs related to hard minerals
and mining, pipelines and coastal construction in the Beaufort, Chukchi and Bering
seas. Numerous problems face engineers in these regions: thermal erosion in the
coastal zone, gas hydrates in the shallow sub-seafloor sediments, sub-seafloor
permafrost and seasonally frozen soils. Changes in temperature cause thaw
subsidence phenomena and frost heave. Movement of the ice creates ice scour and
gouging. In the construction of gravel islands there are problems of limited materials
and the effects of dredging and response to ice and current scour. Foundations are
subject to properties of immature and unstable seafloor silts, seismic risk zones,
sedimentary erosion, soil structure interaction and foundation stability.

Specialised education and training is needed to help engineers adapt basic
engineering principles to resolve these problems, and thus enhance the opportunities
for development. There is the need for long-term studies of nearshore ice and ice
mechanics, field and laboratory investigation of geotechnical properties of sediments
peculiar to the Arctic, design of new coring systems operated from ice platforms,
long-term seismic monitoring nets, mapping and coring of permafrost,

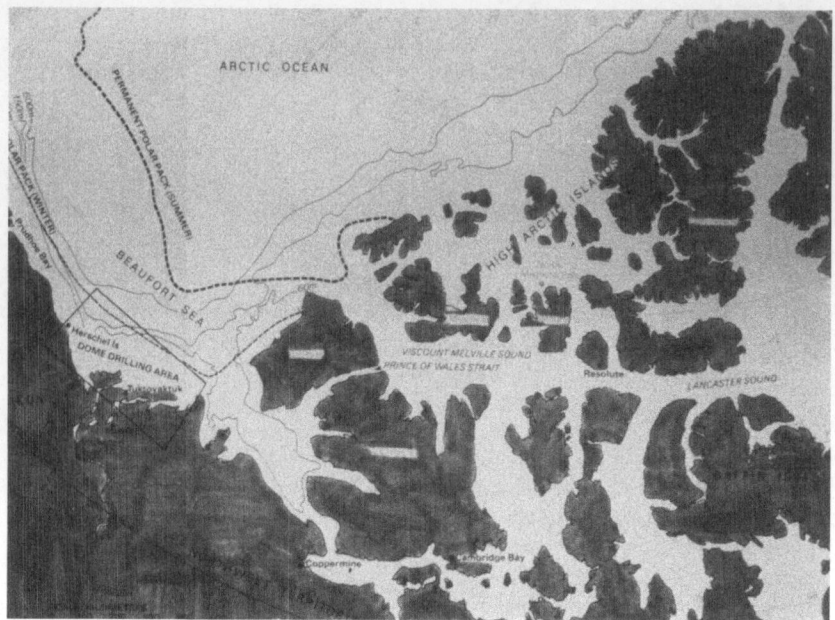

Figure 1 Location map.

instrumentation of well casings, development of under-ice marine placer mining machine, in situ tests on hydrates, and development of predictive capabilities, through research, on these variable parameters.

There is a shortage of skilled engineers to meet this challenge. Young engineers should be encouraged to develop career opportunities in arctic engineering. International cooperation is required in pursuing the many opportunities.

Figure 2 Shell's Seal Island in the Alaskan Beaufort.

Whereas the significant problem of offshore operations in the high Arctic are associated with low temperatures, the problems in the margins of the Arctic are linked to temperature contrasts between the atmosphere and the open ocean (Fig. 3). Four climatic factors of crucial importance to offshore activities are due to the combined effects of cold air and open water. These are drifting icebergs; icing caused by water vapour and seaspray; wind and wave loadings; low visibility.

Nowadays the whereabouts of drifting icebergs can be monitored and the risk of collision is much reduced. Towing is a practical countermission to avoid collisions with platform. Concepts for design of platforms to resist collision are being developed.

Icing has proved a major hazard to fishing vessels in northern waters. Its effect on platforms is less certain. In particular, the relationship between icing and elevation for various geometries and wave conditions is unclear.

Wind and water loadings are caused by large extratropical cyclones or by small "arctic storms". Cyclones are well monitored, but small arctic storms often arrive without warning because they are localised and small. The origin of arctic storms is obscure, but the few observations made indicate that these storms have at least as high wind speeds as those of the extratropical storm. The low air temperature adds simultaneously ice and extra wind loads to the wave current loads, thereby raising the total applied forces to extreme and severe levels. These arctic storms have repeatedly caused the loss of fishing vessels.

Figure 3 Dome's icebreaker *Kigoriak* at work.

Poor visibility is due to the lack of daylight, to frost 'smoke" in the winter period, and to a high incidence of fog in the summer. The conditions affecting visibility are relatively well understood.

HYDROCARBONS (OIL/GAS) POTENTIAL

Drilling commenced offshore on the east coast and onshore in the Mackenzie Delta and Arctic Islands in the 1960s. Potential hydrocarbon reservoirs have been discovered in these areas with the exception of the west coast and Hudson Bay. The main interest is in the search for oil for Canada to achieve self-sufficiency and security of supply. Estimates of the potential recoverable reserves have recently been published by the government agency, Geological Survey of Canada (GSC); see Table 1.

The GSC assesses the potential frontier oil reserve as 24.6 billion barrels. Commercial exploitation, however, depends on many factors: reservoir characteristics, environmental conditions, costs and the fiscal regime. Such locations as those of the Arctic Islands must wait for a favourable economic climate and development of technology. Hibernia oil, however, could be producing by the end of the decade. Average expectations of the GSC for gas reserves in these regions are 232 TCF (trillion cubic feet), of which the Venture one on the east coast is currently the most significant.

ENVIRONMENTAL CONDITIONS

The most prospective sedimentary basins in the Canadian offshore include the Beaufort Sea, the Sverdrup Basin, the Labrador Shelf, and the Grand Banks. Environmental factors govern the design and operation of exploration and production facilities.

Beaufort Sea

Sea ice is the most severe environmental constraint in the Beaufort Sea. The exploration areas of interest are usually ice-covered from October to June.

Table 1
Geological Survey of Canada potential oil and gas reserve estimates

	Average expectancy oil (billion BBL)	Gas (TCF)
Beaufort Sea—delta	8.5	66
Arctic Islands	4.3	80
East coast offshore	11.8	86
	—	—
	24.6	232

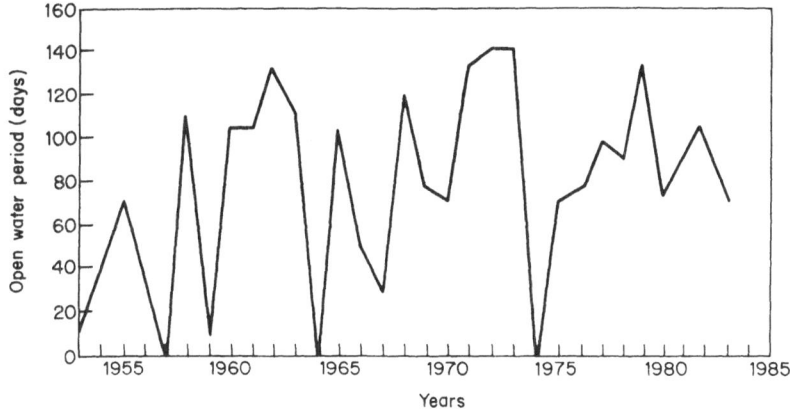

Figure 4   Beaufort Sea — open water days per year (location: 70°10′ N, 133° W).

Freezing-up commences in October and extends seaward some 50 km to depths of approx. 20 m. When the ice breaks up in June or July, open water periods are available, varying from a few days to 140 days (see Fig. 4). Movement of the pack ice usually averages 3–5 km per day generally in an east to west direction. Movements up to 30 km per day have been experienced in a northerly direction. The polar pack therefore generally rotates in a clockwise gyration. The 1983 summer was the most severe in nine years of offshore activity in the Beautfort Sea.

The variations in the ice conditions and limited open water season result in extremely costly exploration drilling. To extend the drilling season specially designed ice-reinforced drillships, support vessels and ice-breakers are used.

Waves, although not excessive, are serious to artificial islands from the action of scour in shallow waters and spray can cause problems of icing to drilling operations. The maximum 100 year significant wave height is 6 m. Unique geotechnical conditions also exist in the Beaufort Sea, including sea bottom gouging or scouring by grounded ice features, subsea permafrost, unstable foundations in certain locations and often a scarcity of construction materials (sand and gravel).

Arctic Islands

The northwestern region of the Arctic Islands, known as the Sverdrup Basin, appears to be the most promising for hydrocarbons. Here the most severe ice problem is the sea ice (mostly multi-year) up to 12 m in thickness.

The shelf falls off rapidly, with most drilling taking place in water depths of 300–360 m. Logistics are also affected by problems of cold temperature, windchill, prolonged darkness and remoteness. A fuller description of arctic ice conditions is given by Sanderson (this volume).

East Coast

The primary environmental problems on the east coast of Canada are icebergs, sea ice and high waves. Icebergs can run aground and thereby represent a hazard to wellheads, subsea templates and flowlines. Other seabed problems are glacial till, boulders, scouring and sediment transport.

As icebergs are the most severe problem they have been and continue to be studied. Estimates indicate that some 20 000 icebergs a year form primarily from the West Greenland glaciers. These drift through Baffin Bay and south along the Labrador Coast. The numbers that survive and cross the 48th parallel near the Hibernia location average about 400 per year. Due to icebergs and sea ice, drilling is severely restricted to summer months only in the Labrador Sea. With the hazard of icebergs, on rare occasions they have been towed to avoid collision with drilling rigs.

The icebergs diminish in mass and draughts with their drift southward and at Hibernia an estimated size is 10 million tonnes. Icebergs drift at an average rate of 4 km per day. No icebergs have been recorded at the Scotian Shelf or in area of Venture.

Sea ice is seasonal in the Labrador Sea, but at Hibernia on the Grand Banks ice only appears every few years and early in the year. Sea ice reaching Hibernia may be up to 1 m thick and rafted into ridges.

Floating drillships and semi-submersibles can be moved out of the path of encroaching sea ice. Fixed structures would have to be designed to withstand both sea ice and icebergs. Sea ice is not a problem on the Scotian Shelf.

EXPLORATION

To carry out offshore drilling in the arctic regions, the following techniques and equipment have been developed: artificial islands; ice pads; caissons; drillships.

Artificial islands

In the Beaufort Sea, although permits were granted in the early 1960s and seismic surveys were conducted, the first offshore drilling was undertaken in 1973 from a gravel island constructed over a two-year period in 3 m of water. To date, 20 islands have been constructed and drilling carried out in 20 m. New ideas have been proposed and others are under construction and planned. All but one of the islands have been expendable and are eroded below sea level by wave and ice action.

Ice pads

In the Arctic Islands region, each year since 1974 drilling has successfully been carried out off thickened ice in water depths to 330 m. The ice pad is built up to 6 m in thickness. In each case, flooding of the first year or multi-year ice was carried out as soon as possible after freeze-up, usually in November. A second well pad for relief well purposes is built nearby after the first pad is complete.

A further innovation introduced is the use of large urethane foam blocks within the ice pad to increase the load carrying capacity and reduce the construction time. The foam is not penetrated by water and can be disposed of by burning in summer. Drilling must be completed in April to permit time for a relief to be drilled if necessary.

Prior to drilling off the ice, measurements of ice movement are made for at least two winters. A lateral displacement of only 5% of water depth can be tolerated by the drilling equipment. Only one well to date has had to be respudded due to excessive ice movement. One test well was actually completed with a seabed completion and subsea flowline to shore.

A further idea being developed by Norwegians is the formation of a gravity structure from artificially produced ice contained within a steel or concrete frame. One advantage of artificial ice over the real stuff is that the former does not suffer from "creep" or erosion and can be used as a platform for production as well as exploration drilling. Artificial ice can be made cheaply and mixed with something like sawdust to provide a very high tensile strength.

Caissons

The advantage of caissons is their flexibility. They can be moved by de-ballasting, and easily placed on another location. The amount of sand and gravel required is much less than that required to make an artificial island. Basically, once the caisson is placed on the berm, sand and gravel are pumped in to fill the space remaining between the berm and surface to stabilise the platform (Fig. 5). The operators tend to use these facilities as the water depth increases, whereas the artificial islands are constructed in shallow water. By using caissons the cost of dredging materials can be reduced by up to 80%; even so, caissons cost around $80 million each.

The first concrete caisson comprising four units of the open box type was installed in 1981 on the Dome "Tarsuit" field and stands in 22 m of water. Some 1.3 million

Figure 5   Artist's impression of Esso Canada's caisson island in the Canadian Beaufort.

Figure 6   Artist's impression of Gulf's arctic caisson in the Canadian Beaufort.

cubic metres of sand/gravel was required to fill the caisson. This compares favourably with Esso's Issungnuk Island, which needed 5 million cubic metres of ballast.

For even deeper waters, the options for designs of caissons are still open. Barges with pre-installed drilling and production equipment which will resemble large gravity caissons are an alternative option. A segment of an oil tanker was equipped with concrete between a double shell hull. It was reinforced to withstand ice and equipped with drilling equipment, and ballasted onto a subsea berm. The caisson known as the "Canmar SSDC" (semi-submersible drilling caisson) performed successfully through the winter of 1982/83 at two locations.

Gulf has also taken the caisson concept for development, and a design in steel has been constructed costing US $140 million. It will be used in the Beaufort Sea. The 33 000 tonne structure will measure 100 m × 100 m in the lower section and 80 m × 80 m in the upper section. Its height will be 20 m and it will be housed in depths of water up to 20 m (Fig. 6). As in previous installations, it will be floated onto a simple foundation. When work is completed, it will be de-ballasted to move on to another location.

Drillships

In parallel to the caisson concept is the development of a protected floating rig. To date, operators have been using specially converted drillships with ice-breaking

Figure 7   A design of a special conical-hulled vessel which is capable of operating in deeper water.

capabilities. In 1976 three reinforced drillships started operating in the Beaufort Sea and a fourth was added later. These have drilled successfully throughout the limited seasons. Some years the season has been extended by using a Canadian Coast Guard vessel and specially built ice-breaker supply vessels. These drillships have operated in water depths ranging from 25 to 75 m. Most of the drilling on the Grand Banks in the region of Hibernia has been carried out by drillships which can move out of the way of drifting icebergs (see Fig. 7). The conical drilling unit called "Kalluk" keeps ice away from the all important drill string while the rig lies anchored with twelve 85 mm lines. The shape of the hull is such that ice will be broken in bending by the downward pressure of the vessel. A shroud on the bottom of the hull deflects the broken ice away from the mooring lines.

OUTLOOK ON DEVELOPMENT

For the development of oil and gas off Labrador, Newfoundland and Canada's East Coast, environmental conditions appear less of a problem, but there are still difficulties. The capsizing of the giant drilling semi-submersible "Ocean Ranger" three years ago has stimulated a major rethink of techniques for development. Icebergs are the main problem and typical fields such as Hibernia, classed alongside the North Sea's Forties in terms of size, lie directly in their habitual paths. A variety of options such as the Norwegian concept are being considered, but it appears unlikely that semi-submersible platforms tied to subsea production equipment will be used until the lessons of the "Ocean Ranger" have been fully evaluated. An additional factor is the need to provide protection for subsea well heads and associated equipment as some of the giant bergs dig trenches along the sea floor.

In the Beaufort Sea production platforms in shallow water could be gravel islands. In deeper water, where islands are not suitable, several concepts are being examined, including steel or concrete caissons set on subsea berms or direct on the seabed. Compliant structures (which could yield slightly) are also being considered. Subsea satellites are possible but the ice depths and effects of sea bottom gouging would necessitate either excavation below the seabed or wellhead protection.

Valuable information will be obtained from the performance of the Beaudril Mobile Arctic Caisson (a large steel floatable caisson) and Global Marines' Concrete Island Drilling System (CIDS) in offshore Alaska, both currently under construction. The use of tankers for transport of the oil rather than submarine pipelines would require different platforms in deep water to permit offshore loading.

In the Arctic Islands, the Cisco field is likely to provide the first oil production, with some 300 million barrels of recoverable reserves. This is, however, unlikely to occur until about 1995. A possible production system comprises subsea completions with individual flowline connections to manifolds and a larger line carrying oil to shore. Several transportation options have been considered, such as ice-breaking oil tankers, submarine tankers, submarine pipelines.

In the Arctic Islands, discoveries have been primarily gas fields. Some of these, such as Drake Point, Hecka and Whitefish, are large. GSC lists discovered gas reserves as nearly 13 TCF. One offshore, Drake Point, a subsea well in 53 m of waters was completed off the ice and flare tested.

Two major transportation studies have been carried out for gas. One, by the Polar Gas Project, evaluated overland and submarine lines to the south. The other, Arctic Pilot Project, involves transporting LNG by ice-breaking tankers. With the low demand for natural gas, there is considerable doubt regarding the development of these reserves for Arctic Islands.

Offshore the East Coast for production of oil from Hibernia, two concepts are being evaluated, one based on a floating production system and the other with a concrete gravity structure. Gas from the Venture field will likely be brought ashore by submarine pipeline with conventional technology for production facilities on platforms.

CONCLUSIONS

The large hydrocarbon potential in the arctic regions offshore Canada will ensure a continued activity of exploration and delineation drilling. Data on ice mechanics,

Figure 8   Dome's Tuk Base in the Canadian Beaufort Sea.

geotechnical and physical environmental parameters have been gathered and are used for design of production systems. Technology for producing and transporting oil and gas is being developed. Exploration awaits the economic viability of these discoveries and further advances in subsea technology.

By the late 1990s, offshore arctic developments could provide Canada with extremely valuable sources of energy and revenue. Most of the developments discussed here should be onstream by then and Europeans, as well as North Americans, will doubtless be involved as customers, participants, constructors and consultants (Fig. 8).

ACKNOWLEDGEMENTS

The author would like to express his thanks to the following organisations who have helped provide background for this presentation: ECOR "Arctic Ocean Engineering" session A18 at the Joint Oceanographic Assembly, Halifax, Nova Scotia, August 1982; Society for Underwater Technology international seminar "Operations in Arctic Waters" London, December 1982; Gulf Canada Resources Inc., Oceanographic Research — Brian Wright, Offshore Engineering Journal — Adrian Cottrill.

24

Ice Conditions in the Arctic

T. J. O. Sanderson, New Technology Division, BP Petroleum Development Ltd,
London, UK

Operations in the Arctic have to cope with seasonal sea ice, old multi-year ice, icebergs and ice islands. The oil industry has carried out a large amount of research to quantify the hazards due to these features and now builds structures to withstand them. Underwater operations in Arctic regions have to recognize the risks of cold and damage by moving ice features but the problems are not insuperable.

INTRODUCTION

Ice conditions in the Arctic are by no means uniform: ice varies significantly in type as well as in distribution, and conditions are strongly dependent on location and time of year. Here I present a general survey of the forms which ice takes in the Arctic and the hazard it presents to offshore operations. In particular, I shall examine the difficulties it poses for diving and other underwater operations.

ICE CONDITIONS

There are three quite distinct types of ice in the Arctic oceans: first-year ice, multi-year ice and icebergs. Their occurrence in the north polar regions is shown schematically in Figure 1.

First-year ice

Each winter season most of the seas north of latitude approximately $58°$ N are frozen (Fig. 1). The notable exception is the northeast Atlantic Ocean (especially the Norwegian Sea) which is kept largely ice-free, owing to the Gulf Stream. The east coast of Canada, on the other hand, experiences more southerly encroachment of first-year ice, as a result of prevailing northerly winds and the cold Labrador Current from Greenland and the Arctic Islands. Newfoundland experiences sea ice and icebergs, yet it is the same latitude as Paris.

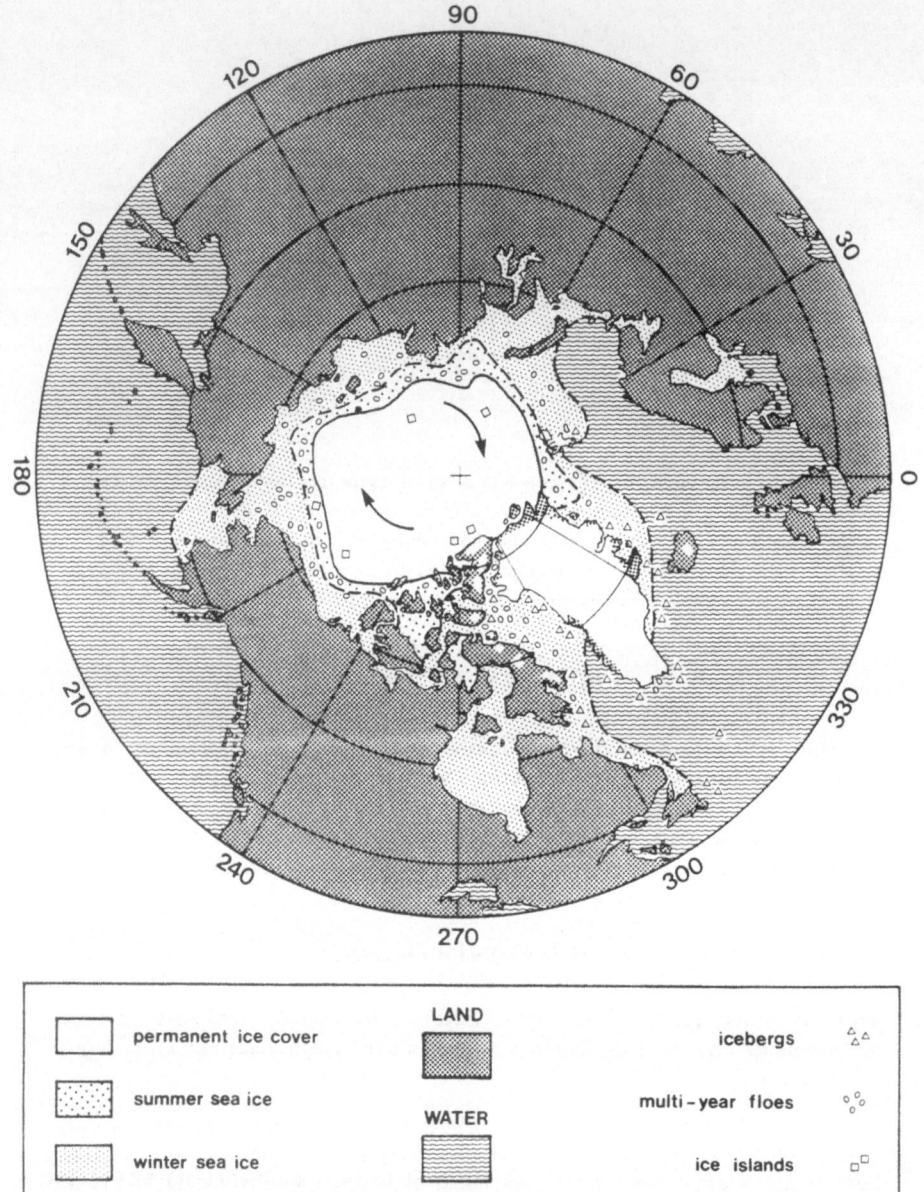

Figure 1 Schematic map of Arctic ice conditions. Dashed lines represent average maximum extent and average minimum extent of sea ice. Permanent ice cover is shown as pure white. Ovals represent multi-year floes, triangles represent icebergs and squares represent ice islands. Representation of these discrete ice features is entirely schematic.

Figure 2   Close-up of compression ridge, Canadian Beaufort Sea, March 1983. The block in
the foreground shows the full ice thickness of 1.5 m.

New first-year ice forms during the winter in Arctic regions and melts during the
following summer. Offshore Newfoundland, it is generally present for several weeks
during February, March and April. In the Beaufort Sea to the North of Alaska and
Canada, it is present for most of the year, disappearing for about two months only,
between August and October.

First-year ice is between 0.3 and 2.5 m in thickness, and typically 1.5 m thick in
the Beaufort Sea. It is generally flat and level, except where movement has caused
ridging (Fig. 2). Such ridges are typically of 2–4 m height above sea-level and
generally have underwater "keels" of depth about 10 m below sea-level (Brooks,
1983). They can be much larger than this, and may reach the seabed in water depths
of 30 m or more.

The material properties of pure freshwater ice are complex, but well understood
(Glen, 1955; Sinha, 1983). It is a creeping material under applied stress, showing
initial immediate elastic deformation, subsequent transient recoverable deformation
and finally settling down to a steady-state constant creep rate. This creep rate follows
Norton's Law, with a power-law coefficient of 3. At high deformation rates,
however, ice is extremely brittle: it fractures very easily and has a lower fracture
toughness than glass (Goodman and Tabor, 1978). These unusual properties make it
a difficult engineering material.

Sea ice has essentially the same properties as pure ice, but with the important
difference that it contains salt water (Weeks and Ackley, 1982). Although the
seawater from which the ice forms has a salinity between 20 and 35 ppt (parts per
thousand), the ice itself excludes salt water as it grows, so that the final consolidated
sea ice cover has a gross salinity of about 5–10 ppt. It is not evenly distrubuted but
concentrated in discrete pockets of brine, generally in the form of vertical cylindrical
channels. The amount of brine is strongly dependent on temperature: it is a complex
thermodynamic balance between the solid ice phase and the concentrated liquid salt

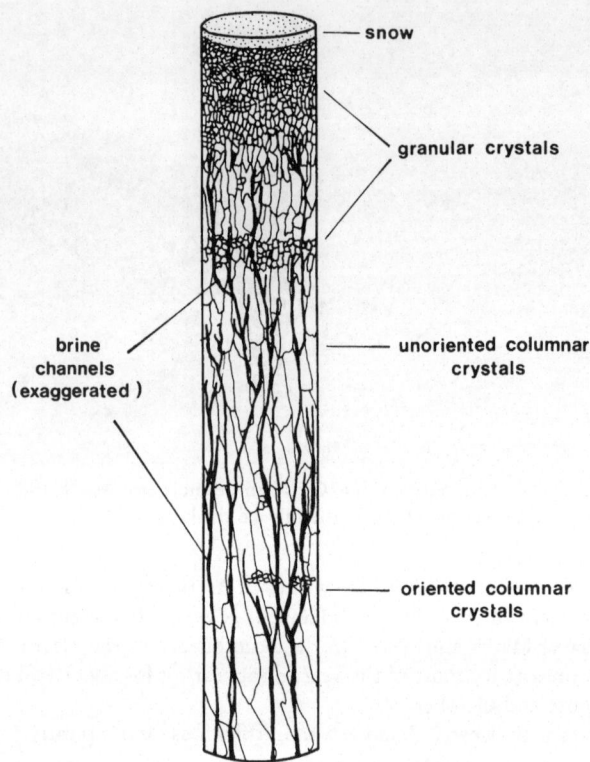

Figure 3 Schematic section through coastal sea ice. The upper layer is of granular crystal structure and lower layers are predominantly columnar with some granular inclusions. Brine pockets are more prevalent in the warmer lower layers.

solution. A schematic diagram is shown in Figure 3. This also shows variation in ice crystal type: in the surface layers, down to 0.20–0.50 m below the air–ice interface, the ice is generally small-grained, equi-axed granular ice (grain-size 1–3 mm), with essentially isotropic mechanical properties. This is ice which has grown quickly during the early ice formation season. Below this layer, when ice is growing more slowly, long columnar ice crystals of typical dimensions 10 × 50 × 200 mm may form. They have predominantly horizontal c-axes, which in coastal areas are also preferentially aligned with seawater currents (Weeks and Gow, 1978). Far from the coast, ice movements and dynamic action seriously disturb crystal growth patterns, resulting in a more complex picture. Columnar ice is anisotropic. However, it is also generally weaker than granular ice, partly as a result of its lower position in the ice cover: it is generally warmer (and therefore softer) and contains a higher proportion of liquid brine. During winter, sea ice may have a temperature of −25 °C at its upper surface and of −2 °C at its lower surface. It is, therefore, much stronger towards the top. Late in spring, the ice becomes quite rotten as it warms up throughout to −2 or −5 °C, and brine pockets enlarge.

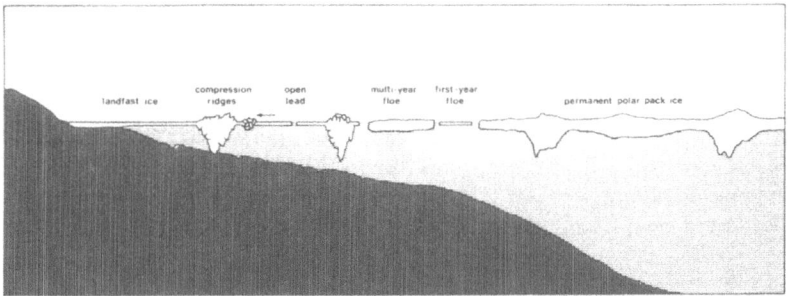

Figure 4 Schematic section of ice conditions in the Beaufort Sea. Landfast ice remains stable, attached to the shore, while the permanent polar pack ice rotates in the Beaufort Gyre. Between them is a zone of intense ice activity.

During most of the winter season, first-year ice close to the coast is effectively stable and landfast. It rises and falls with the ocean tides, but hardly moves horizontally. It is generally flat, level and undisturbed until break-up at the end of winter. The extent of this "fast-ice" zone depends strongly on local geography and regional wind patterns, but it is typically 50-100 km in, for instance, the Beaufort Sea. In the Arctic Islands of Canada, almost all the ice is landfast.

Beyond the fast ice, wave action or large-scale pack-ice movements give rise to moving, disconnected ice flows. The boundary between the fast ice and the pack ice is generally the scene of intense ice action and ridge-building. Where these ridges ground out on the seabed they stabilise the fast ice. The general picture is shown in section on the Beaufort Sea in Figure 4.

Multi-year ice

Further north towards the pole, from approximately 75° N northwards, the ice cover is permanent. It consists of old ice, or multi-year ice (Fig. 5) and is up to 10 or 12 years old (Anon, 1981a).

Figure 5 Surface of a multi-year floe, Kennedy Channel, August 1983. Note the uneven deteriorated surface.

This ice develops from first-year ice by a combination of ridging, rafting, melting and re-freezing (Fig. 6) and is generally 4–6 m in thickness. Multi-year ridges are thicker than this, and may exceed 30 m in thickness (Wadhams and Horn, 1980). The polar pack is not a solid mass, but always contains sporadic patches and leads of open water. It rotates clockwise under wind action of the Beaufort Cyre about a point at roughly 80° N, 150° W, and completes one revolution in about 7–10 years. Its drift speed is normally about 2.5 km/day at its southern rim, but this can reach 25 km/day during spring (Anon, 1981).

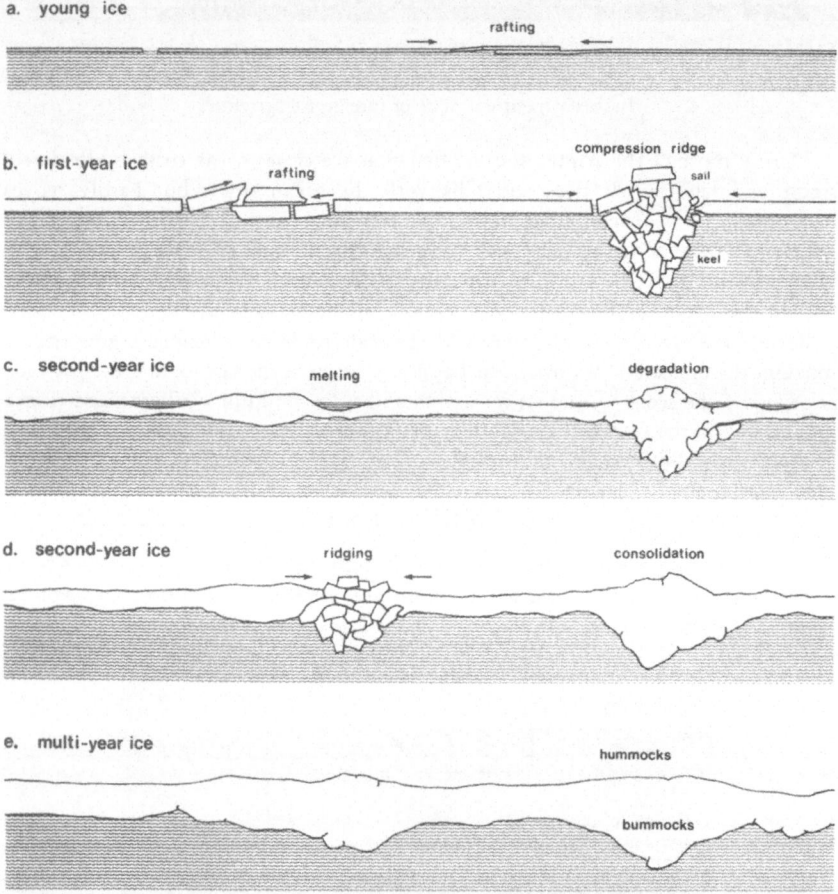

Figure 6 The development of ice cover: (a) Young ice 30 cm thick, early winter. (b) First-year ice, 1.5 m thick, mid-winter/early spring. Ice action causes ridging and rafting. (c) Second-year ice, variable thickness, late summer; surface and bottom melting has occurred. (d) Second-year ice, during its second winter, undergoing refreezing, accretion and further ice action. (e) Mature multi-year ice 4–6 m thick, 7–10 years old containing old smoothed ridges, hummocks and bummocks. This diagram shows normal conditions. Extremes exceed quoted thicknesses.

As a result of its chaotic formation process (Fig. 6), multi-year ice is highly variable in composition: it consists of a conglomerate of lumps of granular and columnar ice with interstitial refreezing of melt water. It is in general rather less saline than first-year ice (1–5 ppt), since brine pockets migrate outwards during the summer along vertical thermal gradients. Multi-year ice is not significantly stronger or weaker than first-year ice (Weeks and Ackley, 1983).

Icebergs and ice islands

Icebergs generally originate from the Greenland Ice Sheet (Fig. 1). This ice sheet is a state of dynamic equilibrium, receiving snowfall accumulation over its entire area, flowing outwards under its own weight, and shedding icebergs at its perimeter. Some smaller bergs come from Franz Josef Land, Svalbard and the Queen Elizabet Islands (Mathisen, 1983). Icebergs vary in diameter from 15 m to more than 200 m and weigh between 1000 000 tonnes and more than 10 million tonnes. A typical iceberg weighs about 2 million tonnes. The motion is principally determined by medium-depth ocean currents, and the area of highest concentrations of icebergs lie along Davis Strait and the east coast of Labrador. Several hundred icebergs each year travel as far south as $48°$ N to the Grand Banks, Newfoundland (Lewis and Benedict, 1981), and cause hazards to shipping and oil company operations (Fig. 7).

Ice islands are an Arctic curiosity: they are large masses of flat freshwater ice which have calved from the ice shelves of Northern Ellesmere Island, notably the Ward Hunt Ice Shelf. They are 60–80 m in thickness and the largest observed was 750 km^2 in area (Koenig et al., 1952). The most famous of these, named T-3, calved in 1946 and was used as a scientific research station for more than 20 years. There are at present at least 40 ice islands in excess of 400 m diameter drifting with the Arctic pack ice (Spedding, 1976). Of these, perhaps ten have diameters of several kilometres and they may periodically be released into the Beaufort Sea. They are the counterpart of East Coast icebergs, but very much rarer.

Figure 7   Iceberg offshore Newfoundland. It is being towed by ice surveillance vessel *Hudson Service*.

HAZARDS

The offshore oil industry has to cope with designing structures to withstand the hazards of these ice conditions. Since the early 1970s intense research activity has been initiated and financed by the industry through two associations: the Alaskan Oil and Gas Association (AOGA) and the Arctic Petroleum Operators Association (APOA). Much of this work is proprietary, and not yet in the public domain. Here I simply show examples of engineering solutions to the problem.

Sea ice loads

A structure generally has to withstand two types of loading from sea ice: static pressure of vast expanses of wind-driven sea ice, and dynamic impact of large current-driven multi-year floes. Both these cases are complicated by the occurrence of ridges. Loads on actual structures have been measured at full-scale (Exxon, 1979; Danielewicz and Metge, 1981, 1982; Blanchet, 1983; Sanderson *et al.*, 1983). Typical design loads lie in the range 50 000–250 000 tonnes.

In certain regions of the Arctic, where the ice does not move very much, these loads can be avoided by means of drilling from the ice itself or from a structure locked into the ice. In the Canadian Arctic Islands, Pan Arctic have carried out exploration drilling from artificial floating ice platforms (Fig. 8). These structures move with the ice and are suitable only where annual ice movement is less than about 5% of the water depth (Hood *et al.*, 1981). They are made from sea ice artificially thickened to about 7 m thickness and have successfully been used in sheltered Arctic Island regions in water depths up to 365 m.

This simple approach cannot be followed in regions such as the Beaufort Sea, where ice may move hundreds of metres per day. Exploration in the shallow waters of the Beaufort Sea has been carried out from simple gravel islands in water depths up to about 20 m. Such an island requires several million cubic metres of fill material. In greater depths than this, floating steel or concrete caisson structures have been ballasted onto sand berms, forming a potentially mobile drilling platform. Tarsiut Island (Gulf/Dome) was built in 1981 and used four independent concrete caissons (Anon, 1981b) in 22 m water depth (Fig. 9). Another elegant and inexpensive solution was Dome/CANMAR's use of an oil tanker (VLCC), strengthened

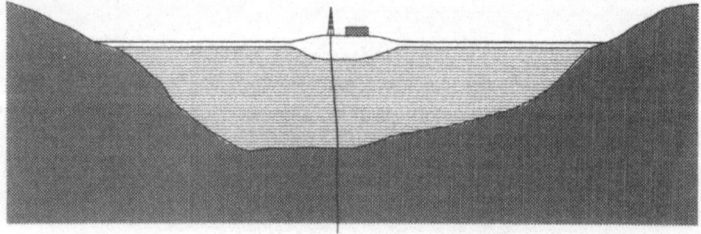

Figure 8 Schematic diagram of a floating artificial ice platform as used in the Canadian Arctic Islands. They are about 7 m thick, made of artificially thickened sea ice, and can cope with lateral movement of about 5% of the water depth.

Figure 9 Tarsiut Island (Gulf/Dome), a concrete caisson-retained sand island in 22 m water depth in the Beaufort Sea, June 1982.

internally and then placed on a sand berm in 31 m of water (Fig. 10). Around this structure (Cottrill, 1982) artificial grounded ice pads have been constructed by bulldozing and water-spraying. They afford protection to the structure.

During this year and next year, more ambitious mobile conical drilling units and caisson structures will come into service (Cottrill, 1983).

Iceberg loads

Impact of an iceberg or ice island weighing several million tonnes and travelling at several knots presents a severe problem to an offshore structure (Dunwoody, 1983). No-one yet knows what load such an impact would impose on a rigid structure. Preliminary estimates are typically in the range 0.5–1 million tonnes (Cammaert

Figure 10 The Single Steel Drilling Caisson (SSDC) at Uviluk, Canadian Beaufort Sea, operated by Dome/CANMAR. It is surrounded by an artificial grounded ice pad.

and Tsinker, 1981). It is possible, but extremely expensive, to design a structure to such a specification: it would have to be huge and heavy.

Other solutions are available. Where iceberg impact is expected to be rare (say, every few years), a mobile floating structure can be used for operations: it simply disconnects and moves off station if an impact is threatened. This method is already used for exploration drilling in Hibernia, offshore Newfoundland. The other feasible solution is to absorb impact energy gradually by means of a "fender" structure, which fails, slides or deforms on impact (Anon, 1982) allowing energy to be dissipated over an appreciable distance.

Scouring

Large ice features such as multi-year ridges, first-year ridges and icebergs may frequently ground out and gouge channels into the seabed. This is a serious risk to any subsea installation, particularly export pipelines. Gouges on the seabed have been measured and mapped (though they are often difficult to date) (Pilkington and Marcellus, 1981; Lewis and Barrie, 1981) and statistics suggest that a pipeline should be buried to about 5 or 6 m depth for the Beaufort Sea, and to approximately 10 m for offshore Labrador and Newfoundland.

DIVING AND UNDER-ICE OPERATIONS

Underwater operations in arctic environments have three novel problems to cope with: the cold, access through solid static ice cover to the water, and damage caused by moving ice features.

Cold

Although Arctic waters are colder than, for instance, the North Sea, they cannot be below about $-2\ ^\circ$C, the freezing point of typical seawater. However, these temperatures do make use of wet suits marginal, except for very short periods (Jenkins, 1975). The problems of cold water can be overcome fairly straightforwardly by use of the variable volume dry suit or conventional warm water hosing from the surface. There are some minor engineering problems to be overcome with the use of such equipment in cold conditions.

At deeper diving depths (150 m plus), the cold ambient water conditions exacerbate the problems of respiratory heat loss when using oxy-helium mixtures (Nuytten, 1973). Air temperatures may, of course, be well below water temperatures, perhaps $-40\ ^\circ$C, and also windy. This means that divers' access to the water should always be sheltered and heated. Heating also helps avoid mechanical problems with surface-based regulators and life-support equipment (McLaren and Frobel, 1975).

Access through ice

Where sea ice cover is statutory and solid, diving may be carried out from the floating ice as a base. The ice may be anything between 50 cm and 15 m thick.

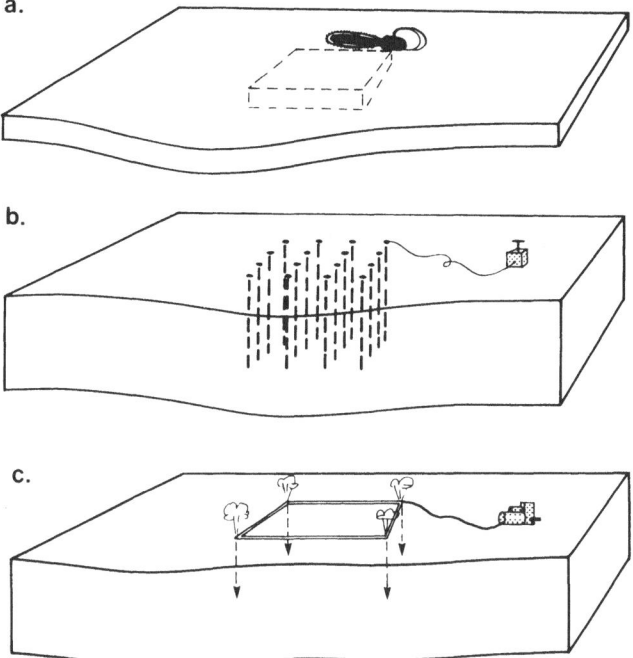

Figure 11 Methods for cutting an access hole through ice: (a) cutting; (b) blasting; (c) melting. See text for fuller explanation.

There are three ways of cutting through the ice (Jenkins, 1975; Jess, 1983) (Fig. 11):

(i) *Cutting*. If the ice cover is quite thin (up to about 1 m thick) a hole can be cut using mechanical means such as handsaw, pick, ice chipper or chainsaw.

(ii) *Blasting*. Thicker ice can be blasted using, for instance, seismic explosives. This is remarkably efficient. A clean diving hole can be made through even very thick ice (up to at least 15 m), by drilling a series of holes in an even lattice, packing with explosive and then simultaneously detonating. Holes should be of about 5 cm diameter, and drilled vertically to within 10–20 cm of the ice–water interface at the bottom. Holes should be spaced in a rectangular lattice at intervals of 30 cm, and 1 inch (2.5 cm) sticks of explosive have been found to be suitable (Jess, 1983). After explosion some floating blocks of ice need to be cleared from the hole, but most of the ice is scattered or vaporized. The hole may then be covered after use and kept open by clearing once per day.

(iii) *Melting*. If, for environmental or safety reasons, explosives are not appropriate, melting can be used to cut a block from the ice. The scheme is shown in Figure 12, and uses a steam generator to pass steam through a square frame made of 2 m lengths of 2.5 cm diameter water-pipe. Guidelines are required to stabilize it as it melts its way through the ice. Typical melting rate is about 1 m/h. The

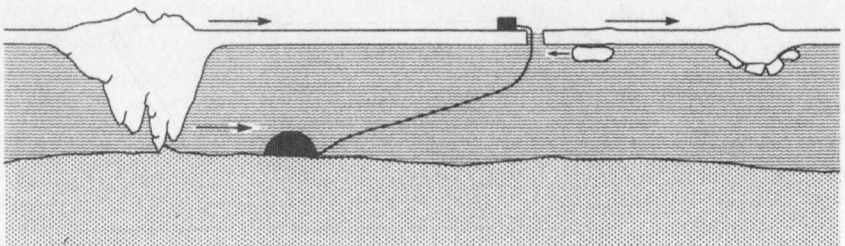

Figure 12   Risks of diving under moving ice: motion away from worksite, damage to
communication lines, and scouring by deep ice features.

free-floating cube can then be lifted and removed by crane (Jess, 1983). It might
weigh 5–10 tonnes.

In general, where a major operating structure is involved in Arctic waters it will not
be necessary to dive from the ice: access should be possible from a moon-pool on a
ship (Anon, 1977) or from a leg of a semi-submersible (Anon, 1981c) or fixed
platform. Access can be made directly from beneath the ice level.

Moving ice

If the ice cover is moving at a rate exceeding a few metres per hour, then operations
under ice are hazardous.

Surface support

If surface support is carried out from the ice or from a vessel moving with the ice,
there is the obvious problem of movement relative to the area of subsea operation. In
general, operations in moving ice should be carried out from a fixed structure with
access beneath the mobile ice. In deep waters, diving may be possible from
submarine.

Damage to communication lines

Ice in motion is subject to impact, splitting and crushing. Any line passing through
mobile ice runs the risk of damage or destruction. There is also a risk of buoyant
rubble trapped beneath consolidated ice moving up into any access hole, damaging
lines or even sealing it off (Fig. 12).

Scouring

Any operation using remotely operated vehicles or subsea habitats placed on the
seabed must take account of the risk of impact from major ice features grazing the
seabed (Fig. 12). This is normally a remote risk, but becomes considerable if
equipment is left for a long period on the seabed.

CONCLUSIONS

Underwater operations beneath Arctic ice cover differ in several respects from standard underwater operations. If these differences are not properly appreciated then operations may be hazardous. However, the problems are essentially minor, and not insuperable. If properly approached, underwater operations in the Arctic should be no more hazardous than elsewhere in the world.

REFERENCES

Anon, 1977. Rough weather diving system, *Ocean Industry*, February, 62–63.

Anon. 1981a. Arctic ice: how it moves, the forms it takes, in *Offshore Engineer Arctic Supplement*, August, 21–26.

Anon. 1981b. The building of Tarsiut, *Beaufort*, 1(2), 8–11.

Anon. 1981c. Semi-rub granted ice classification, *Offshore Engineer*, September, 181.

Anon. 1982. Platform will absorb iceberg impact, *Ocean Industry*, December, 87.

Blanchet, D. 1983. Interpretation of Tarsiut Instrumentation, Winter 81–82, Load Results and Analysis, *Report No. 7(b) under APOA Project No. 197* (Restricted).

Brooks, L. D. 1983. Statistical analyses of pressure ridge keel definitions and distributions, in *POAC '83 Conference, Helsinki, 5—9 April 1983, Proceedings*, Vol. 1, pp. 69–78.

Cammaert, A. B. and Tsinker, G. P. 1981. Impact of large ice floes and icebergs on marine structures, in *Proceedings of the symposium: Production and Transportation Systems for the Hibernia discovery, February 16—18, St. Johns, Newfoundland*, Petroleum Directorate, Government of Newfoundland and Labrador, pp. 189–203.

Cottrill, A. 1982. Oil tanker turned drilling caisson brings revolutionary concept to Beaufort Sea, *Offshore Engineer*, August, 17.

Cottrill, A. 1983. Beaufort pioneers pile on pressure as production looms, *Offshore Engineer*, Arctic Review, January, 29–33.

Danielewicz, B. W. and Metge, M. 1981. Ice forces on Hans Island, August, 1980, *APOA Project No, 180* (Restricted).

Danielewicz, B. W. and Metge, M. 1982. Ice forces on Hans Island, 1981, *APOA Project No. 181* (Restricted).

Dunwoody, A. B. 1983. The design ice island for impact against an offshore structure, in *Offshore Technology Conference, 1983, Houston, Paper No. OTC 4550*, pp. 325–330.

Exxon Company, USA. 1979. *Technical seminar on Alaskan Beaufort Sea gravel island design*, Presented by Exxon Company USA, Anchorage, Alaska, 15 October 1979.

Glen, J. W. 1955. The creep of polycrystalline ice, *Proceedings of the Royal Society of London, Ser. A*, 228, (1175), 519–538.

Goodman, D. J. and Tabor, D. 1978. Fracture toughness of ice: a preliminary account of some new experiments, *Journal of Glaciology*, 21 (85), 651–660.

Hood, G. L., Masterton, D. M. and Watts, J. S. 1981. Installation of a subsea completion in the Canadian Arctic Islands, *Journal of Canadian Petroleum Technology*, October–December, 41–52.

Jenkins, W. T. 1975. Diving in the harsh Arctic environments, *Ocean Industry*, 10(4), 363–374.

Jess, P. E. 1983. Personal communication (Dome Petroleum Ltd., Calgary).

Koenig, L. S., Greenaway, K. R., Dunbar, M. and Hattersley-Smith, G. F. 1952. Arctic ice islands, *Artic*, 5(2), 66–103.

Lewis, C. F. M. and Barrie, J. V. 1981. Geological evidence of iceberg groundings and related seafloor processes in the Hibernia discovery of Grand Bank, Newfoundland, in *Proceedings of the symposium: Production and Transportation Systems for the Hibernia discovery, February 16—18, St Johns, Newfoundland,* Petroleum Directorate, Government of Newfoundland and Labrador, pp. 189–203.

Lewis, J. K. C. and Benedict, C. P. 1981. Burial perameters: an integrated approach to overdesign, in *Proceedings of the Symposium: Production and Transportation Systems for the Hibernia Discovery, February 16–18, St John's Newfoundland,* Petroleum Directorate, Government of Newfoundland and Labrador, pp. 189–203.

Mathisen, J.-P. 1983. Ice and environmental data from the Arctic, in *POAC '83 Conference, Helsinki, 5—9 April 1983, Proceedings,* Vol. 1, pp. 158–171.

McLaren, P. and Frobel, D. 1975. Under-ice scuba techniques for marine geological studies, *Geological Survey of Canada,* Paper 75–18.

Nuytten, P. 1973. Deep helium diving in the High Arctic, *Offshore Technology Conference, 1973, Dallas, Paper No. OTC 1779.*

Pilkington, G. R. and Marcellus, R. W. 1981. Methods of determining pipeline trench depths in the Canadian Beaufort Sea, in *POAC '81 Conference, Quebec, August 1981, Proceedings.*

Sanderson, T. J. O., Child, A. J., Duckworth, R., and Westermann, P. H. 1983. *BP total ice load programme, Tarsiut Island, 1983,* BP Petroleum Development Limited Report to programme participants (restricted).

Sinha, N. K. 1983. Creep model of ice for monotonically increasing stress, *Cold Regions Science and Technology,* 8, 25–33.

Spedding, L. G. 1976. Ice island count southern Beuafort Sea, *APOA Project No. 99.*

Wadhams, P. and Horne, R. J. 1980. An analysis of ice profiles obtained by submarine sonar in the Beaufort Sea, *Journal of Glaciology,* 25(93), 401–424.

Weeks, W. F. and Ackley, S. F. 1982. The growth, structure, and properties of sea ice, *USACRREL Monograph 82—1.*

Weeks, W. F. and Ackley, S. F. 1983. Recent advances in understanding the structure, properties and behaviour of sea ice in the coastal zones of the polar oceans, in *POAC '83 Conference, Helsinki, 5—9 April 1983, Proceedings,* Vol. 1, pp. 25–41.

Weeks, W. F. and Gow, A. J. 1978. Preferred crystal orientations along the margins of the Arctic Ocean, *Journal of Geophysical Research,* 84(C10), 5105–5121.

25

Development and Operation of ROVs

H. R. Talkington, Director Engineering and Computer Sciences, Naval Ocean Systems Center, San Diego, CA, USA

A series of undersea remotely controlled vehicles has been developed. Some of these vehicles are described and details given of the part one played in the rescue of the PISCES III.

INTRODUCTION

Although divers and manned submersibles can, properly equipped, perform almost any conceivable mission in the ocean, in many cases they are limited in depth and undersea operating time, and perform their operations at great cost and some risk to their personnel. Thus a program was initiated at our Center for the development of tethered, unmanned, remotely controlled work systems. These systems have proven to be very reliable ocean engineering tools to the oceanographic investigator, the offshore industry, and the Navy. The continuing research at the Center assures that each generation of such systems is significantly improved in operational capabilities, while operational costs and shipboard support requirements are substantially decreased.

A series of undersea remotely operated vehicle types has been developed, from the small SNOOPY to the larger, deeper operating CURV. The following will describe the characteristics of these types of undersea vehicles and will present an example of a particularly difficult operation, the rescue of the PISCES III.

CABLE-CONTROLLED UNDERWATER RECOVERY VEHICLE (CURV II)

The improved CURV II is an unmanned tethered submersible capable of operating to 4000 feet. It is the successor to CURV I, which recovered the H-bomb off the coast of Spain in 1966. In configuration, CURV II is typical of most unmanned vehicles; it has an open rectangular framework to support the sensors and tools, two horizontal propulsion motors to drive and steer the vehicle, one vertical motor for close vertical control, and buoyancy of approximately 25 pounds. The vehicle is 6.5 feet high, 6.5 feet wide, and 15 feet long, weighs 3000 pounds in air, and operates at submerged speeds to 4 knots. The sensors include an Applied Research Lab active/passive sonar, acoustic altimeter and depthometer, compass, two Hydroproducts television cameras with lights, and an EG & G 35-mm still camera with strobe. One major feature of all surface-powered vehicles is that their bottom time is restricted only by the ability of the surface support craft to remain on station.

The CURV II system consists of the vehicle, control cable, and control console. Although it often operates from the YFNX 30 surface support ship, the system can be air transported to operate from any surface ship available. It is primarily used to support test operations on the Center test ranges.

CABLE-CONTROLLED UNDERWATER RECOVERY VEHICLE (CURV III)

CURV III (Fig. 1), a greater-depth version of CURV II, is capable of operating at depths to 7000 feet at a submerged speed of 4 knots. It has been modified for emergency operations to 10 000 feet. The CURV III system consists of the vehicle, control cable, and control console. Although it often operates from the YFNX 30, the system is designed so that all major operational components can be disassembled, air transported to a work site, and installed on any surface craft that has adequate

Figure 1   CURV III.

deck space. The vehicle normally carries a hydraulically operated claw for attaching and recovering items, such as ordnance, from the ocean floor. For special tasks, the claw is removed and replaced by a variety of grasping, cutting, or working tools. The vehicle also contains the necessary equipment for searching, locating, and documenting the lost item. Control of the vehicle and monitoring of operations are performed in the control van. The vehicle is 6.5 feet high, 6.5 feet wide, 15 feet long, and weighs 4500 pounds in air. Its instrument suite includes a Straza 500 active/passive sonar with transponder interrogation capability, acoustic altimeter and depthometer, compass, two Hydroproducts television cameras with lights, and an EG & G 35-mm still camera with strobe.

CURV III is a versatile underwater vehicle that can be readily modified to accommodate a wide variety of underwater tasks. It has demonstrated its search and recovery capabilities off the west coast as well as in the Atlantic Ocean, most notably during the 1973 rescue of the PISCES III submersible off Ireland.

SNOOPY

SNOOPY is the smallest in a series of lightweight, portable, unmanned undersea vehicle systems. It is capable of carrying a TV camera with a 250 W mercury vapour light source into the sea environment. As such, it is capable of replacing a diver in many tasks for which observation or surveillance is required. The system uses hydraulic fluid pressure of 1500 p.s.i. and delivers one-fifth horsepower to each of two thrusters for underwater propulsion. SNOOPY has two unique features: all propulsion power is sent from the surface by hydraulic lines, and an automatic depth-keeping capability is provided by a variable-buoyancy chamber and a depth-feedback system. A small, electrically powered grabber is mounted on the forward end for implanting or retrieving lightweight objects. The vehicle weighs 50 pounds and operates to depths of 100 feet.

ELECTRIC SNOOPY

ELECTRIC SNOOPY (Fig. 2), the successor to SNOOPY, differs from its predecessor principally in its propulsion scheme and its depth capability of 1500 feet. This vehicle uses three one-quarter-horsepower, oil-filled, pressure-balanced electric motors for thrust in the horizontal and vertical directions. This approach allows the use of a small-diameter tether cable. Power (AC), along with multiplexed control signals, is sent down the cable and converted to variable-DC motor-drive voltage through motor controllers at the vehicle. Twin pressure hulls house all vehicle electronics in addition to a television camera and a Super-8 photo camera, thus providing a streamlined and responsive vehicle. The Super-8 photo camera provides intervals of action footage or a large number of individual-frame pictures. The vehicle is 24 inches wide, 40 inches long, and 18 inches high. It weighs 155 pounds in air and is made neutrally buoyant by a rectangular slab of syntactic foam with a density of 36 pounds per cubic foot.

Figure 2   ELECTRIC SNOOPY.

NAVFAC SNOOPY

NAVFAC SNOOPY (Fig. 3) is a small, remotely controlled vehicle system which was designed and fabricated at the Center at the request of the Naval Facilities Engineering Command for ocean construction work. Its primary uses are optical survey of proposed undersea construction or implantment sites, surveillance and documentation of diver operations, and general undersea inspection and documentation.

The vehicle's design depth is 1500 feet. It employs four hydraulically powered thrusters for horizontal and vertical excursions. The three horizontal thrusters are controlled by a three-axis proportional joystick for integrated forward, reverse, turning, and lateral vehicle motion. The vehicle thruster control employs automatic depth and altitude holding circuitry with manual override. A small scanning sonar subsystem provides target localization and limited area search. A television camera is used for real-time viewing, and a Super-8 movie camera provides color photographic documentation. Vehicle instrumentation data, which consist of heading, depth, and altitude, are relayed to the surface by a time-division multiplex system. All instrument data are digitally displayed at the control console, and an additional "sense indication" display is provided for vehicle heading. The vehicle power, control signals, video signal, and instrument data are multiplexed onto a single coaxial tether, which is of small diameter for reduced hydrodynamic drag and ease of

Figure 3   NAVFAC SNOOPY.

handling. This vehicle weighs approximately 500 pounds in air and is 48 inches long, 36 inches wide, and 28 inches high.

SUBMERSIBLE CABLE-ACTUATED TELEOPERATOR (SCAT)

As orginally configured, SCAT was built to act as a test-bed demonstration vehicle primarily for the purpose of evaluating head-coupled television. It has recently been reconfigured as a light duty, inspection/work vehicle capable of operating to depths of 2000 feet.

SCAT can be considered an intermediate vehicle between SNOOPY and CURV. Tethered by a multiconductor cable, SCAT carries a black and white television camera with one 250 W quartz iodide lamp and a 50-frame 35-mm camera and strobe mounted on a pan-and-tilt assembly. Sensor capability also includes a sonar system, depth sensor and compass. A specially designed electrohydraulic system consisting of an electric motor, fixed displacement pump, relief valve, and reservoir drives hydraulic motors connected to ducted thrusters and provides maneuverability in both horizontal and vertical directions. In addition, five two-way servo valves are being added to the system to allow for various work functions, which may include a rudimentary manipulator or special purpose tools. While seated at the control console, the SCAT operator views on a conventional TV monitor the image relayed by the submersible's television camera mounted on a remotely controlled pan and tilt unit. The entire SCAT system (vehicle, control console, power distribution unit, main cable, and power generator) is designed for transportation by commercial aircraft and can be accommodated by their handling systems and cargo spaces. SCAT has a dry weight of 975 pounds and is 39 inches wide, 71 inches long, and 52 inches high.

SOLID ROCKET BOOSTER DEWATERING SYSTEM

The Center was asked by the National Aeronautics and Space Administration (NASA) to design, develop, and test a solid rocket booster (SRB) dewatering system. The prototype dewatering system consists of a tethered, unmanned vehicle (nozzle plug); a control console; and handling, deployment, storage, and support subsystems. The dewatering system will be used to recover the expended SRB cases of the space shuttle system after they have been jettisoned and are floating in the sea.

The nozzle plug (NP) vehicle docks and locks itself into the nozzle of the partially flooded booster case, which is floating in a spar mode with the nozzle 100–125 feet below the surface. Water is forced out of the SRB with compressed air from the surface support ship. When sufficiently dewatered, the SRB goes into a "log" mode, and a sealing bag on the NP vehicle is inflated to seal the nozzle. A hose is then deployed and the remaining water is forced out through the nozzle plug. Prime power for the system is provided by a 440 V, 400 Hz, three-phase generator aboard the support ship. This power is supplied to the vehicle through a 600 feet umbilical. The umbilical also contains a 1.5 inch air line, through which air to dewater the SRB is supplied at 150 p.s.i. Vehicle thrust is provided by six 5.5 horsepower hydraulically powered thrusters, four horizontal and two vertical. The horizontal thrusters are controlled individually so the vehicle can move in any horizontal direction without requiring a yaw maneuver. This feature allows the operator to follow the SRB motions during the docking maneuver. The vehicle is equipped with a TV camera and lights and a compass for navigation. Horizontal control can be tied into the compass at the option of the operator for automatic horizontal hold. The vehicle is 14 feet high; the main body is 30 inches in diameter. Total vehicle weight in air is 3400 pounds.

RESCUE OF PISCES III

On 29 August, 1973, the deep submersible vehicle (DSV) PISCES III sustained flooding of the after machinery compartment and sank to the ocean floor in 1375 feet of water some 100 miles southwest of Ireland. Trapped inside the main sphere of the vehicle were pilots Roger Mallison and Roger Chapman; about 72 h of life support remained.

The PISCES III, owned and operated by Vickers Oceanics, Ltd., was under charter to perform tasks associated with the laying of a telephone cable from Ireland to Nova Scotia. Initial notification of the accident was received at the Naval Ocean Systems Center (NOSC), San Diego, at 0445 Wednesday, with a request for information on the status of the CURV III. Vickers had appealed to the US Navy for assistance as backup to their primary plan to raise the distressed submersible with the aid of her sister DSVs PISCES II and V, which were being airlifted to the accident scene.

The CURV III is the third in a series of unmanned remotely controlled vehicles developed by NOSC primarily for use in recovery of test ordnance. However, like her predecessor, the CURV I, which placed the lines to recover the hydrogen bomb off Palomares, Spain, in 1966, the CURV III has been called on to respond to emergency situations.

The CURV III with associated equipment was transferred from its primary support ship to the dock at the Naval Air Station, North Island (San Diego), where all components were placed on pallets in preparation for aircraft loading by 1430 Wednesday. Two US Air Force C-141 Starlifter aircraft transported the vehicle, its support equipment, and crew direct from North Island to Cork, Ireland, arriving at 1930 Thursday, 30 August. At Cork, the equipment was unloaded from the aircraft, transferred to a barge at the dock, and taken to the Canadian cable-laying ship *John Cabot,* which was lying to about ten miles down the river, since low tide prevented her approach to the dock. By 0645 Friday, with all CURV equipment and crew embarked, the *John Cabot* proceeded to the accident site, arriving at about 1930 Friday.

A recovery attachment device, a large toggle bolt manufactured by Vickers, was provided to the US Navy crew for use on the CURV. Since the after hatch of the PISCES III was open, the easiest place to make an attachment for lifting the submersible was through this hatch. Also, most of the submersible was covered by fiberglass fairing, which could not sustain the loads needed for lifting. Since the vehicle had sunk when the after-machinery compartment flooded, it was sitting tail down with the hatch opening in a perpendicular plane. An 8 inch (circumference), double-braided nylon line was attached to the toggle bolt, then taped (with low strength masking tape) to the CURV frame, and taped at intervals all along the tether cable.

At 0900 Saturday, September 1, a Vickers representative requested that the CURV be sent to the PISCES III to attempt to attach the heavy lift line. At 0942, the CURV was over the side ready to dive. At 1030, it arrived on the bottom at a 1500 feet depth and commenced a 360° sonar scan. A large sonar target was detected at 240 yards. The bottom current was estimated at about 0.5 knot. The sonar target was closed, classified as PISCES III, and the toggle bolt placed by 1040. It was ascertained that the toggle bolt was secure by observing it via the television camera.

Then a strain was put on the lift line by the *John Cabot,* the tool holder was ejected from the CURV manipulator, the masking tape ties broke as the lift cable separated from the CURV tether, and the lift of the PISCES III was begun. By 1300, the submersible had been raised to the surface (Fig. 4), and an additional line was being attached by swimmers so that it could be held horizontally for the egress of the personnel. By 1320, the men had climbed out and were transferred, in good condition, via rubber boat to the Vickers Voyager. Although 70 h had passed since the start of the initial dive and the life support system was designed for 72 h duration, it was estimated that the two men could have safely remained within the

Figure 4   PISCES III rescue.

submersible for an additional 12–15 h. Both men in the PISCES III were experienced in diving as well as submersible piloting, and they exercised considerable restraint during their wait for rescue by relaxation, sleeping, and taking other measures to reduce their metabolic rate, thereby conserving the oxygen supply and the carbon dioxide absorbent.

The water was pumped out of the after machinery compartment, and the PISCES III was then returned to the Vickers Voyager, thus completing not only the rescue of the two pilots but also the salvage of their submersible. The *John Cabot* returned to Cork where the CURV system was off-loaded, transferred to the airfield and the two C-141 aircraft, and returned to the NOSC San Diego home base.

26

One-Man Submersibles

S. B. Boulton, Technical and Operations Director, HMB
Subwork Ltd. Great Yarmouth, Norfolk, UK

Diving operations carried out in support of drilling or associated operations are described. The problems of working in the seas off Nova Scotia and Labrador are recounted; in the former precautions are necessary to prevent equipment freezing up and in the latter pack ice is a major problem.

INTRODUCTION

The operations referred to in this paper are those carried out from the rig supply vessel *Balder Cabot*. The *Balder Cabot* is in fact used as a dual purpose vessel, having a diving moonpool immediately aft of the accommodation through which diving operations take place. Diving operations in this instance means diving services carried out by the *Mantis* one-man tethered submersible in support of drilling or associated operations. An air diving spread, complete with DDC is also carried on board to enable drilling operations to be supported at or near the surface should the need arise. When diving operations are not taking place, the *Balder Cabot* is used purely as a rig supply vessel, and serving this dual purpose role enables extremely cost effective operations to be carried out.

The *Mantis* submersible spread takes up the deck space on the *Balder Cabot* immediately aft of the bridge, and placed in obvious close proximity to the diving moonpool. In fact, two *Mantis* submersibles are deployed on board, one of which is a dual purpose ROV capable of manned or unmanned operations. The complete spread, including the air diving equipment, takes up a space approximately 30 ft × 30 ft, which is quite economical, and leaves a large amount of deck space to be used in the rig supply role.

It is necessary to have two *Mantis* submersibles on board because the *Balder Cabot*'s area of operations stretches from Nova Scotia to the northern waters off Labrador, and other rescue facilities are normally not available. The second *Mantis* must, therefore, be on instant readiness when the first *Mantis* is involved in diving operations to be of assistance should the first *Mantis* require assistance.

It is realised that, in the true sense of the word, the area of operations described does not in fact cover polar operations. Polar operations are, therefore, taken to mean conditions one would normally expect to encounter if operating in polar regions, and are in fact the worst conditions in which submersible operations can safely take place when such operations are carried out from a surface vessel.

As was mentioned earlier, diving operations from the *Balder Cabot* are carried out in support of drilling or associated operations. To this end, the *Balder Cabot* is a DP ship, the DP system performing satisfactorily in all conditions in which *Mantis* can safely dive. No problems have ever been encountered with the DP system which have prevented diving operations from taking place.

Initially, the moonpool on the *Balder Cabot* was fitted with top and bottom doors. The bottom doors were removed at an early stage, before any diving operations commenced and a coming was manufactured at deck level, around the moonpool opening, with a top hatch fitted. This coming was to a height of approximately 2 ft 6 in. Unfortunately, even though pressure relieving holes were let into the top cover, pressure caused by surging in the moonpool caused the hatch securing latches to spring and the door to lift. A further problem was that the pressure build-up also forced water through the pressure relieving holes, forcing spray and water to the height of the after bridge, and completely saturating the after deck. This problem could be lived with during the summer months, but caused untold problems during the winter months, when all the spray turned to ice — literally to bridge height, and covering all the equipment in close proximity with ice several inches thick.

The problem has now been solved to a great extent by removing the top moonpool door completely, and raising the coming around the moonpool a further 18 inches. This has drastically reduced most of the moonpool spray effect, and to a lesser degree, the amount of water spilling over the moonpool sides and onto the deck.

During the early days of *Balder Cabot* operations, no shelter was provided for the submersibles, and they were therefore subjected to the same conditions as the rest of the support equipment. This situation was remedied after the first winter at sea, and a garage provided to house the submersibles when not in use. This is obviously a considerable improvement, because, although not outfitted to tropical standards, it at least ensures that the submersibles are afforded some protection from the extremes of weather, and obviously makes working conditions a little more practicable and comfortable. However, all support equipment, such as umbilical winches, shot weight winches, and cranes, must obviously remain *in situ*, and are therefore always subjected to the extremes of weather encountered in these operating conditions.

This then sets the scene for *Balder Cabot* diving operations. Unfortunately, the *Balder Cabot* always seems to get the worst of both worlds. She operates off Nova Scotia during the winter months when weather conditions are atrocious, and off Labrador during the summer months when the weather and sea conditions are in many instances only marginally possible for diving operations, and in many cases diving cannot take place at all. The two areas of operations are entirely different in their environmental and climatic conditions, so I shall attempt to describe the conditions for each area separately.

NOVA SCOTIA OPERATIONS

Operations off Nova Scotia normally take place from October to May, which covers the worst of the winter weather, with very little good weather operating conditions to compensate. One of the worst enemies during the winter months is the effect of ambient spray, by which is meant that spray which comes over the side of the ship with an intensity which varies with ship, wind and tide directions, and the relevant speeds of these three factors. Often in transit, much water also comes onto the deck, and it is possible at times for the deck, and obviously the diving equipment sited there, to be half submerged. The free water is not too much of a problem, but the spray for this free running water commences to freeze rapidly at approximately $-9\ ^\circ$C, especially if accompanied by winds in excess of 20 knots.

Although ice several inches thick is often evidenced on equipment, this is usually only during the colder periods, when temperatures may drop to $-30\ ^\circ$C or more, and are normally only experienced when the ship is alongside, or passing through waters close to shore, especially when pack ice is also in evidence. Offshore, the temperature rarely drops below $-15\ ^\circ$C, and, providing pack ice is not in evidence, it is just about possible to dive in these conditions.

Problems are obviously experienced with personnel, and dress in these conditions is of paramount importance, because a temperature of $-15\ ^\circ$C, often accompanied by a moderate wind, reduces the chill factor to an almost unbearable degree, and personnel cannot be exposed to these conditions for prolonged periods, even with thermal type clothing. The submersibles themselves have many ice associated problems, and precautions must be taken against these. When the submersibles are stored in the garage, then there is normally sufficient heat to prevent freezing conditions. However, when diving operations are imminent, and the submersibles brought out of the garage to diving stations, precautions must obviously be taken.

The manipulators work on seawater hydraulics, and if any delay to launch is envisaged, then the manipulators must be flushed through with pure antifreeze. The air ballast system must be treated in like fashion — all the water blown out of the soft tanks and replaced with antifreeze but, in this case, the antifreeze must be aluminium-compatible to prevent corrosion, as the inside of the soft tanks are not anodized.

The air bottles themselves have a regulator fitted to control air flow. Over this regulator is fitted a rubber boot, which must be filled with glycerine to prevent freezing with the consequent loss of air ballast until the seawater thaws out any ice.

All the 24 VDC thrusters are oil-filled, and the shaft seals on the larger 120 V DC boost thrusters are oil-filled also. To retain this oil, shaft seals are obviously necessary and, again, obviously, these seals are in contact with the water on the external sides. If this water freezes, even though we are talking of minute quantities only, the ice between the ceramic and carbon seals is sufficient to score the seals and then cause thruster leaks. Thrusters must therefore be turned over slowly by hand before being tested with power on.

Apart from these problems, the submersibles are quite robust, and suitable by material selection to operate in temperatures of $-55\ ^\circ$C. This, of course, is of no consolation to the pilot, who often has to climb into the submersible with an internal temperature of 0 °C — it cools down very quickly once the heater has been removed. Good selection of clothing is essential for pilots operating in these conditions, and one of the most critical factors with the *Mantis* submersible is to keep the pilot's feet warm, as his feet rest in the after-dome, which, being aluminium, is a natural heat sink. The best solution to this problem has been to issue pilots with "moon boots" and pilot comfort is now not a problem.

Even though the pilot often commences the dive with an internal temperature of 0 °C, due to a combination of pilot's body heat and the amount of electrical and electronic equipment within the vehicle, the internal temperature quickly rises to between 18 and 19 °C, and remains steady at this temperature, a comfortable temperature in which to operate. In fact, due to the thermal clothing worn by the pilot, the internal temperature often appears to be too warm, and pilots complain more about this discomfort rather than being too cold.

There is not a lot that can be done about the formation of ice on support equipment, or the submersible's hull. Fortunately, offshore conditions are not normally severe enough to cause any great problem, and normally any ice experienced can be removed by ice mallet or by steam heat or hot water. Care must obviously be exercised when using ice mallets, especially when used on the submersible itself. Any ice adhering to the submersible quickly disperses once the vehicle is in the water and, once deployed, the cold conditions do not present any problems to the submersible's operating capabilities.

Once on diving station, with the *Balder Cabot* on DP, the ship is positioned such that the minimum effect of the weather is felt, with little or no spray coming over the ship's side, thus during actual diving operations, the surface support equipment is generally kept reasonably clear of ice. For the spray to commence freezing usually requires an ambient temperature of −9 °C, backed by a wind of 20 knots or more, usually from the north or northeast. Diving would not normally take place in these conditions with winds in excess of 25 knots, and would certainly cease if the wind speed increased to approximately 30 knots.

Although the submersible itself is capable of operations to −55 °C, the support equipment, in particular the winches, have a minimum operating temperature of −15 °C. Fortunately, conditions when diving operations are possible are rarely encountered below −15 °C, and this is normally the cut-off point for diving operations. One problem which normally affects the support equipment is the hydraulic oil. In low temperatures the oil thickens, and it is essential to open all the bypass valves in the systems to ensure that the oil is allowed to circulate freely until normal working temperatures are reached.

Working off Nova Scotia, little pack ice is normally encountered throughout the winter, in the area where drilling and hence diving operations take place. Problems with pack ice normally occur only when the ship nears land, and several times the *Balder Cabot* has had to battle her way through to berth, fortunately sustaining only minor damage.

Unfortunately, the extreme conditions experienced alongside obviously mean that diving operations — when training is normally carried out — cannot take place for the majority of the winter months. Also, the ice normally forms so thickly and quickly on the support equipment, especially when approaching the shore through pack ice, that it is almost impossible to shift, even with steam cleaners. The ice forms again almost immediately, so that to even paint equipment becomes a real problem. However, the garage means that the submersible can be maintained to a high level. Also, even though winches and the rest of the equipment cannot be painted, and even though the equipment is often awash with seawater, especially when steaming, liberal coatings of special greases and wax oils means that even if not pristine in appearance, the equipment is at least first class in condition and operation.

These then are the main problems associated with operations off Nova Scotia. Fog, especially when freezing, can obviously be a problem at times, but generally not to the extent that prohibits diving operations.

LABRADOR OPERATIONS

The operating season off Labrador is always short, usually from late May to early October, but prevailing conditions determine the actual length of season. Forward planning is almost an impossibility, as no two seasons are the same. Some seasons see little or no pack ice; in other seasons the pack ice is in evidence for the majority of the time. It is the same with icebergs. In some seasons, the icebergs are few and diminish rapidly. Other seasons see icebergs throughout the period, many of them huge in size, which brings its own set of problems.

The problem with pack ice is that it is always shifting, and the severity of the problem is never known until arrival at dive site. It is not uncommon to arrive at a dive site reported clear by the Coast Guard to find the ice again closed in, and diving impossible. Although pack ice is a problem to the submersible, it presents even greater problems to the surface ship, as the DP system cannot be used in these conditions.

Pack ice can do irreparable damage if sucked into thruster ducts and, even in conditions of minimum pack ice, it is a constant threat to the DP taut wire system. In such conditions, a close watch must be kept on the weather, and in particular on wind speed and direction. At the first hint of a possible clearing of the pack ice, the ship must immediately proceed to the dive site, get the submersible into the water as quickly as possible, and carry out the subsea work.

Standby vessels must keep a constant look out for any tendency for the ice to start closing in, and the submersible must be recovered at the first hint of this happening. Pack ice backed by even a moderate wind, and especially with any current flowing in the same direction can quickly close in, and these factors must always be considered as part of pre-dive planning.

Icebergs present different problems altogether, and standby vessels are constantly in attendance to tow them clear of diving or drilling operations. I believe the minimum distance to which an iceberg is allowed to approach drilling or diving operations to be 1 mile before the operations must be aborted and the vessels moved clear, but icebergs in fact often approach much closer than this on occasions. I think it is something to do with the ship's Master's contention that distances are always more difficult to judge at sea.

Large icebergs can be a positive nuisance even when not threatening diving operations. Many large icebergs can scour the bottom several metres deep and this scouring can often reduce visibility to practically nil when dispersed by bottom current.

Another phenomenon is that sea mist is a constant problem, especially when many icebergs are in evidence, due to the reaction of the cold sea surface and warm air conditions, which can make team changes and supply runs by helicopter extremely difficult. The *Balder Cabot* is often completely hidden in these conditions and depends upon her team changes and supplies being on passed from near at hand drilling ships, or by going alongside. Even the drilling ships often have only the tip of the drilling derrick sticking out of the mist, but this at least enables the helicopter to locate the heli-deck.

Taken all round, the conditions even in summer can be pretty dismal and depressing, and it is often a rare occasion when the sun breaks through the deck level. These then are the conditions experienced with the *Balder Cabot*.

We have also carried out diving operations from the drill ships *Ben Ocean Lancer* and *Pelerin* off the Labrador coast and, although not ideal, it is certainly more conducive to diving operations from vessels of this nature than from the *Balder Cabot*. The biggest single problem when working from drilling ships is icebergs, as it is obviously not easy to up-root everything and move off station just because an iceberg is threatening, and this is when the obligatory mile shrinks to about 100 m. These then are the everyday problems encountered in operations. I take off my hat to the gentlemen who are going to solve the problem of actually recovering the oil should drilling operations discover oil in commercial quantities.

27

Submarine Navigation

M. G. T. Harris, Royal Navy Office of the Captain SM, H.M.S. *Neptune*,
Faslane, Helensburgh, Dumbartonshire, UK

The author reflects on his experience of navigating a submarine to the North Pole,
below the ice-cap. He concludes that it is safer and more comfortable to undertake such
a journey than to travel on the surface, provided the submarine has an energy source
that does not require surfacing and modern navigational equipment.

My subject is navigating a submarine beneath the ice-cap and I must emphasize that
my views are those of a deep sea sailor. My credentials are that I commanded the
British nuclear-powered submarine HMS *Sovereign* (Fig. 1) on a voyage to the North
Pole in 1976. The *Sovereign* is a vessel of 4200 tons displacement, 83 m long with a
draught of 9 m. She can dive to depths of about 150 m and has a maximum speed of
over 25 knots. She has a crew of about 100.

We were going to the Arctic and it is useful here to note that the natural main
deep water highway into the Arctic is from the North Atlantic — broad and deep
with no shoals. The other ocean entrances, the Bering Strait, and Bering and
Chukchi seas, are all shallow and, with the ice on top, more hazardous simply
because of the small vertical distance between the seafloor and the base of the ice,
especially in winter.

Before encountering the ice-cap myself for the first time, like many people I
thought of it as a vast ice-covered lake, together with some huge icebergs of the sort
that sank the *Titanic*; I hoped that I would not meet any of the latter! One imagines
the ice as a flat topped low wall rising out of the ocean, continuing the whole way to
the pole, and it makes one think that the same thing will be true of the base of the
ice — just a huge ice-covered duck pond. Of course, it is not like that at all, but this
partially explains the failure of the first attempt to reach the pole by submarine, that
of Sir Hubert Wilkins in 1931. He converted an old American submarine, even
putting a 'moon pool' in her, re-christened her *Nautilus*, and gave her a flat top
which he called the sled deck (Fig. 2). On this, he put a cushioned arm so that
Nautilus, having dived, could slip along the smooth bottom of the ice. He also
designed long vertical ice drills so that when he needed more air for charging
Nautilus' batteries, he could stop and drill an air hole up through thick ice.

So, if the bottom of the ice is not as Sir Hubert Wilkins envisaged it, what is it

Figure 1   HMS *Sovereign* (Copyright, Royal Navy).

Figure 2   Sir Hubert Wilkin's *Nautilus*.

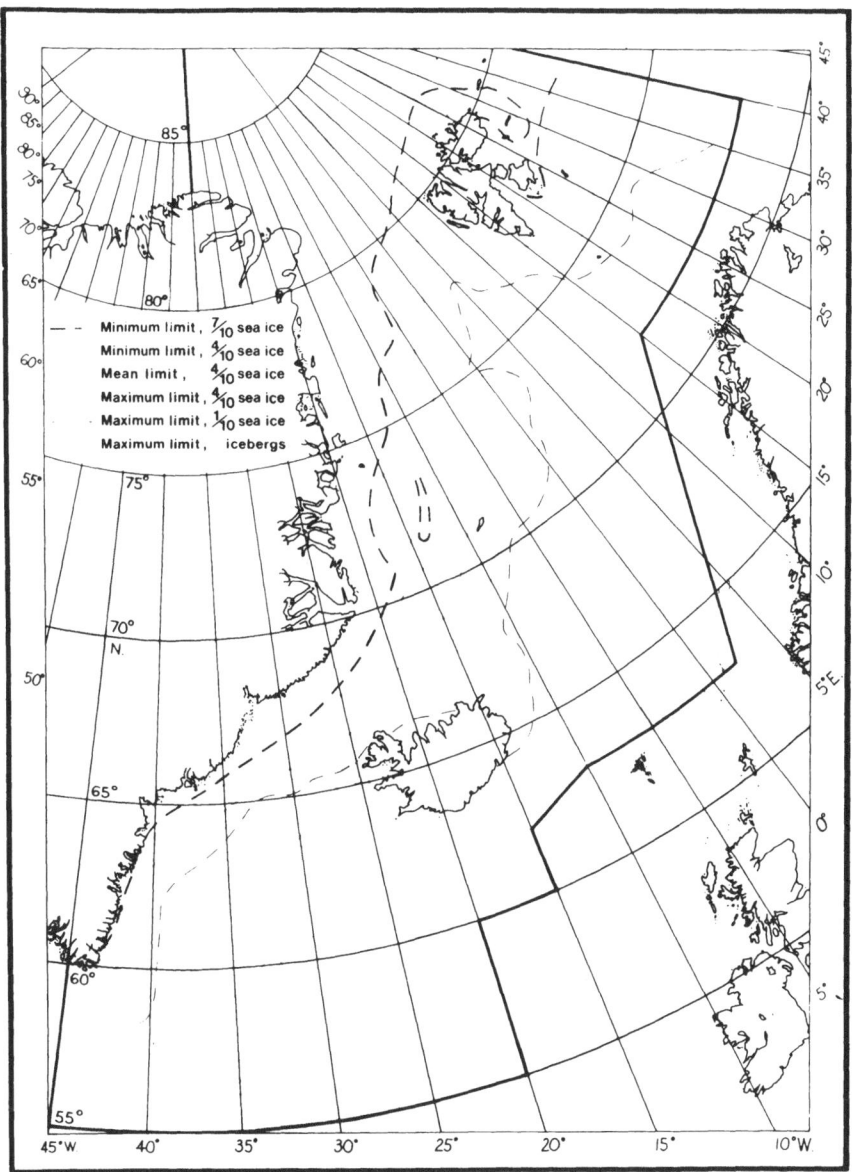

Figure 3 Limits of the ice cap in the Greenland Sea.

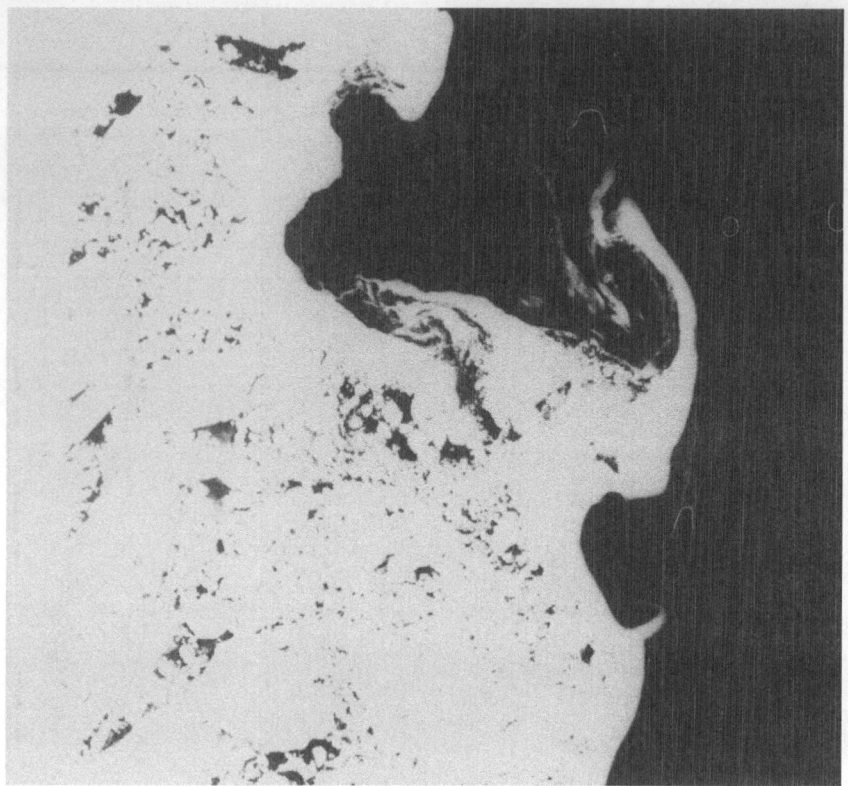

Figure 4   Satellite photograph of ice edge.

like, from the point of view of the mariner or, should I say, submariner? First of all the ice edge. Looking at the Atlantic ice edge, one notes first of all that it does not follow along a parallel of latitude but takes a generally diagonal shape (Fig. 3), reaching up to Spitzbergen, where there is always an area of open water; this was so well known to the whalers of old that it came to be known as Whaler's Bay. It is a bay, but in the ice, not in the land. The cause for this odd shape is simply the pattern of ocean currents to be found in the Greenland Sea, with the cold East Greenland current supporting the formation of ice whereas the north-going warm Norwegian Atlantic current causes Whaler's Bay to remain open; and its neighbour, the Irminger current finally snuffs out the Greenland current's efforts. Figure 4 is a satellite photograph of the ice edge (79° 30′ N 3° E), taken to the West of Spitzbergen; the edge is anything but clearly defined. Bear in mind also that, especially in heavy weather, of which there is a lot in these latitudes, the edge in any one place does not remain static but can drift at a rate of several miles per hour. I experienced a 20 mile shift overnight while I was up there.

Now what does the ice edge or, more properly, the marginal ice zone, really look

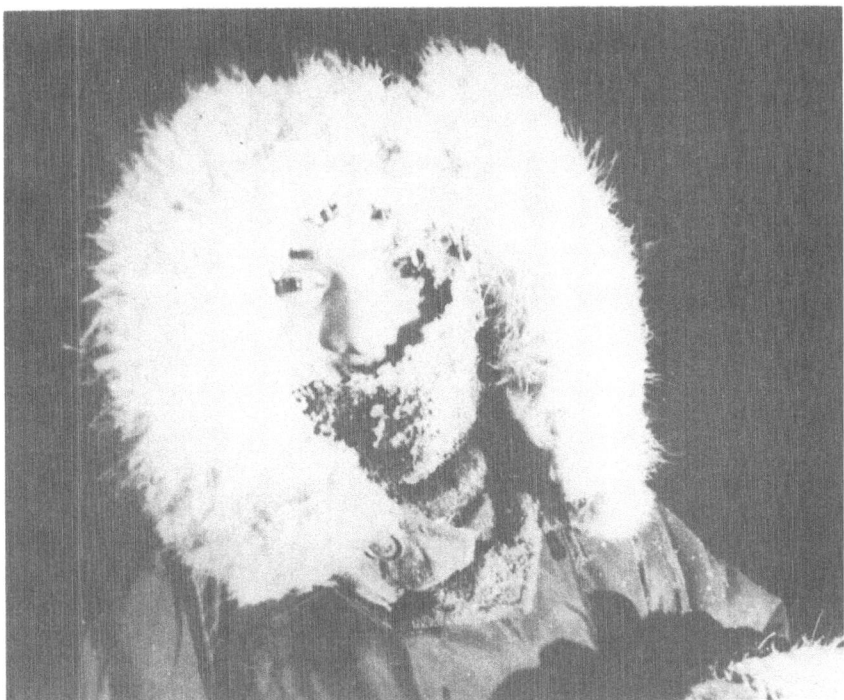

Figure 5   An Arctic lookout.

like to the mariner? Since ice does not show up well on radar, the marginal ice zone is not a pleasant place to be at night and it is arguable that these latitudes are not a pleasant place to be at all — on the surface anyway! So — and here follow some key points — while the mariner in the Arctic might be peering through the gloom (Fig. 5) at 2 knots, the submariner will be going at 15 knots in perfect comfort and safety.

Furthermore, the submarine just keeps going when she meets the consolidated ice of the main Arctic pack. Indeed instead of calling it by that threatening description, we submariners refer to it as the "ice canopy", which is just what it is — a canopy to protect us from the weather. The high Arctic is a perpetually calm ocean.

Until the advent of the nuclear powered submarine, the overwhelming disadvantage of the ice canopy was that it prevented the submarine from replenishing her atmosphere in order to charge the batteries and also for the crew to continue to breathe. Now we have an energy source that does not require huge amounts of oxygen and that can provide plenty of power to maintain a breatheable atmosphere by driving machinery that produces oxygen and also gets rid of unwanted and noxious gases, such as carbon dioxide and carbon monoxide. Fuel cell technology could provide a similar advantage but is not yet in general use.

So now let us imagine ourselves under the main ice canopy. It stretches ahead of us

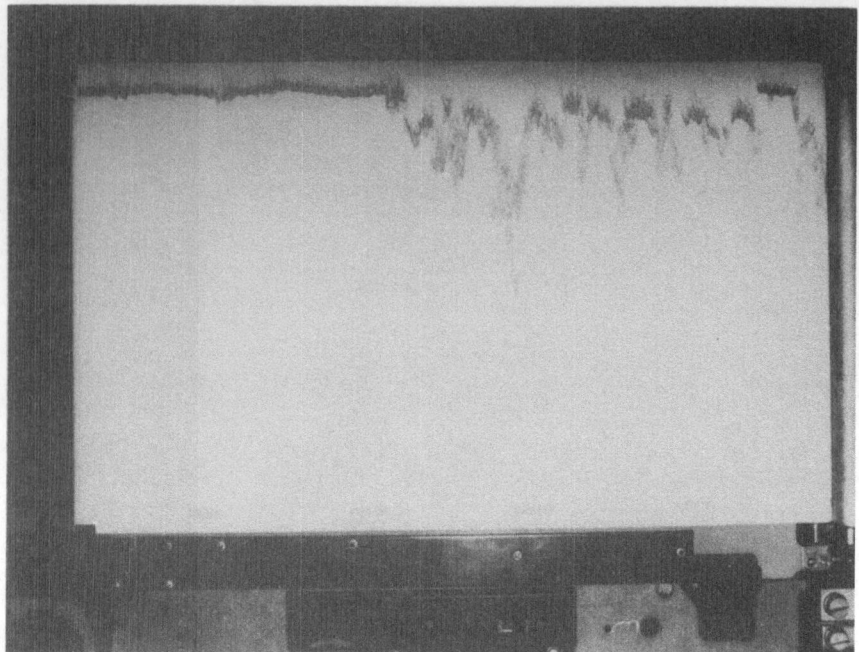

Figure 6   Under-ice profile with polynia.

for over 1000 miles. What is it like? Well, it is no more like an ice-covered duck pond than the marginal ice zone is like an edge. For a start, sea ice forms differently from freshwater ice, and the arctic ice canopy has been there for many years. It suffers partial melting and refreezing in the summer months and its borders extend southwards in the northern winter. Great stresses are placed on it by virtue of its huge size and the different weather systems and currents which act upon it. Cracks are opened in the canopy by divergent wind patterns. Because the air temperature is much lower than the freezing point of water, ice forms immediately on top of the exposed seawater, when the main ice is put under pressure again, causes the new thin ice in the crack to be crushed upwards into ridges and downwards into keels.

Patterns build up into a landscape, or "ice scape", typical of the Arctic ice: a 2 or 3 m thickness of undisturbed ice with ridges reaching some 14 m above it but seldom more than 4 m, while below the keels can extend down to 45 m but, again, not often more than 25 m. These keels are important to the submariner, since he must keep below them to avoid a collision; as long as he keeps out of the known iceberg infested areas, he can steam happily along forever with the uppermost point of his submarine at about 35 m depth.

Before leaving the ice itself, I will mention surfacing through ice, should one wish to do so. What one needs is a fresh crack or a polynia since the 2–3 m of undisturbed ice is too strong to break through. This is easier than, perhaps, it sounds. In broad

Figure 7   HMS *Sovereign*, surfaced in a polynia.

terms, one tends to come across something suitable every mile or so. How do you identify such a place? By placing an ordinary echo sounder on top of the submarine and pinging upwards against the ice as you go along, a characteristic trace is obtained but, beneath a polynia, a hard, flat echo is produced which is instantly recognisable (Fig. 6). The technique is to turn back and hover beneath the new ice of the polynia, then rise slowly to rest the top of the submarine against the ice — come up too fast and you risk damage. Then you surface normally and the submarine

Figure 8 Using a rod strain-meter to measure deep penetration of the long ocean well into the Arctic.

pushes her way up through the ice (Fig. 7). I mention this surfacing through ice not because a submarine needs to bob up all the time to see where she is, but to illustrate a submarine in the polar ice and more seriously to demonstrate that a submarine can not only navigate freely beneath the ice, but also come up, perhaps to conduct experiments (as here) with a rod strain-meter to measure deep penetration of the long ocean swell into the Arctic (Fig. 8).

How do we navigate? First, I will mention two methods that can be used in the polar regions but not by diving submarines, as they require a person or an aerial to be above the ice: satellite navigation and the sextant, the latter needing an artificial horizon and the use of special low temperature refraction tables, found in the Mariner's Handbook. Now for under-ice systems, and these are pretty ordinary items of navigational equipment: the sonar and the ship's inertial navigation system, or "SINS" for short. A perfectly ordinary ship's electromagnetic log is adequate but one must remember that it measures the speed and distance run but, just as importantly, it feeds a speed input to both the gyro and the SINS. It must therefore be accurate and have been recently calibrated, otherwise navigational accuracy will be fatally undermined before you start.

Next the compass. The magnetic compass may be discounted because of the position and drift of the magnetic pole, which makes its use in the Arctic impracticable. Interestingly enough, this problem clearly intrigued an early proponent of submarines in the Arctic, Professor Anschütz-Kämpfe, who published a plan to go by submarine to the North Pole in 1901. Although his plan came to nothing, he foresaw the directional difficulties and later developed the gyro compass, which he began producing in 1908 and which is in general use today. A gyro compass for use in the Arctic should clearly be optimised for high latitude work. Even so, as the pole is approached, any gyro compass will lose its north-seeking quality, so if navigating by gyro compass only, it is best to stay more than 200 miles or so from the Pole. If you must go near the pole, then the gyro needs to be put into a directional, or free-spinning mode, and aligned to a meridian. Once aligned, it will take you through the pole and over the other side until you reach a low enough latitude to use it as a compass again. Obviously, this is not a convenient state of affairs and the way to avoid it, if you must go close to the pole, is to have a ship's inertial navigation system. For navigation, SINS, gives a continuous output of latitude and longitude as well as the heading, thus taking the place of a compass, and it gives a local vertical reference as well. Also, modern equipment capable of working correctly in the vicinity of the pole is important here. SINS is simply a gimballed stable platform which houses three gyros with their axes mutually at right angles, so that between them they can sense any rotational motion of the platform and thus of the ship. Accelerometers are also fitted to the platform to detect any acceleration in the east/west or north/south axes. By doubly integrating these accelerations and feeding them into a computer which already has information on the Earth's shape, gravity and spin, the SINS system produces changes in latitude and longitude. So, as long as the original position is known, SINS will always give the correct position. There is an interesting development being pursued by British Aerospace in the United Kingdom now; the ring laser gyro is smaller, cheaper and has no moving parts, and that includes no gimballes.

Although, as I have said, satellite navigation requires an aerial to be above the ice, there are two existing radio aids that can be used by an aerial below the ice. These are Omega, a very low frequency (10–14 kHz) world wide radio system, and Loran C, a low frequency system at 100 kHz. However, the submarine must slow down and come into shallow water to receive signals from them and so they are not ideal.

The last navigational aid I wish to discuss briefly is sonar. It seems to me that if you wish to navigate in the Arctic, you will probably be wanting to go somewhere, say an oil terminal, rather than just be wandering about. Such a terminal will presumably be in shallow water which is, in general, poorly charted in the Arctic for obvious reasons. Thus an acoustic homing beacon at or near the terminal would make good sense. This would mean, of course, having an acoustic direction finder fitted to the submarine. A scanning active sonar set may also be worthwhile to detect the continental shelf, which it may not be very good at because of the water structure, and for detecting the hard ice of possible surfacing areas, at which it will be good for the same reason. If, despite my advice, you wish to go into iceberg territory, such a set will be necessary for iceberg avoidance.

I hope that I have shown that a modern submarine is a most practical vehicle in which to travel around the Arctic as, indeed, it is for any of the oceans of the world. Not only does the Arctic ice itself present no barrier but the hazardous marginal ice zone and the foul weather to be found in the North Atlantic Ocean and Greenland Sea can be avoided by going beneath it.

I spent, in all, some two weeks under the ice, dressed most of the time in shirtsleeves in warm, comfortable conditions and, without being in a hurry, I reached the pole three weeks after sailing from Plymouth and got back again in two. It gave me great pleasure to have on board Captain George Nares's telescope to take to the pole. He led an expedition to try to reach the North Pole exactly 100 years before me and although he failed, the Strait between Ellesmere Island and Greenland bears his name. I am sure that many people will join me in hoping that we shall soon see submarine merchant ships in the shape of tankers, plying under the pole. After all, it is less than 3000 miles from Prudhoe Bay to Sullom Voe, about the same to Bergen and only another few hundred miles to all the major North European ports.

ACKNOWLEDGEMENTS

I wish to acknowledge the assistance of Dr Peter Wadhams of the Scott Polar Research Institute, Cambridge, and also of the Director, Arctic Submarine Laboratory, Naval Ocean Systems Center, San Diego.

28

Underwater Navigation and Positioning Systems

K. Vestgård, Manager, Development Department, SIMRAD Subsea A/S, Horten, Norway

Advantages and disadvantages of long baseline navigation and supershort baseline navigation are considered. The two systems may be used in under-ice navigation; the final choice of system configuration depends on requirements for simplicity in use, accuracy and navigational range as well as on the type of navigational operation.

INTRODUCTION

The existing principles of acoustic underwater navigation systems fit directly into the under-ice navigation applications. Obviously, the systems have to be designed for the arctic environments with respect to low temperature and acoustic reflections.

Operation in low temperatures is a requirement for the parts of the system being exposed to open air and seawater. Environments with high acoustic reflection have to be handled by optimising the geometrical position between the transmitter(s) and receiver(s), and by discriminating the unwanted reflected signals from the main signals by proper signal filtering and by narrow beam transmit and receive technique. The following gives a short introduction to the two main principles of acoustic underwater navigation and how these two principals can be used for navigation under ice-covered areas.

EXISTING UNDERWATER NAVIGATION SYSTEM

The two main principles of acoustic underwater navigation in use are: long baseline navigation (LBL) and supershort baseline navigation (SSBL).

Long baseline navigation (LBL)

The long baseline navigation system is illustrated in Figure 1. A number of transponders are deployed at the sea bottom. By measuring the slant range between

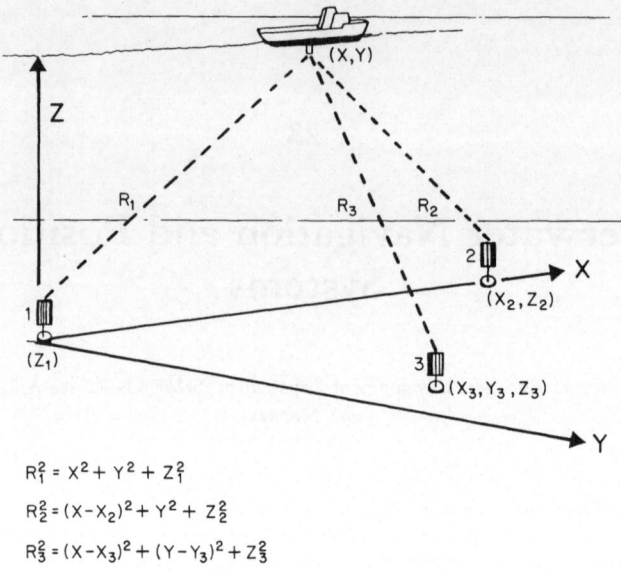

$$R_1^2 = X^2 + Y^2 + Z_1^2$$

$$R_2^2 = (X - X_2)^2 + Y^2 + Z_2^2$$

$$R_3^2 = (X - X_3)^2 + (Y - Y_3)^2 + Z_3^2$$

Figure 1   Long baseline positioning.

each transponder and the ship, and knowing the baseline length between the transponders, the ship's position can be calculated.

Before the actual navigation takes place one has to calibrate the transponder array — that means to establish the relative baselines between the transponders. Earlier, this was carried out by a surface survey that could last for hours. Today transponders are equipped with low power microprocessors. These transponders can call up each other and measure the relative position between themselves. That means that grid calibration can be carried out within seconds, which is a great saving of time compared with the survey method. The communication between the transponders and the ship is carried out by a telemetry link transmitting the necessary commands and data.

A drawback with the existing LBL systems is the use of a ship-mounted wide beam transducer. This is necessary to cover all transponders. A wide beam transducer is highly sensitive to reflected signals and to acoustic noise generated by thrusters and propellers on the ship. Another drawback is the high number of acoustic signals that has to be successfully transmitted in order to obtain a reliable position fix.

Supershort baseline navigation (SSBL)

The supershort baseline navigation system is illustrated in Figure 2. The position of the surface ship is obtained by a combined range and bearing measurement between the SSBL transducer and the reference transponder on the sea bottom. The position of the submarines are obtained in the same way. Due to the bearing measurement principle, the SSBL transducer is much more complex that the LBL transducer. The

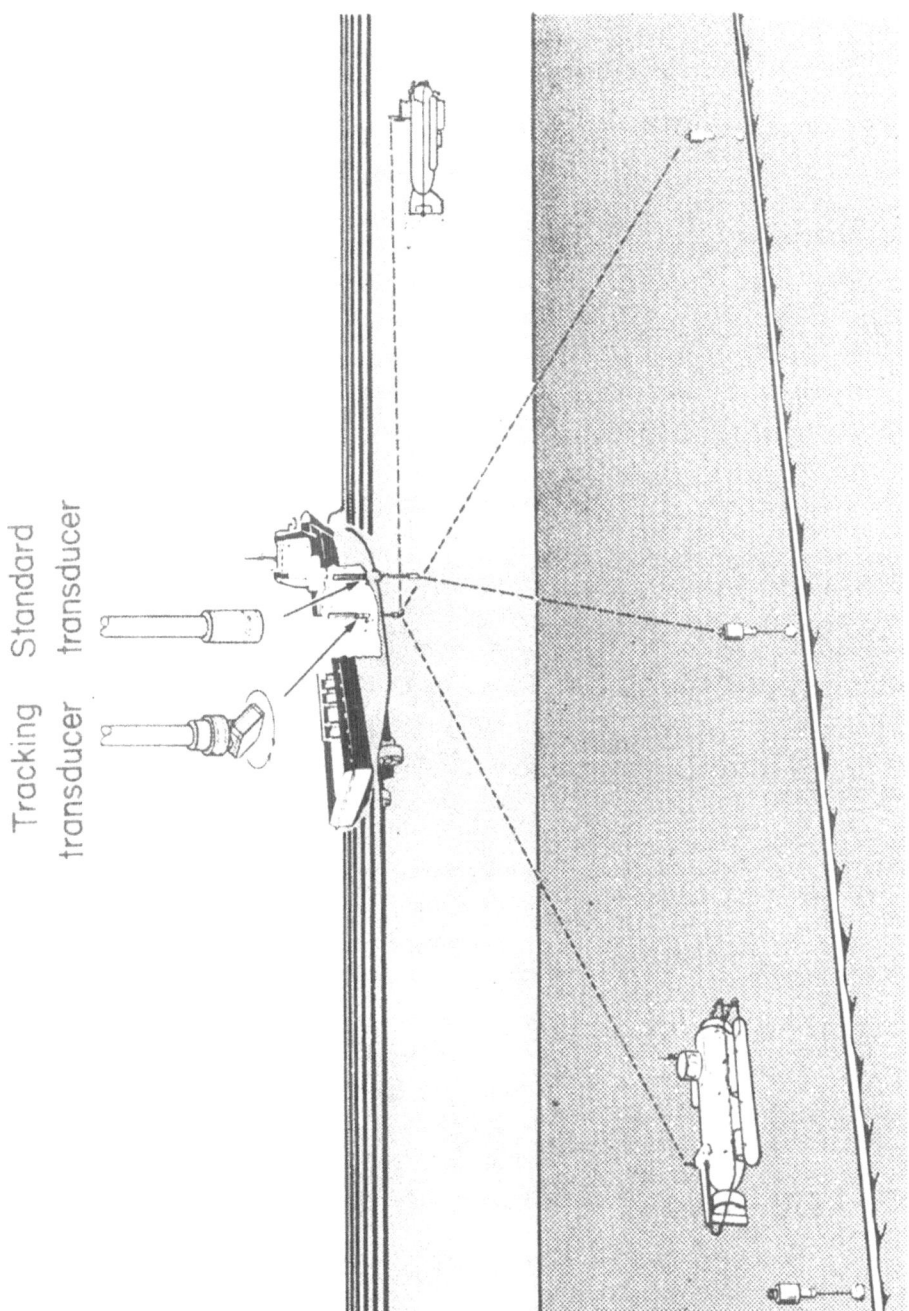

Tracking   Standard
transducer   transducer

Figure 2   SIMRAD HPR SSBL acoustic navigation system.

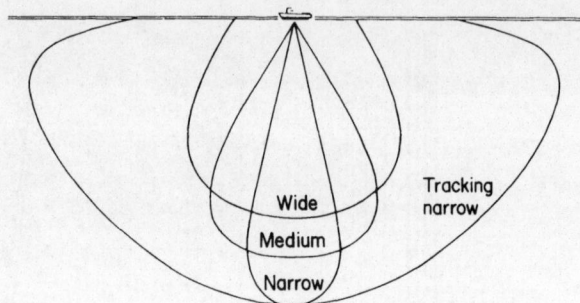

Figure 3   Operation area for the wide, medium, narrow and tracking narrow transducer.

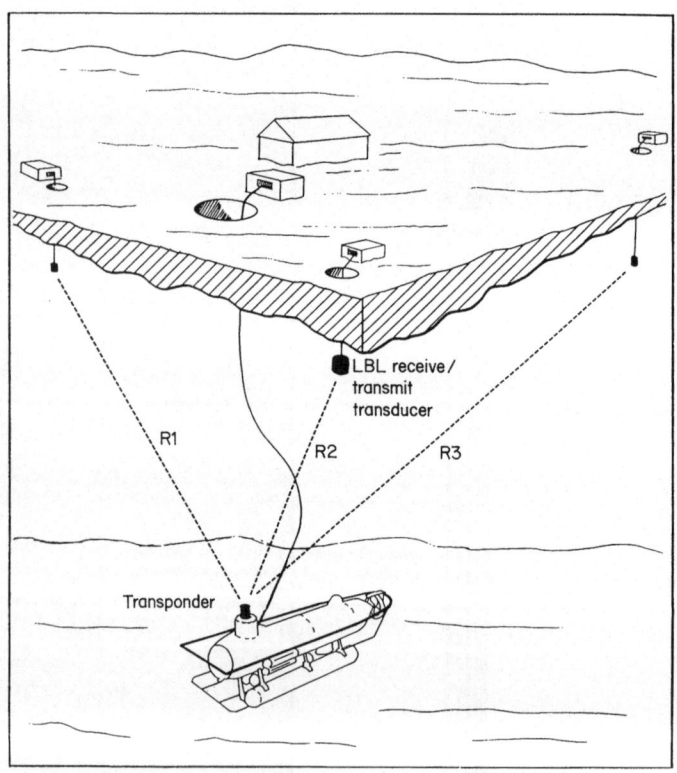

Figure 4   Under-ice navigation with LBL system.

main advantage of the SSBL principle is, however, its simplicity in use, requiring only one reference transponder and no transponder grid calibration. Another important advantage is that this system allows narrow beam signal detection and beam steering to be used.

Use of a narrow beam gives a better accuracy and higher noise immunity, but with a fixed beam the operation is limited to an area covered by the beam itself. This can be avoided by using beam steering technique. The signal from each element is processed and added together in a way which allows the beam to be steered to any required direction. In this way, the whole underwater sphere below the vessel is covered, allowing narrow beam tracking of one or more transponders independent of their position with respect to the ship. Figure 3 illustrates operating areas for different transducer types.

Another advantage of the narrow beam tracking transducer is its capability to reject reflected signals arriving outside the beam pattern as illustrated in Figure 4. The narrow beam is significantly reducing the amount of noise detected by the system. The system illustrated is working with a 30 degree narrow beam, which gives an increase of 12 dB in S/N ratio. This has proved to be a very important system feature when working in offshore environments with extreme noise from propellers and thrusters.

UNDER-ICE NAVIGATION WITH LONG BASELINE (LBL) NAVIGATION SYSTEM

The LBL system described earlier is turned upside down, with the transponders hanging down through the ice floor instead of being anchored to the sea bottom. The principle is illustrated in Figure 4. The transponders are lowered down to a depth where free line of sight is obtained between the LBL transducer and the submarine (or other object) to be positioned.

Before the actual navigation takes place the transponders array must be calibrated. The calibtration can be carried out in two ways:

(i) Measuring the distance between the transponders using an appropriate range/radio navigation system,
(ii) Measuring the distances by using acoustic measurements.

The submarine to be positioned carries out a simultaneous interrogation of three or more transponders. Each transponder replies with its own coded signal. By measuring the time between interrogation and reply, and using the information about relative position between the transponders, the position of the submarine can be calculated.

Advantages

Positioning accuracy will normally be within 2–3 m using a transponder array of 2–3 km distance between the transponders.

Disadvantages

The system requires deployment of a minimum of three transponders through the ice. Calibration of the array must be carried out before navigation takes place.

The system requires successful transmission of a high number of acoustic signals for each position fix.

Use of a wide beam transducer makes the system sensitive to the acoustic signal being reflected from the ice-covered environments.

UNDER-ICE NAVIGATION WITH SUPERSHORT BASELINE (SSBL) NAVIGATION SYSTEM

The SSBL navigation system can be configured in two ways:

(i) The SSBL transducer is lowered through the ice floor. The transducer is tracking the transponder mounted on the submarine or other objects to be positioned.

(ii) The SSBL transducer is mounted on the object (submarine, structure) itself, which is to be positioned. The transducer is measuring the position with reference to a transponder hanging down from the ice floor, anchored to the sea bottom or mounted on a subsea structure.

In the first example, the SSBL transducer is lowered down to a depth obtaining free line of sight between the SSBL transducer and the transponder(s), Figure 5. Since the SSBL system establishes the position by measuring relative range and direction between the SSBL transducer and the transponder, compensation of transducer heading and inclination is needed. Heading and inclination sensors are mounted inside the transducer, and automatic compensation is carried out by the system's navigation computer. A number of transponders can be positioned simultaneously allowing a complicated subsea operation to be displayed.

Advantages

The SSBL principle is simple in operation. Only one SSBL transducer and one reference transponder is required in order to track a submarine. There is no need for a transponder array requiring array calibration. The system can track transponder(s) using a narrow acoustic beam. This provides increased signal to noise ratio. Signals reflected from the environments arriving outside the beam are significantly reduced and will not disturb the direct path main signal. The system requires only two successful acoustic signal transmissions for each position fix.

Disadvantages

The accuracy of the system is a function of the distance. Accuracy around 0.5% of distance can be obtained. This means 5 m accuracy at a distance of 1 km. The system requires input from a heading sensor. A high quality heading sensor is required when operating in arctic areas.

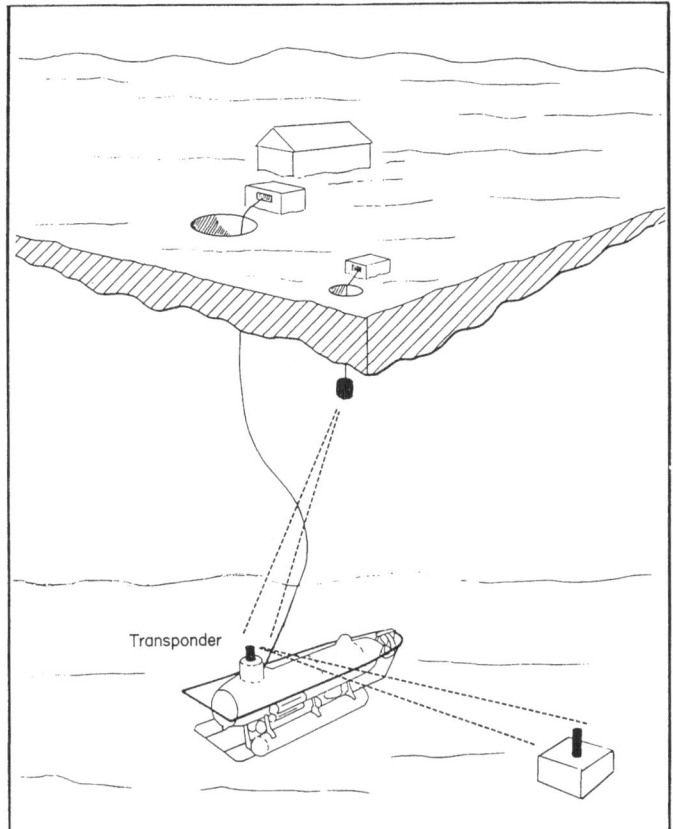

Figure 5   Under-ice navigation with SSBL narrow beam tracking system.

SUMMARY

There are two main principles of underwater navigation: long baseline navigation (LBL) and supershort baseline navigation (SSBL). Both systems can be used for acoustic under-ice navigation. The choice of system will depend upon the requirements for simplicity in use, accuracy, and navigational range. The final system configuration will depend upon the type of navigational operation (submarine tracking, structure installation, etc.) and has to be decided upon after a detailed discussion between user and system supplier.

29

The Development of a Submarine Freighter

J. Chappuis, Technical Director, Ateliers de Constructions Mécaniques de
Vevey, Switzerland
and
F. Abels, Manager, Ingenieurkontor Lübeck GmbH Lübeck, Federal Republic
of Germany

The transportation of heavy loads by means of a special heavy load submarine-carrier is
proposed and the design requirements for such a submarine-carrier outlined. Four
major conventional carrier types are envisaged, two of which have been further
investigated. The systems are described in detail.

INTRODUCTION

In a few decades there has been a tremendous development of subsea technology; this
is generally well known and it is unnecessary to emphasize the fact here. This
progress has been made under the pressure of various activities when they were
extended to the subsea world, for instance scientific research, military research,
commercial exploration of mineral wealth, etc. But, today, the most active and
demanding of these activities is very probably petroleum exploration and
production, when considering only civilian applications.

For some years, petroleum activities have been undertaken in the Arctic area.
This orientation has produced a new surge of technological development in general
and more specifically in subsea technology. In fact, it can be said that these
developments are still in full progress and that we can expect to witness in the future
the use of what can still be considered today as exotic means. In this paper, we will
present a subsea technological line of development, which is the direct result of this
trend, dealing with the transportation of heavy loads.

DESCRIPTION OF THE PROBLEM

If we consider the arctic petroleum quest in northern areas, we have to distinguish
several regions, each one with its own particularities. We can, very roughly, make a
distinction between at least two main areas:

(1) areas such as northern Norway, Labrador, Nova Scotia, South West Alaska, etc.,
    where the winter ice-shelf disappears in summer, possibly with some remnant
    drifting ice, or still more serious icebergs;

(2) areas such as North Slope in Alaska or the northern Canadian archipelago, where multi-year ice-shelf or at least high density drift ice still prevails even in summer.

Hereafter, we shall deal with the conditions prevailing in these last areas where very audacious exploration programs have reached a quasi-standard operational status, particularly in the Canadian archipelago. Routine drilling from the ice-shelf in water up to 300 m deep has been described on several occasions.

A certain number of gas and oil fields have already been found, for instance, located in the area included between Melville Island and the Elles Ringes islands (Fig. 1). Work is being carried out to increase the knowledge of the true economic potential of such areas. It is already known that the oil and gas will not necessarily be found in giant or very large pools, but may occur in geologically disconnected pools of a relatively limited size. Such facts have to be taken into account when projecting the data into a production stage. In fact, any production scheme in such areas will have to deal with very difficult problems, resulting not so much from the geological features of the fields, and not necessarily from the fact that products will have to be extracted from sea-bed wells. Much more, the extreme remoteness of the geographical locations, the very difficult climatic conditions, the permanent ice-shelf on the sea and the permafrost on the ground are amongst the most challenging conditions for achieving an economically viable production.

Whatever technical solutions are considered, some fixed structures will have to be located at or near the well heads on the sea-bed. They will be conceived to perform standard operations such as well control, gas separation, polling or several wells together in a manifold system, reinjection for enhanced recovery, etc.

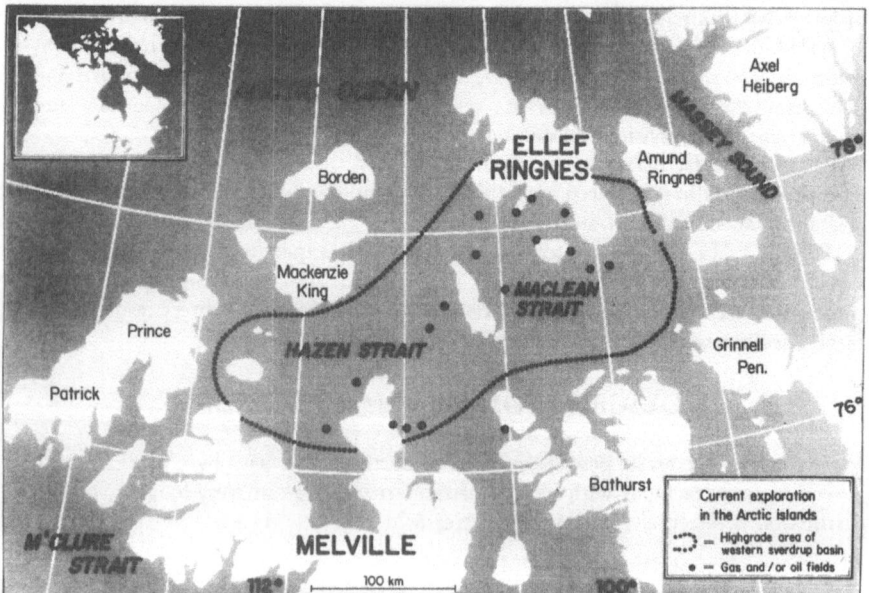

Figure 1   Exploration zone on Melville and Elles Ringes islands.

It is already known that at least some of these systems are quite large and heavy items, typically several tens of meters long and weighing hundreds or possibly thousands of tons in air. When completed in an industrial zone and shipped to some northern base, they will have to be transported from the summer boundary of the permanent ice-shelf to the well locations, and deposited on the sea-bed underneath the ice-shelf. This is obviously not an easy task, and quite unusual transportation systems may be expected.

A feasibility scheme discussed with northern operators has led to the concept presented below, by means of which the final transportation leg could be carried out on a subsea route by means of a special heavy load submarine-carrier.

It sometimes happens that, once a new procedure has been devised for some specific task, uses other than those first intended appear advantageous. We do think that this may reasonably be the case with this subcarrier, which could lead to easier handling of heavy subsea loads than provided for now in other special schemes.

STATE OF THE ART

Submarines and submersibles play a more important part today than ever before. It is well known that military nuclear-powered submarines routinely operate under ice in the north pole region. But commercial submarines and submersibles also operate in the northern areas under severe conditions.

Design objectives to be imposed on this submarine freighter for offshore operations in ice-covered areas include a medium cruising range; thus it is advantageous to base development, design and construction of such a freighter on the modern conventional submarines which are provided with the well-proven lead–acid batteries. During the past 25 years, more than 100 conventionally driven submarines have been designed and built (or are under construction still) in the Federal Republic of Germany, so that wide-spread experience in this field has been gained, including the design of new submarines by using well-proven elements on one hand and the integration of new developments on the other hand. A description of new developments in German submarine design is given in Abels (1978).

As far as commercial projects are concerned, a decisive change in the importance and duties expected of subsea craft in the offshore range has taken place during the past 20 years. From the original function of a subsea craft as a demonstrator for technical possibilities and a special scientific vehicle, this change has turned it into a craft which must prove itself when employed in offshore regions as a subsea work vehicle as well as a subsea research vehicle. As many as 100 submersibles were constructed worldwide, but not all of them were actually operated successfully. Their size varied between 2 and 30 tons and their maximum diving depth varied between 30 and 10 000 m. Meanwhile, submersibles have proved themselves as superb work vehicles, and it may be expected that in the future they will be fulfilling additional tasks for the offshore industry.

The use of a submersible has become an economical factor, when being compared with divers who start working from the water surface. Advanced working

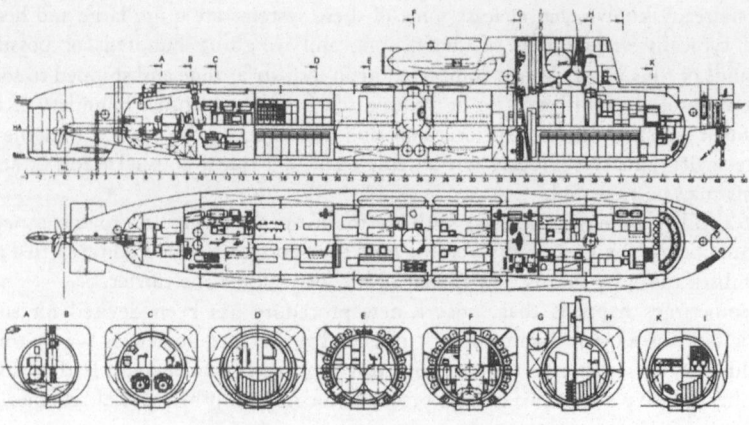

AUTONOMOUS SUBMARINE, TOURS 520

Figure 2   Autonomous submarine TOURS 520/500.

submersibles are equipped with diver lockouts and require saturation diving systems which usually are accommodated on the support ships. Nevertheless, most of the submersibles are dependent on surface support ships, which is a disadvantage for many offshore tasks. A compilation of most of the commercial submersibles and submarines is shown in Trillo (1982).

Trends in development have lead to autonomous submarines of long-term underwater endurance which are independent of a support ship. These units are equipped for the underwater inspection of pipelines and cables, for installations and repairs, for core drilling tasks and for oceanographic and geophysical duties. Furthermore, these units are provided with diver lockout systems which ensure even greater flexibility, and, in some cases, auxiliary submersibles too. The vessels may operate independently for several weeks and can run over long distances of up to 3000 miles. One example is the design of TOURS 520/500, which is shown in Figure 2.

For the transportation of arctic crude oil and natural gas throughout the year, the Electric Boat Division of General Dynamics performed studies and concept designs for a Subsea Transport System (Veliotis and Reitz, 1981). The latest developments lead to large submarine tankers for LNG transportation of 60 000 tons with displacements of 700 000 and 830 000 tons.

Even 70 years ago, commercial cargo-carrying submarines and submarine tankers had been developed, constructed, and successfully employed in Germany. Although the conditions were quite different, their duty, too, was the transportation of goods and fuel in submerged operation over long distances, if required. The first cargo-carrying submarine transported high-quality deadweight cargo of 780 tons in 1916. These submarines had a length of 65 m, a displacement of 1800 tons, and speeds of 9.5 knots on the surface and 7.5 knots when submerged for a period of 2 h. Their total cruising range was 14 000 nautical miles at 9.5 knots.

During the 1940s, several cargo-carrying submarines and submarine tankers had been developed and constructed. Moreover, existing military submarines were converted into transport submarines for the maintenance of the connection with the Far East. The Type 20 design had a displacement of 2700 tons and a load capacity of 800 tons. The box-shaped cross-section was a typical feature of these transport submarines, of which 30 units had been ordered. A detailed description of all submarines planned and built in Germany is given in Rössler (1981).

Although a conventional propulsion system is provided for the projects introduced here, the air-independent propulsion systems should also be mentioned, as demands for operation under the ice often return to an autonomous energy supply and an air-independent propulsion system because of the large cruising ranges and long periods of mission. Nuclear propulsion seems to be the most favourable solution, but it is very expensive for civil applications. Its development cost is high if intended for commercial use, and it is available to a few potential users only. Moreover, it is subjected to very stringent demands on safety and environmental protection.

There is a great variety of air-independent propulsion systems between nuclear and conventional systems. The German Federal Navy decided to require an air-independent propulsion system for the future Class 208 submarines, so that, for many years, any air-independent systems which have been deemed possible have been investigated in the Federal Republic of Germany and, integrated into submarine designs, they have been compared with one another and valued. Comparison has been made between the conventional propulsion system consisting of diesel generator sets, lead batteries, and electric propulsion motor, and the propulsion systems and energy sources as follows:

batteries
   improved lead storage battery (Pb)
   sodium/sulphur storage battery (Na/S)
   lithium/thionyl chloride storage battery (LiSOCl$_2$)
   lithium/peroxide storage battery (Li/H$_2$O$_2$)
closed cycle engines
   diesel engines
   stirling engines
gas-turbine propulsion plants
   Walter turbine
   Brayton turbine
fuel cells

Depending on given conditions, a great number of criteria was involved in the evaluation, examples of which are as follows: cruising range, diving depth, development costs, development time, operating costs, development time, operating costs, infrastructure, service, maintenance, etc.

In our experience, the following propulsion plants can solve the propulsion problems of this project and should be accurately investigated in detail.

   closed cycle gas-turbine
   closed cycle diesel engine
   fuel-cell

The German Federal Navy has decided in favor of the fuel cell, and intense development work is being done at present. In 1985, the first fuel cell system having a capacity of about 100 kW will be operated in a shore testing installation under submarine conditions. On completion of all trials and tests, these fuel cell systems may be installed in underwater vessels. But it is already feasible to plan the design of an underwater vessel with conventional propulsion in such a way that it is "fitted for" installation of a fuel cell system during a midlife conversion at a later date.

A brief description of a fuel cell system is given as follows: Fuel cell propulsion systems are independent of outside air, working on hydrogen and oxygen. In this system, the chemical fuel energy is directly converted to electric power at a high efficiency without any exhaust gas. The system is almost maintenance-free and is highly reliable. Beside a possible high overload, the increase of the efficiency rate at partial load is a big advantage. Food storage is a different matter. Oxygen is suitable to be stored in liquid phase. As for hydrogen storage, detailed investigations are necessary to decide whether gaseous storage in pressure vessels or storage in a metal hydride is better.

Development work in the field of air-independent propulsion is described by Nohse (1982).

BASIC OBJECTIVES

As indicated above, the aim of this study was to prove the feasibility of a submarine transfer of well production modules known as "templates" or "manifolds". Characteristics of such objects have been studied and published with relation to sea-bed offshore projects, generally located outside the Arctic geographical area (Jones, 1981).

Large size (several tens of metres in each direction in plane view) and heavy weights in air are the determining aspects of such objects, even if not necessarily associated with large oil fields. We can conceive of reducing or even cancelling the apparent weight in water, using natural buoyancy or additional compensating tanks attached to the structure. However, in its basic objectives, it has been assumed that the submarine system should be able to handle apparent loads of at least several hundred tons under water, possibly to be extrapolated to several thousands of tons in alternatives. The purpose of this last objective was to prove the feasibility of an extension of such a means of transportation to fulfill other purposes, either during exploration phases or during early production stages of offshore oil projects (see "Barge Shuttle Solution" below).

As a first target, only a relatively short distance is needed for the submerged leg of the trip, estimated not to exceed 200 m. In fact, this covers areas located under the multi-year ice-shelf, or within the heavily ice-infested marginal zone. Typically this corresponds, in the north Canadian exploration program, to the distance between an intermediate unloading point near the boundary of the permanent ice-shelf and proved well locations.

The main purpose of the system is not to carry out long-haul surface missions in light single-year ice or in light drift ice, as more conventional transportation means can and are already used in those cases. Therefore, if some reinforcement against accidental ice impact is a must, no ice-breaking capability as such is requested; regulations may dictate the minimum ice strength capability to be used as a basic design parameter.

High submerged speed is not important in that case, and no strong depth currents are anticipated. Figure 3 shows some typical numerical data available for the areas concerned.

Potentially productive oil wells have already been drilled out under the ice at a sea-bed depth of 300 m. Consequently, the ship must be able to operate at least to such a depth.

Of course, both legs of the trip, northward and the return trip, as well as the unloading procedure at the target location, have to be carried out without any intermediate resupply from outside, neither for the propulsion system, nor for the life support system.

In the first stages of analysis, a possible means of unconventional propulsion could be closed gas or deisel engines, or fuel cells. These would considerably extend the range with extremely limited on-board energy storage space and, in some instances, they do not require outboard dumping of spent combustion by-products — all extremely attractive factors. Nuclear propulsion has been ruled out as being completely outside the power and economic ranges needed for such an undertaking.

Interest in the use of such unconventional propulsion means has been revived recently and some test applications are under way at the present time (Santi and Fowler, 1983).

A more detailed approach shows that, if closed systems have been built and/or are presently developed and perfected, no system has yet reached the stage of continuous

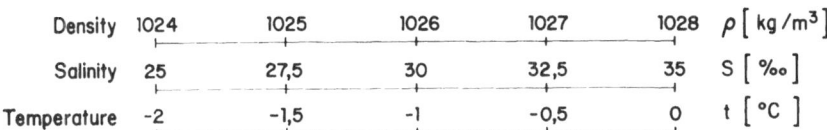

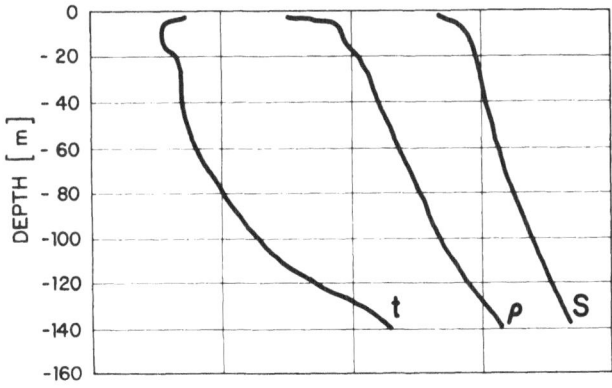

Figure 3   Data for water in Arctic Regions.

and dependable industrial operations, at least at the level of output required. In this case, industrial operations means long-term operation with very low maintenance and extremely low failure probability, even of minor failures. This is an essential requirement for a system which is to be operated completely isolated from any form of access during routine trips of 400 m under the ice, and which is supposed to stay in northern latitudes, far away from comprehensive maintenance facilities during, most likely, multi-year campaigns.

These considerations finally lead to the choice of the proven and highly reliable conventional electric propulsion, with advanced storage batteries, as used in non-nuclear military submarines. In any case, closed-circuit systems could be introduced or retrofitted at later stages, as soon as considered sufficiently reliable and proven on less stringent applications. In fact, the basic philosophy of the project has been to examine its feasibility using exclusively existing and proven technologies. Thus it is believed that the difficulty of the mission in itself, the safety and reliability which would necessarily be requested not only for its industrial acceptance but also for crew safety and environment protection are such that the need for exotic or unsufficiently proven systems would be a serious obstacle.

Whatever technical solutions are chosen, the final probability of incidents or failures can, however, never be assumed to be zero. Some types of failures have to be assumed and must be introduced into an operational risk and reliability analysis at later design stages. This can be carried out by means of one of the methods presently available (Aldwinkle and Pomery, 1983) to define at the very least the most statistically probable boundary incident or accident. This also means that the preliminary design concept has to incorporate a certain number of safeguards to provide for such boundary events and this represents, in fact, non-negligible technical limitations.

Additionally, a preliminary approach to other aspects of the problem (e.g. guidance, position fixing and tracking under the ice) has had to be made to ascertain the existence of reasonable solutions, again, whenever possible, based on existing technologies.

Finally, the basic numerical parameters retained for the study are the following:

Weight to be lifted and in the basic solution	min. 500 tons
Main dimensions of the object	min. 30 × 15 m
Operating depth	300 m
Submerged range, only one leg of the trip with load, no intermediate resupply	400 m
Submerged speed	5/6 knots

THE SUBMARINE CARRIER

Based on the above requirements, different solutions are possible for this transportation task. Apart from future-oriented solutions of the problem, different alternatives were worked out as preliminary concept designs, consisting of conventional structural elements. The advantage will be that system-proven components can be utilized immediately. At least four major conventional carrier types can be envisaged:

(1) A *Submarine carrier* diving to a *limited depth* of 50 m with an installed winch for lowering the load down to 375 m max–buoyancy with air.
(2) A *submarine carrier* of the bathyscaphe principle, diving to the full depth of 375 m with the load–buoyancy with a liquid of low density.
(3) A *submarine carrier* diving to the full depth of 375 m with the load–buoyancy with air.
(4) A *submarine barge carrier* diving to the full depth of 375 m with the load–buoyancy with air.

At the present stage of design work, the carrier systems presented under proposals 3 and 4 appear as most promising solutions. They will now be described in more detail.

The submarine carrier

The design of a submarine carrier for operation under ice has to take into consideration not only all of the design principles for conventional submarines running world-wide, but also the special requirements of under-ice operation under arctic conditions. These are:

Navigation, communication and control under ice over long distances.
Snorkel operation at special points with drilling through ice.
Maneuvrability and controllability within 6 degrees of freedom.
Loading and unloading procedures.
Compensating, trimming, ballasting and deballasting.
Number of crew and accommodation.
Diving depth and pressure tight compartmentation.
Safety and rescue.

With respect to these principles, dealing with the systems has led to different proposals for alternatives defined in the following as types A, B and C. For all three submarine carriers the following commonly applies:

All pressure hulls have a diving depth of 375 m and a collapse depth of 600 m.
Conventional propulsion plant with lead–acid batteries and DC propulsion motor for submerged cruise, diesel-generator sets for charging the batteries during surface or snorkel condition.
Control and monitoring so that, as far as possible, engine rooms can be operated unmanned.
Equipment for surface and underwater navigation, detection and communication, which is adapted to the mission area.
Satic and dynamic maneuvering facilities on all 3 axes.

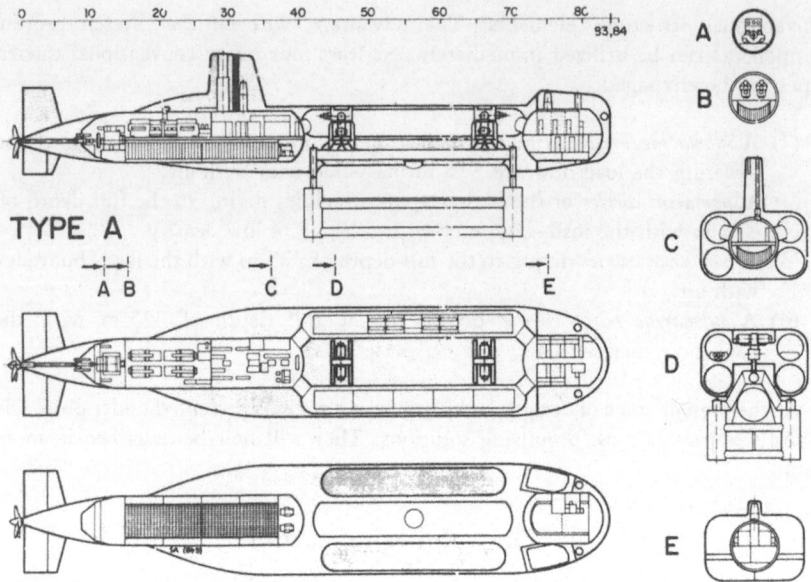

Figure 4   Submarine carrier type A.

Accommodation of crew acording to the presently used naval standard.
Utilization of any submarine-proven components.
Application of regulations of any classification society required.

Main data for submarine carrier type A

Single pressure hull type with central cargo bay, see Figure 4.

Length	85.3 m
Breadth	15.0 m
Height	16.5 m
Surface displacement	2950 m^3
Submerged displacement	3550 m^3
Payload	400 t
Max. speed submerged	9.5 knots
Transit speed submerged	5.0 knots
Cruising range submerged	200 nautical miles
Total cruising range	9300 nautical miles
Crew	8 persons (also type B and C)
Mission time	15 days (also type B and C)

Navigation (also valid for types B and C)
   Surface: Loran C, satellite navigation, radar, horizontal stabilized gyro,
optronic, log, echo-sounder and plotting table.

Submerged: Horizontal stabilized gyro, doppler log, sonar (passive/active), scan sonar, echo-sounder, plotting table, submerged TV-camera with monitor and recorder.

Communication (also valid for types B and C)
External: UHF/VHF antenna, HF antenna, underwater telephone.
Internal communication.

Propulsion motor	880 kW/80 r.p.m.
Battery	720 cells with 380 t weight
Diesel generator	4 × 880 kW/1800 r.p.m.

Main data for submarine carrier type B

Double pressure hull type with central cargo bay, see Figures 5 and 6.

Length	78.3 m
Breadth	16.0 m
Height	15.4 m
Surface displacement	3100 m³
Submerged displacement	3720 m³
Payload	400 t
Max. speed submerged	9.0 knots
Transit speed submerged	5.0 knots
Cruising range submerged	210 nautical miles
Total cruising range	9700 nautical miles
Propulsion motor	2 × 450 kW/100 r.p.m.

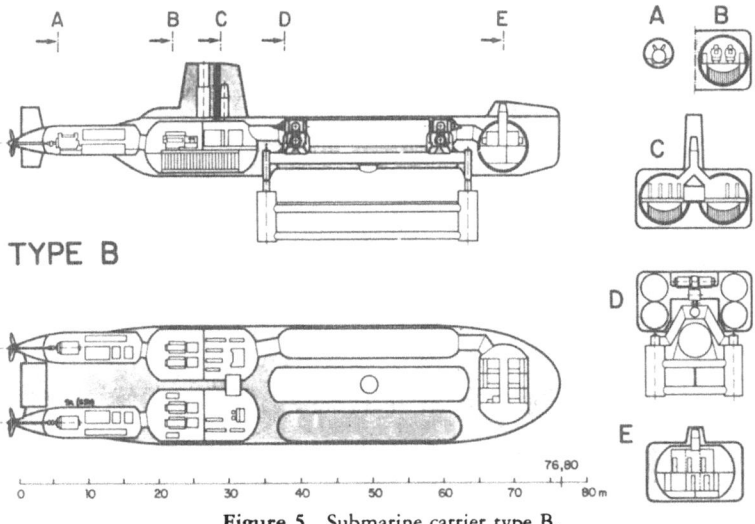

Figure 5 Submarine carrier type B.

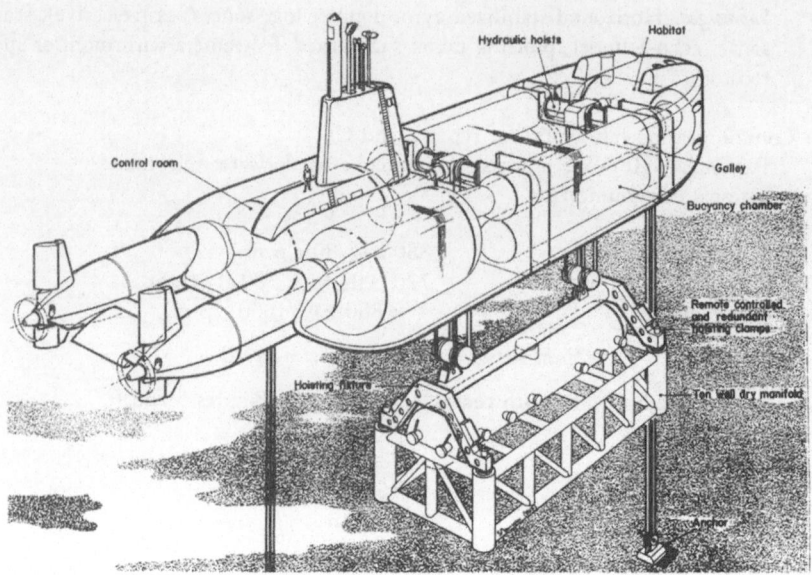

Figure 6   Submarine carrier type B, lowering a 400 ton load.

Battery	720 cells/380 t
Diesel generator	4 × 880 kW/1800 r.p.m.

Main data for submarine carrier type C

Single pressure hull type *without* cargo bay, see Figure 7.

Length	56.0 m
Breadth	15.0 m
Height	16.5 m
Surface displacement	2930 m^3
Submerged displacement	3530 m^3
Payload	400 t
Max. speed submerged	9.0 knots
Transit speed submerged	5.0 knots
Cruising range submerged	220 nautical miles
Total cruising range	10200 nautical miles
Propulsion motor	880 kW/80 r.p.m.
Battery	720 cells/380 t
Diesel generator	4 × 880 kW/1800 r.p.m.

The submarine barge carrier

For the submarine barge carrier, all of the statements for the three solutions of submarine carrier types A, B and C are also valid. Unlike in submarine carriers, the compensating tanks are arranged within a separate barge. It is conceivable that,

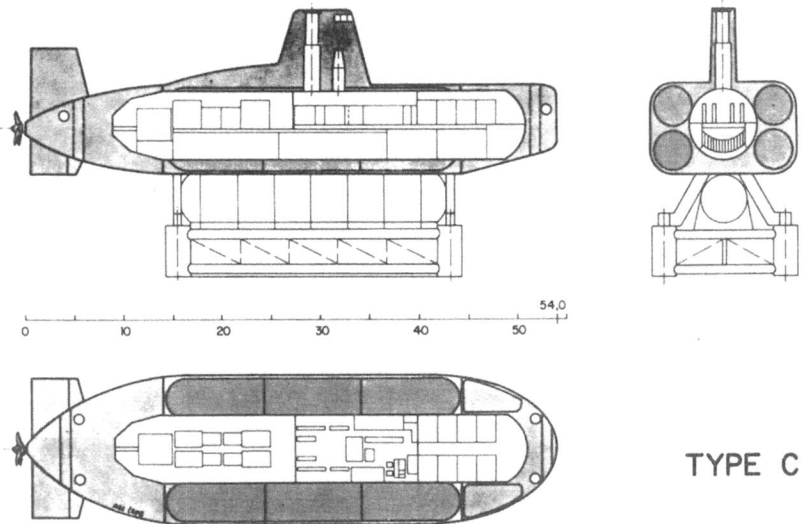

Figure 7   Submarine carrier type C.

TYPE C

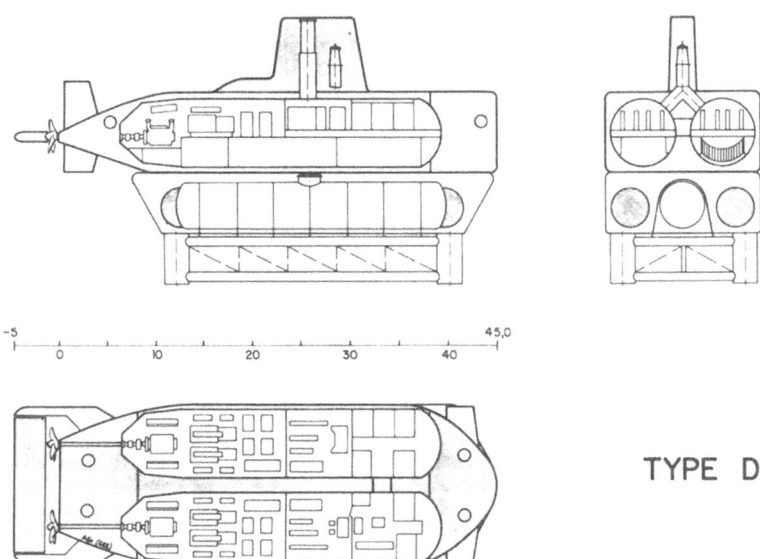

Figure 8   Submarine barge carrier.

TYPE D

corresponding to the different manifolds, the barges will be equipped with specific connecting elements, thus achieving weight compensation.

Main data for submarine barge carrier

Single pressure hull type with separate barge, see Figure 8.

Length	49.5 m
Breadth	16.0 m
Height	15.4 m
Height with barge	21.4 m
Surface displacement	
without/with barge	2300/3100 m^3
Submerged displacement	
without/with barge	2650/3450 m^3
Payload	400 t
Max. speed submerged	5.0 knots
Cruising range submerged	195 nautical miles
Total cruising range	8800 nautical miles

The other data are the same as for submarine carrier type B.

Preliminary conclusions

The design was based on rough assumptions of the largest and heaviest manifold of four hypothetical ones to be transported. The cargo bay and the hoisting frame (and consequently the carrier types) should not be considered as being optimized regarding minimum height, minimum drag, minimum weight and manifold integration into cargo bay, the solutions shown could be carried out by numerous shipyards with no particular difficulty.

Four concept designs were initially envisaged. Two have not been further investigated: the submarine carrier with limited depth, and the submarine carrier with bathyscaph principle. The two remaining alternatives are designs which were further investigated. In both designs we assume that the manifolds were connected rigidly and integrated to a large extent to the carrier system.

Compared with the three other concepts, the submarine barge carrier presents a much larger flexibility regarding the transport of variable cargo, using an autonomous barge. The more costly components required for hoisting might be transferred in a reasonably cheap barge specially designed for an exceptional load without any modification to the autonomous carrier submarine, which would mate with various barge sizes. In addition, the three integrated designs which are attractive because of their good performance and maneuvering are also designed for a maximum and limited cargo buoyancy. This does not apply in the case of the barge. It is not our intention to favor one particular design, as each concept mentioned could be worthy of more detailed examination provided that complete design data are available.

LOAD HANDLING PROCEDURE

We have to consider that the task of transporting and, more particularly, loading and unloading large loads with a submarine, is not as simple a procedure as it may appear, particularly when the relative weight of the load in water is important with relation to the displacement of the submarine. By definition, a floating vessel has a large positive buoyancy reserve, in fact, generally very much larger than any load to be taken on board. A reasonable weight added to, or subtracted from, the vessel is automatically compensated by a relatively small variation in depth of submergence. There are few, if any, longitudinal problems of stability, and trimming for good balance (pitch attitude) is easy. Transverse stability is more easily upset; vertical and transverse load distribution (roll attitude) must be balanced carefully.

On the other hand, a submerged submarine is a vessel in exact hydrostatic equilibrium with surrounding water, with roughly the same pitch and roll stability. But submarines are sensitive with respect to load variations over their length. This means that internal adjustment measures must be taken by length-wise shifting of trim or compensation water or of a trim weight by smaller submersibles. Vertical movement is, in fact, practically in indifferent equilibrium. Without active corrective measures, any vertical impulsion will tend to be quite extended. When in movement under water, a stable trajectory is reached by hydrodynamic effect of the forward and aft hydroplanes; ballasts or compensating tanks are mostly used for adjustment of the mean neutral buoyancy at a given depth, and of course for the diving and emergency operations. To be able to dive, a submarine on the surface can have only a small buoyancy reserve, typically about 15 % of the displacement.

These characteristics require a specially adapted scheme for loading and unloading when those loads are a considerable fraction of the total displacement (in our case, approx. 20%). The best practical solution is to have them hanging in water under the hull, where they are to be lifted from (and deposited on to) the sea-bed by means of suitable winches, with a close adjustment for pitch equilibrium. This is the procedure adopted for this project, as shown in Figures 9 and 10.

Several successive steps are needed, best described by means of sequence sketches. A short summary of these sequences, with reference to the corresponding sketches, is given below. This is, of course, only one of the scenarios which can be used, but it is a reasonable one, showing some of the special procedures specifically needed for such a task.

Loading

L1   At a convenient location, near the beginning of the transportation leg, the load is deposited on the sea-floor, directly from the surface ship or barge used for travel from manufacturing center, or is simply sunk if floated for that travel.

L2   Transport submarine is located directly above the load, by on-board remote location equipment, sonar, video or acoustic transponders. It is hooked to pilot lines released remotely from the load and lifted by buoys.

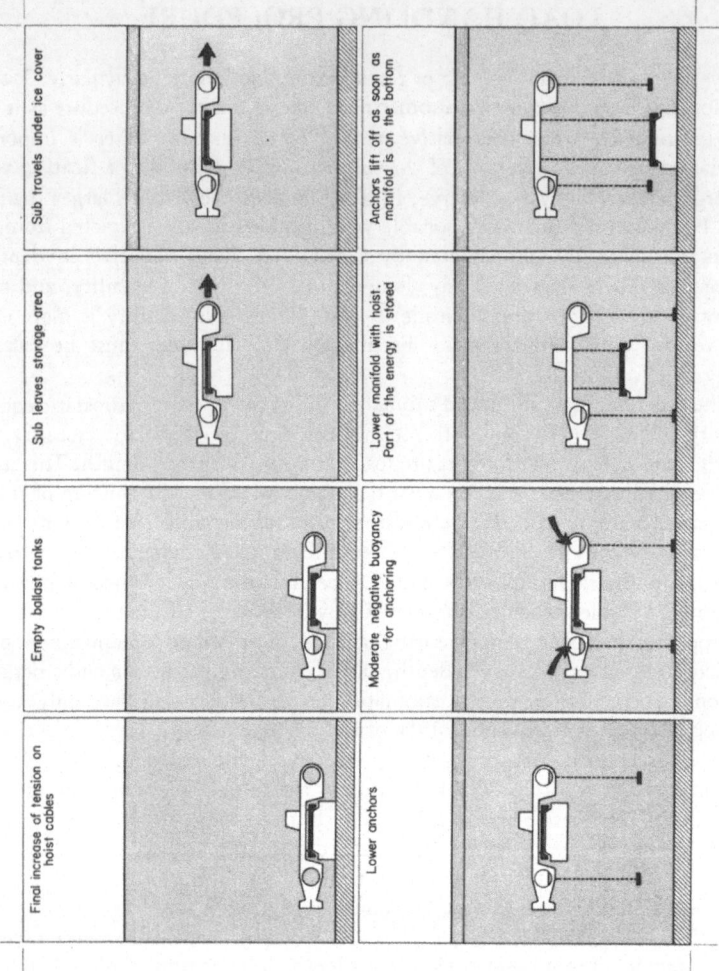

Figure 9   Sequential sketches of loading procedure.

L3   A connected frame, needed for a balanced geometric matching between various types of loads and the submarine holding points, is hoisted down from the submarine to the load, guided by the pilot lines. This frame is remotely and mechanically locked to the load.

L4   In its turn, the submarine, in a state of slight positive buoyancy, is maneuvered down over the load and mechanically locked onto the connecting frame.

L5   Ballast and compensating tanks are adjusted and trimmed to bring the submarine to its initial operating depth, in a neutral state of buoyancy.

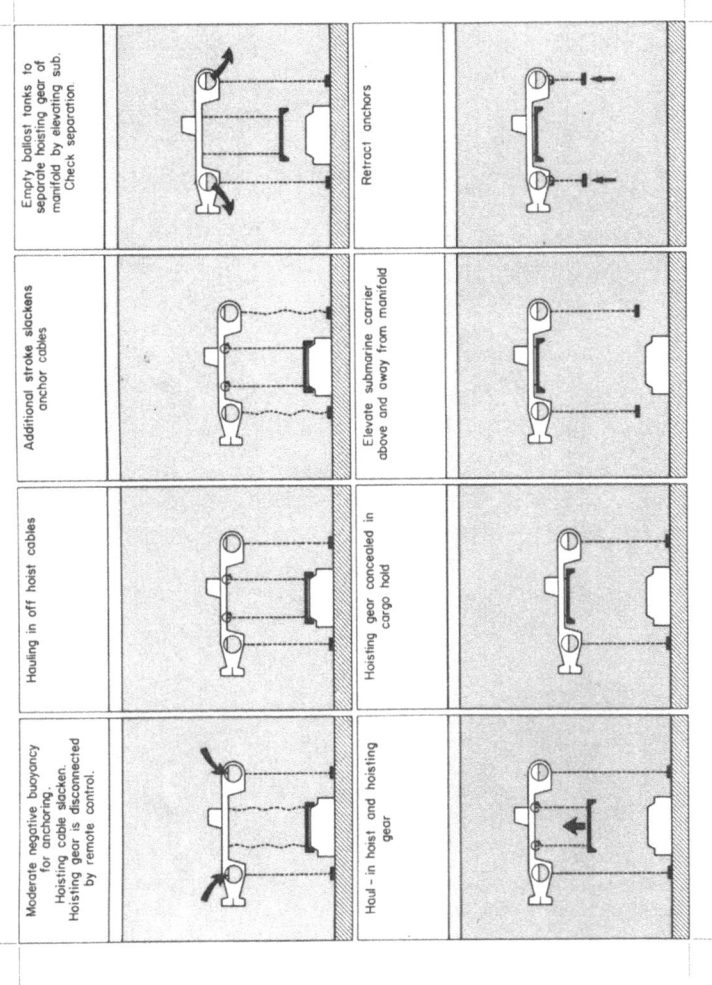

L6   During travel, correct attitude and trajectory are dynamically controlled in the usual way, in addition to the action of hydraulic thrusters needed because of the slow travel speed.

Unloading

U1   The submarine is positioned where the load is requested, slightly above the sea-floor, using conventional location equipment. It is anchored to the sea-bed by means of weight anchors.

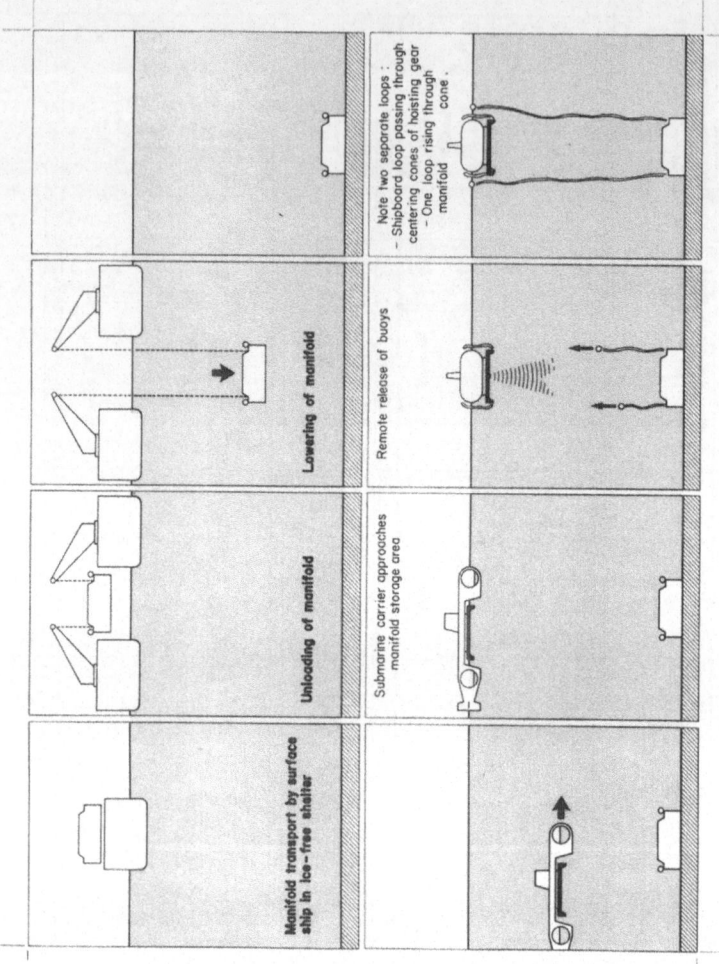

Figure 10   Sequential sketches of unloading procedure.

U2   Load with connecting frame is hoisted down to the sea-floor until the submarine with its weight anchors is, in fact, "floating" above the load.

U3   Anchoring on the weights is to be restored by decreasing the buoyancy of the submarine so that some slack can be given to the main hoisting cables and the locking mechanism of the frame can be released.

U4   Adjusted positive buoyancy allows lifting of the frame and retracting of the weight anchors to reach the normal empty travel configuration.

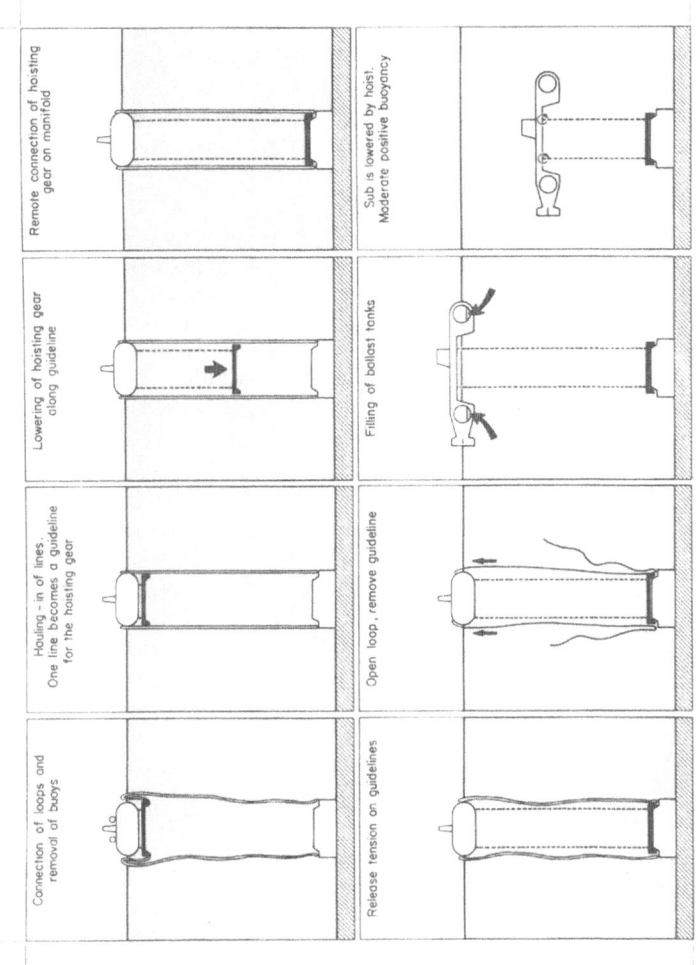

BARGE SHUTTLE SOLUTION

In many technical feasibility studies, operational limits such as maximum weights, dimensions, or performances such as speed or range of operation, are quite unstable. This also applies in our particular case. It has become rapidly evident that if we assume a heavy submarine haul capacity available, it could be potentially used for various purposes, possibly a larger range of loads, in the thousands rather than hundreds of tons. For instance, a sufficiently large submarine loading capability could be thought of as an alternative to other extremely expensive solutions, for initial production stages on small and remotely located oil fields. It may not be needed to approach the very large capacity of some earlier schemes, using nuclear propulsion to meet the goal. These are the types of considerations which lead to the examination of the larger alternatives mentioned in "Description of the problem" above.

Load volume, capacity required for ballasts, for compensating tanks and energy production or storage, increase tremendously, as would the cost of a corresponding specialized vessel. Some of these problems can be at least partially circumvented by the use of a barge shuttle submarine system. In this case, the submarine in itself remains relatively limited in size, used mainly as a propulsion, guidance and control unit; it incorporates only a fraction of the total volume needed for ballasts, energy storage, and has no on-board direct load carrying capability.

This "tug submarine" is associated to one, or several, special-purpose barges, in which are grouped the total load-carrying function, part of the buoyancy compensation tanks and some of the energy storage (either electric, or possibly other batteries). In one of the possible configurations, the "tug submarine" will ride and be mechanically locked onto the barge during the actual transportation phase, as shown in Figure 11. The concept is not new and has already been proposed. Within the framework of our project, we have made a preliminary examination of its practical feasibility.

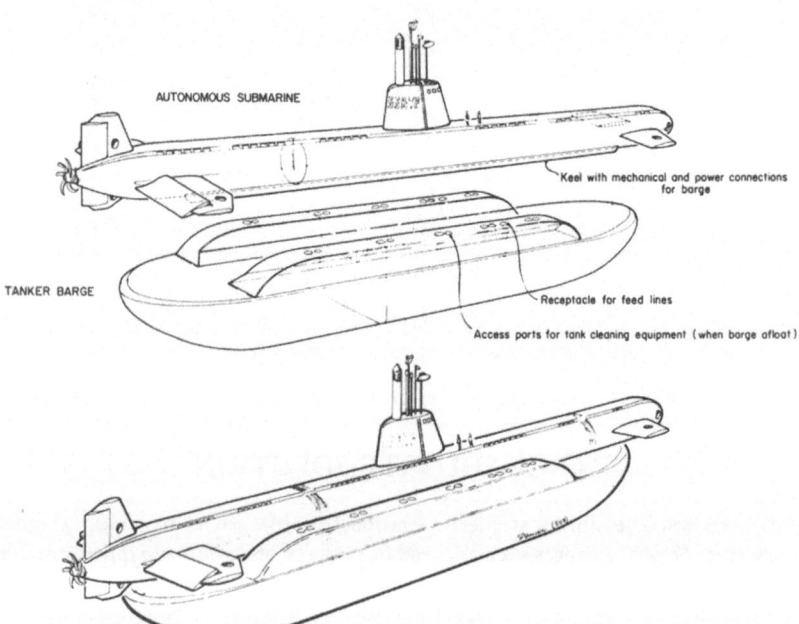

Figure 11   Conceptual design of a barge shuttle submarine tanker.

It has been found that, on the basis of still relatively modest target values,

load carrying capacity	min. 10 000 tons
operating range	min. 100 m
speed submerged	9 to 10 knots

the concept is feasible, again using mostly conventional and proven technological bases.

It has to be considered that the concept opens the way for a relatively wide range of possible barges capacities for a given size of "tug submarine". Therefore, the values given above are certainly not upper limits, and reasonably larger targets should be feasible. Additionally, it can be conceived that a barge can be made and adapted for different purposes, not only load retrieval and deposition, but possibly for tasks such as exploration, or laying of special submarine equipment, for instance. Moreover, barges are cheaper to build than complete submarines, so that several different bulk-carrying or several multipurpose specialized barges would be serviced by a single submarine. Some more or less independent on-board operational equipment can be incorporated into a barge, which, consequently, does not need to stay hooked to the "tug submarine" to perform its basic task on the sea-floor. In fact, several industrial submarine schemes include various operational modules, such as power production modules, service modules, which can also (and this is not the only possibility) be built as barge modules to be transported, deposited, serviced or retrieved by a "tug submarine".

In this early stage of the feasibility study, we had to limit our projections to the fact that the barge—submarine scheme is a practical solution which can quite effectively extend the large load-carrying potential of a submarine system.

SAFETY FEATURES

Safety and rescue are important factors of the development objectives and design principles for submarines. Today, safety and rescue means have a high standard and these have been described recently by Gabler (1975) and Abels and Nießen (1983). A submarine is at least as safe as a surface ship and rescue may be possible down to collapse depth.

First of all, on-board safety is achieved by the application of the rules and regulations of the classification societies. Although these designs are based on the regulations of Germanischer Lloyd, the regulations of all the other classification societies are, of course, applicable. Over and above these requirements, there are additional safety and rescue means on board. The main items dealing with safety and rescue are summarized as follows:

Surfacing
Hydrodynamic surfacing by forward and aft hydroplanes.
4 forward and aft vertical thrusters for vertical speed of 1 knot.

Quick blowing of main ballast tanks by high-pressure air or by hydrazine gas generators.

Air pressure volume for 6 times blowing of main ballast tanks.

Volume of main ballast tanks 15–20% of surface displacement.

Freeing of compensating tanks, buoyancy tanks and trim tanks.

Quick dropping of load.

Strength and tightness

Dimensioning of pressure hull for collapse depth of 600 m, diving depth of 375 m, safety factor of 1.6; maximum depth in operation area 250 m, safety factor of 2.4.

Double hull design, i.e. pressure hulls are protected by outer shell with steel structures.

Pressure hull material of high strength steel with high static and dynamic strength properties.

Double shut-offs for all pressure hull penetrations, as for pipes, cables and openings.

Test pressure for tightness of all shut-offs and valves and fittings at seawater pressure equal to collapse depth of pressure hull.

Strengthening of bridge fin for breaking through ice by blowing of different tanks.

Maneuverability and controllability

Good maneuverability and controllability at high and low speeds by rudders, hydroplanes, thrusters and internal ballasting and trimming by tanks.

Good hydrostatic and hydrodynamic stability.

Automatic, remote-controlled or hand-operated boat handling.

Redundancy

Of all main systems and equipments, for example:

Double propulsion motor with 2 windings.

Carrier type B double shaft design with 2 propulsion motors and 2 propellers.

4-part main batteries.

4 diesel-generator sets.

6 thrusters for horizontal and vertical speed.

Blowing of main ballast tanks by high-pressure air or hydrazine gas generators.

3 high-pressure air compressors.

4 hydraulic stations.

4 piston bilge pumps.

Cross connections between the main power supply lines.

Automatic and manual control of all systems.

Redundancy in the navigation and communication systems.

Under-ice navigation

Acoustic transponder arrays on the sea bottom.

Active/passive sonar and doppler sonar.

Detection and collision avoidance sonar.

Floating wire antenna for radio signal reception.

Inertial navigation system.

Life-support facilities

Access trunks for free escape and free ascent.

Life-raft container

Emergency signal buoy.

Built-in breathing-air system and breathing-air supply

Water and provisions.

Bilge pumps, boat fan and lighting.

Emergency switchboard.

CO_2-absorption and O_2-supply.

Life-jackets and rescue suits.

Acoustic pinger.

Breaking through ice and rescue

Blowing of main ballast tanks or freeing of other tanks under ice and breaking through ice by bridge fin.

Positioning the submarine under ice and drilling through ice by rotating snorkel.

Rescue by Deep Submergence Rescue Vehicle (DSRV) of US Navy or by rescue sphere (Gabler, 1975; Abels and Nießen, 1983).

REFERENCES

Abels, F. 1978. New Developments in German Submarine Design, in *International Naval Technology Exposition 78, Conference Proceedings*, Interavia SA, Geneva, pp. 175–188.

Abels, F. and Nießen, E. 1983. The pressure-tight bulkhead in the submarine, The Royal Institution of Naval Architects, London, Symposium on Naval Submarines, Paper No. 5.

Aldwinckle, D. S. and Pomery, R.V. 1983. A rational assessment of ship reliability and safety. *Transactions of the Royal Institution of Naval Architects*, London, 125.

Gabler, U. 1975. *Sicherheit und Rettungseinrichtungen von Unterseebooten, Jahrbuch der Schiffbautechnischen Gesellschaft*, Springer-Verlag, 69. Band, 1975, s. 9–17. (Safety and rescue means, English translation available.)

Jones, M. 1981. *Deep water oil production and manned underwater structures*, Graham & Trotman, London.

Nohse, L. 1982. Submarine propulsion conventional and outside-air-independent. *Naval Forces*, No. IV.

Rössler, E. 1981. *The U-boat. The Evolution and Technical History of German Submarines*, Arms and Armour Press, London/Melbourne.

Trillo, R. L. (ed). 0000. *Jane's Ocean Technology*, Jane's Yearbook, London.

Veliotis, P. T. and Reitz, Sp. 1981. A Submarine LNG Tanker Concept for the Arctic, *Gastec '81 LNG/LPG Conference and Exhibition*, Hamburg.

———. 1983. The design and operation of underwater vehicles, in *Proceedings of Subtech 83*, SUT, London, November 1983, Santi and Fowler Papers 2.3 and 8.2.

List of Authors

Dr. Fritz Abels
Manager
Ingenieurkontor Lübeck GmbH
Postfach 1620
D – 2400 Lübeck 1
Federal Republic of Germany

Dr. John Adolfson
FOA 58 Naval Medicine Division
Naval Diving Center
S – 130 61 Hårsfjärden
Sweden

Mr. Paul Arthur Berkman
Graduate School of Oceanography
Narragansett Bay Campus
University of Rhode Island
Kingston, Rhode Island 02882-1197
USA

Professor Arnoldus Schytte Blix, Ph.D.
Department of Arctic Biology
University of Tromsø
Box 635
N – 9001 Tromsø
Norway

Mr. S.B. Boulton
Technical Director
HBM Subwork Limited
Bessemer Way
Harfreys Industrial Estate
Gt. Yarmouth, Norfolk
Great Britain

Mr. P.J. Butler, Ph.D., F.I.Biol.
Professor of Comparative Physiology and
Head of Department
Department of Zoology and
    Comparative Physiology
The University of Birmingham
P.O. Box 363
Birmingham B15 2TT
Great Britain

Mr. Jakie Chappuis
Technical Director
Ateliers de Constructions Mécaniques de
    Vevey S.A.
CH – 1800 Vevey
Switzerland

Dr. W.A. Crosbie
Employment Medical Adviser
Health and Safety Executive
Employment Medical Advisory Service
1 Long Lane
London SE1 4PG
Great Britain

Mr. Geir Wing Gabrielsen, Ph.D.
Department of Arctic Biology
University of Tromsø
P.O. Box 635
N – 9001 Tromsø
Norway

Professor Per-Ola Granberg, M.D.
Head
Section of Endocrine Surgery
Karolinska Sjukhuset
President, The Nordic Council for Arctic
   Medical Research
Box 60500
S – 104 01 Stockholm
Sweden

Mr. Ron Goodfellow FICE, MASCE
President of the Engineering Committee
   on Ocean Resources (ECOR)
Vice President of the Society for
   Underwater Technology (SUT)
Goodfellow Associates Ltd.
61 Southwark Street
London SE1 1SA
Great Britain

Captain M. G.T. Harris
Royal Navy
Office of the Captain SM
Third Submarine Squadron
H.M.S. Neptune
Faslane
Helensburgh, Dunbartonshire
   G84 8HL
Great Britain

Dr. Philip A. Hayes
RAF Institute of Aviation Medicine
Farnborough, Hampshire
Great Britain

Professor A.M. House, M.D.,
   F.R.C.P.(C)
Director
Center for Offshore & Remote Medicine
   (MEDICOR)
Faculty of Medicine
The Health Sciences Centre
Memorial University of Newfoundland
St. John's, Newfoundland A1B 3V6
Canada

Mr. John Kanwisher, Ph.D.
Woods Hole Oceanographic Institution
Woods Hole, Massachusetts 02543
USA

Professor William R. Keatinge
Department of Physiology
London Hospital Medical College
University of London
Turner Street
London E1 2AD
Great Britain

Dr. Lorne A. Kuehn
Director Science and Technology
(Human Performance)
Research and Development Branch
National Defence Headquarters
National Defence
Ottawa, Canada K1A 0K2

Mr. Lauri Laitinen, M.D.
Head
Department of Clinical Physiology
Central Military Hospital
Box 5
SF – 00281 Helsinki 28
Finland

Mr. Peter Lomax, M.D., D.Sc.
Professor of Pharmacology
Department of Pharmacology
UCLA School of Medicine
University of California
Los Angeles, California 90024
USA

Dr. Henry J. Manson
   M.B., Ch.B., F.R.C.P.(C)
Assistant Director
Center for Offshore & Remote Medicine
   (MEDICOR)
Faculty of Medicine
The Health Sciences Centre
Memorial University of Newfoundland
St. John's, Newfoundland A1B 3V6
Canada

Professor Academician
Vladimir A. Negovsky
Laureate of State Prizes USSR
Head of Laboratory of General
    Reanimatology of the USSR Academy
    of Medical Sciences
25th October str. 9
Moscow, 103012
USSR

Mr. Phil Nuytten
President
Can Dive Services Ltd.
1367 Crown Street
North Vancouver, B.C. V7J 1G4
Canada

Mr. R. Pralong
Forces Mortrices Neuchâteloises S.A.
Departement Hydro-Vision
Les Vernets
CH – 2035 Corcelles
Switzerland

Surgeon Vice Admiral Sir John Rawlins
KBE, OHP, FRCP, FFCM, FRAeS
Past President
The Society for Underwater Technology
Wayhouse
Standford Lane
Bordon, Hants, GU35 8RH
Great Britain

Professor Louis Rey, Ph.D.
Distinguished Professor of
Arctic Science and History.
The University of Alaska-Fairbanks
President, Comité Arctique International
La Jaquière
CH – 1066 Epalinges
Switzerland

Dr. T.J.O. Sanderson
Cold Regions Offshore Research
    Programme, EXT
BP Petroleum Development Limited
Britannic House
Moor Lane
London EX2Y 9BU
Great Britain

Mr. Benno Schenk
Laboratorium für hyperbare Medizin
    und Tauchforschung
Department für Innere Medizin
Universitätsspital Zürich
Rämistrasse 100
CH – 8091 Zürich
Switzerland

Mr. Seppo Sipinen, M.D.
Department of Clinical Physiology
Central Military Hospital
P.O. Box 5
SF – 00281 Helsinki 28
Finland

Mr. Kristjan Sveinsson
Seacaptain
Salvage Vessel Godinn
Björgunarfelagid h/f
Sudurlandsbraut 6
105 Reykjavik
Iceland

Mr. H.R. Talkington
Director
Engineering and Computer Sciences
Naval Ocean Systems Center
Department of the Navy
San Diego, California 92152
USA

Dr. Stein, Tønjum
Head of Diving Division
Norwegian Underwater Technology
   Center (NUTEC)
P.O. Box 6
Gravdalsveien 255
N – 5034 Ytre Laksevåg / Bergen
Norway

Mr. Bruce E. Townsend
Diving Officer
Department of Fisheries & Oceans
Western Region
Freshwater Institute
501 University Crescent
Winnipeg, Manitoba R3T 2N6
Canada

Surg. Capt. Leif Vanggaard, M.D.
Chief of Defence
Royal Danish Navy
Postbox 202
DK – 2950 Vedbæk
Denmark

Mr. Karstein Vestgaard
Manager
Development Department
SIMRAD Subsea A/S
Strandpromenaden 50
P.O. Box 111
N – 3191 Horten
Norway

Dr. L.E. Virr
Head of Research
Admiralty Marine Technology
   Establishment
Experimental Diving Unit
c/o HMS Vernon
Portsmouth PO1 3ER
Great Britain

Dr. P. Wiesner
Werk Druckkammertechnik
Drägerwerk A.G.
Postfach 150 149
Auf dem Baggersand 17
D – 2400 Lübeck-Travemünde 1
Federal Republic of Germany